*Fish and Fisheries*
*of Tropical Estuaries*

# CHAPMAN & HALL FISH AND FISHERIES SERIES

Amongst the fishes, a remarkably wide range of fascinating biological adaptations to diverse habitats has evolved. Moreover, fisheries are of considerable importance in providing human food and economic benefits. Rational exploitation and management of our global stocks of fishes must rely upon a detailed and precise insight of the interaction of fish biology with human activities.

The *Chapman & Hall Fish and Fisheries Series* aims to present authoritative and timely reviews which focus on important and specific aspects of the biology, ecology, taxonomy, physiology, behaviour, management and conservation of fish and fisheries. Each volume will cover a wide but unified field with themes in both pure and applied fish biology. Although volumes will outline and put in perspective current research frontiers, the intention is to provide a synthesis accessible and useful to both experts and non-specialists alike. Consequently, most volumes will be of interest to a broad spectrum of research workers in biology, zoology, ecology and physiology, with an additional aim of the books encompassing themes accessible to non-specialist readers, ranging from undergraduates and postgraduates to those with an interest in industrial and commercial aspects of fish and fisheries.

Applied topics will embrace synopses of fishery issues which will appeal to a wide audience of fishery scientists, aquaculturists, economists, geographers and managers in the fishing industry. The series will also contain practical guides to fishery and analysis methods and global reviews of particular types of fisheries.

Books already published and forthcoming are listed below. The Publisher and Series Editor would be glad to discuss ideas for new volumes in the series...

## Available titles

1. **Ecology of Teleost Fishes**
   Robert J. Wootton
2. **Cichlid Fishes**
   **Behaviour, ecology and evolution**
   Edited by Miles A. Keenlyside
3. **Cyprinid Fishes**
   **Systematics, biology and exploitation**
   Edited by Ian J. Winfield and Joseph S. Nelson

4. **Early Life History of Fish**
   **An energetics approach**
   Ewa Kamler
5. **Fisheries Acoustics**
   David N. MacLennan and E. John Simmonds
6. **Fish Chemoreception**
   Edited by Toshiaki J. Hara
7. **Behaviour of Teleost Fishes**
   Second edition
   Edited by Tony J. Pitcher

# Fish and Fisheries of Tropical Estuaries

Stephen J.M. Blaber

CSIRO Division of Marine Research
Queensland
Australia

**CHAPMAN & HALL**

London · Weinheim · New York · Tokyo · Melbourne · Madras

**Published by Chapman & Hall, 26 Boundary Row, London SE1 8HN**

Chapman & Hall, 2–6 Boundary Row, London SE1 8HN, UK

Chapman & Hall GmbH, Pappelallee 3, 69469 Weinheim, Germany

Chapman & Hall USA, 115 Fifth Avenue, New York, NY 10003, USA

Chapman & Hall Japan, ITP-Japan, Kyowa Building, 3F, 2-2-1 Hirakawacho, Chiyoda-ku, Tokyo 102, Japan

Chapman & Hall Australia, 102 Dodds Street, South Melbourne, Victoria 3205, Australia

Chapman & Hall India, R. Seshadri, 32 Second Main Road, CIT East, Madras 600 035, India

First edition 1997

© 1997 Stephen J.M. Blaber

Typeset in 10/12 Photina by Acorn Bookwork, Salisbury, Wiltshire
Printed in Great Britain by T. J. International Ltd., Padstow, Cornwall

ISBN 0 412 78500 5

∞ Printed on permanent acid-free text paper, manufactured in accordance with ANSI/NISO Z39.48-1992 and ANSI/NISO Z39.48-1984 (Permanence of Paper).

# Contents

# Series foreword

Among the fishes, a remarkably wide range of biological adaptations to diverse habitats has evolved. As well as living in the conventional habitats of lakes, ponds, rivers, rock pools and the open sea, fish have solved the problems of life in deserts, in the great deeps of the sea, in the extreme cold of the Antarctic, in warm waters of high alkalinity or of low oxygen, and in habitats like estuaries where such physical factors are characterized by changing all the time. Along with these adaptations, we find the most impressive specializations of morphology, physiology and behaviour. For example, we can marvel at the high-speed swimming of the marlins, sailfish and warm-blooded tunas, air-breathing in catfish and lungfish, parental care in mouth-brooding cichlids, and viviparity in many sharks and toothcarps.

Moreover, the fish in our oceans, estuaries, lakes and rivers are of considerable importance to the well-being of the human species in the form of nutritious, diverse and delicious food. Sustainable exploitation and responsible management of our global stocks of fishes must rely upon a detailed and precise insight of their biology and their role in aquatic ecosystems.

The Chapman & Hall *Fish and Fisheries Series* aims to present timely volumes reviewing major thematic aspects of the science of fish and fisheries. Most volumes will be of interest to research workers in biology, zoology, ecology, physiology and fisheries, but an additional aim is for the books to be accessible to a wide spectrum of non-specialist readers ranging from undergraduates and postgraduates to those with an interest in industrial and commercial aspects of fish and fisheries. Moreover, it is the intention to focus interdisciplinary work in relevant social and economic spheres upon fish and fisheries and their cultural role.

Stephen Blaber's book on the fish and fisheries of tropical estuaries is the 22nd book to be published in Chapman & Hall's *Fish and Fisheries Series* since its inception in 1989. This book continues the series' tradition of high scholarship, peer review, and a world-wide geographical scope in addressing fish and their exploitation by humans.

Estuaries are defined as areas open to the sea where salinity is reduced from freshwater land drainage. In this volume Dr Blaber justifies broadening the scope to include coastal lagoons, estuarine lakes and shelf areas with strong estuarine influence. He describes the fish that live in tropical estuaries, analyses the critical features of their ecology, evaluates their role in estuarine food webs and describes their fisheries. The book includes case studies and examples from well-studied tropical estuaries located world-wide including many in Australia, India, South-east Asia, South Africa, West Africa, East Africa and Latin America.

Dr Blaber has reviewed a large global literature on tropical estuaries, as well as summarizing for the first time an extensive grey literature scattered among many reports and local periodicals. More than three quarters of the world literature on estuarine fishes covers temperate species, but it has proven difficult to generalize findings about their ecology from temperate regions to the tropics. As in other areas of fish and fisheries research, a more fundamental synthesis concerning estuarine fishes encompasses science from both tropical and temperate regions (e.g. Pauly, 1994).

All estuarine fish face a difficult physical challenge surviving in a dynamic regime of tidal, daily and seasonal changes in depth, salinity, oxygen, temperature and turbidity. But for those fishes that have evolved suitable physiological adaptations, estuarine environments bring a rich reward of food, breeding and nursery areas. In the tropics, exceptionally high rates of turnover make estuaries very productive environments. Although coral reef fisheries (e.g. Polunin and Roberts, 1996), and tropical lakes (e.g. Pitcher and Hart, 1995) have attracted much attention, in fact estuaries are the most economically important among tropical aquatic habitats. Moreover the book makes the case for their conservation status as falling among the most threatened and modified of aquatic environments.

Dr Blaber hopes that his book will encourage others to discover more about the ecology, biodiversity and responsible harvest of tropical estuarine fishes. Like many other books in the *Fish and Fisheries Series*, Dr Blaber's book evidently fills an empty niche and so I am confident that readers will soon wonder how they did without such a useful compendium and a stimulating source of insight and ideas.

**Professor Tony J. Pitcher**
*Editor, Chapman & Hall Fish and Fisheries Series*
Director, Fisheries Centre, University of British Columbia,
Vancouver, Canada

## Literature cited

Pauly, Daniel (1994) May apply in tropics, but not here! in *On the Sex of Fish and the Gender of Scientists*, Chapman & Hall Fish and Fisheries Series, volume 14, pp. 3–7.

Pitcher, Tony J. and Hart, Paul J.B. (eds) (1995) *Impact of Species Changes in African Lakes*, Chapman & Hall Fish and Fisheries Series, volume 18, 604pp.

Polunin, Nicholas V.C. and Roberts, Callum M. (eds) (1996) *Reef Fisheries.* Chapman & Hall Fish and Fisheries Series, volume 20, 478pp.

# *Preface and acknowledgements*

Most information on subtropical and tropical estuarine fishes is in scientific papers and a very extensive grey literature. I have felt for some time that a book was needed to synthesize the wealth of data and bring it to as wide a readership as possible, especially in developing countries where access to scientific libraries and their journals is usually severely limited. For this reason in particular, in this book the coverage of the ecology of tropical estuarine fishes, their distribution, their interrelationships with the environment, and their fisheries has been kept broad. Although considerable detail is provided on some aspects, often reflecting my own interests, other topics are covered more briefly. However, in all cases I have attempted to include sufficient references to allow the reader to research the literature further. This has resulted in a large bibliography that I hope will assist all those involved in any way with the fishes of this fascinating, dynamic and often threatened environment. Fish biologists from colder climes, many of whom act as consultants or supervise students in the tropics, may also find the book a useful source of comparative information and a valuable reference point. My own research experience has been largely in the subtropics and tropics of the Indo–West Pacific and the text inevitably reflects some bias towards this area. Nevertheless, many of the attributes and relationships of fishes and estuaries in this, the region with the world's highest diversity of fish species, are mirrored in the other major biogeographical regions.

It is hoped that the book will encourage research workers and students to discover more about the ecology of estuarine fishes in the tropics. Tropical estuaries are among the most modified and threatened of aquatic environments as well as supporting essential fisheries. As full an understanding as possible of their fishes is vital if they are to continue to provide a sustainable source of food for burgeoning populations. Furthermore, the recognition of the intrinsic value of these species-rich and diverse wetland habitats for conservation has led to some being declared World Heritage Areas. But the pressures from a wide array of human activities, ranging from industry to recreation, are not likely to diminish, and the maintenance of biologically healthy estuaries and their fishes depends more than ever on biologically well-informed management. It is hoped that the information collated in this book will further this end.

I should like to thank the many colleagues in many countries who have contributed so much to the research and ideas in this book. It is unfortunately impossible to list them all, but I am particularly grateful to Brian Allanson, Chris Brown, Mike Bruton, Digby Cyrus, Burke Hill, Tim Martin and Alan Whitfield in South Africa; to David Brewer, David Milton, Ian Potter, Nick Rawlinson, John Salini and Jock Young in Australia; and to Josephine Pang, Ong Boon-Teck and Philip Wong in Sarawak, for their intellectual input over the years, as well as their companionship in the field. At Chapman & Hall I am very grateful to Nigel Balmforth, whose enthusiasm and encouragement helped shape this book, and to Martin Tribe and Chuck Hollingworth, who worked magic to transform the manuscript into a book. The writing of this book would not have been possible without the forbearance and assistance of Tessa, Lucy, Gail and Helen.

Stephen J.M. Blaber

*Chapter one*

# *Introduction*

Estuaries, especially those in the subtropics and tropics, are fascinating places for the biologist. Not only are they the focal point of a lot of human activity, but their study requires the bringing together of scientists from many disciplines: hydrologists, sedimentologists, chemists and geologists as well as botanists and zoologists. Within zoology, the research needed on fishes requires taxonomic, physiological and ecological approaches and there are important gaps in knowledge in all these areas. In few other environments are animals from such a wide range of taxa so closely associated: annelid worms, prawns, crocodiles, birds, hippos and of course fishes – all may form part of the overall tropical estuarine community, often with functional ecological links. Estuaries are rigorous environments and many of the species living there may be operating at or close to their physiological limits, yet species diversity and productivity are high. From a fisheries point of view, compared with other environments, there is a greater complexity of factors that have to be taken into account when managing a tropical estuarine fishery. This complexity not only includes all the above points, but also socio-economic considerations. Fish and fisheries in tropical estuaries are part of an intricate web of relationships that extends beyond the borders of the estuary.

Many of the world's great estuaries are in the tropics: the Amazon, Orinoco, Zaire, Zambezi, Niger, Ganges and Mekong are all very large and receive drainage from enormous catchments. Despite their size, however, and the diversity of their faunas compared with their temperate counterparts, much less is known about their fishes.

Most scientific research is conducted in industrialized developed countries, nearly all of which are in cold or temperate regions. Although there has been somewhat of an upsurge in tropical estuarine fish research in the last 20 years, it still remains true that the amount of data, especially detailed ecology, available for tropical estuarine fishes, is little compared with that for higher-latitude estuaries. To illustrate this point, a search of the *Aquatic Science and Fisheries Abstracts*® database from 1978 to the end of 1995 reveals that approximately 1800 publications dealt with fishes in temperate estuaries, whereas there were only 600 for tropical estuarine fishes. The latter also include a considerable number from the warm temperate parts of the south-eastern USA and southern Australia, hence the real number is fewer than 600.

In addition to this, most books or reviews about estuarine fishes have focused primarily on temperate areas and use temperate examples. There are a number of such excellent texts that can be referred to for information on estuarine fishes, such as Green (1968), Perkins (1974), McDowall (1988), and Day *et al.* (1989). Without in any way disparaging the importance of the temperate work, it is unfortunate that many of the generalizations that can be drawn from the wealth of data about temperate estuarine fishes are very hard to apply in tropical or subtropical situations – hence for the researcher or student in warmer climes trying to interpret data based on theory from temperate regions, there has all too frequently been much head scratching and doubts about whether his or her data are correct, whereas in reality it is usually a case of, to misquote from Pauly (1994), "it may apply in temperate regions – but not here!"

As the amount of data about tropical estuarine fishes has grown, so has the need for a book about the fishes of subtropical and tropical estuaries that attempts to draw some of the threads together.

This book is about subtropical and tropical estuarine fishes; their biology and ecology and how their life histories are attuned to the diverse and changing nature of tropical estuarine environments; and their often long-standing relationships with humans, involving fisheries exploitation, conservation and management, all in the face of unprecedented pressures on the environment from other human activities. Because much of the research on fishes in tropical estuaries has been done in developing countries, often for important fisheries purposes, and not necessarily for straightforward scientific reasons, frequently with limited or sometimes non-existent scientific facilities, many of the results are only to be found in the 'grey literature' of annual reports, reports to aid or funding organizations, unpublished conference papers and articles in newsletters. Such data are often important and may be the only information on a particular area or species, and those that are relevant have been fully drawn upon in this book to support and extend results from the primary literature. While much of this 'grey' literature included in the References may be hard to obtain (one possible argument for excluding it), the loss of data resulting from its exclusion is far worse.

It is assumed that the reader has a basic knowledge of fish anatomy and biology as found in many excellent texts such as Alexander (1967), Norman and Greenwood (1975), Wootton (1990), Rankin and Jensen (1993), Bone and Marshall (1994) and Jobling (1995). Other more specialized texts are referred to in the relevant chapters. Books on the fishes of adjacent tropical habitats that are particularly useful include Welcomme (1979) and Lowe-McConnell (1987) on fresh waters, Longhurst and Pauly (1987) for a good coverage of most marine habitats and Sale (1991) for coral reef fishes.

Although the correct taxonomic identification of tropical marine and estuarine fishes is vitally important, it is often no simple matter. The large number of species and lack of adequate field keys or guides frequently hamper research, particularly in developing countries where scientists seldom have access to good library facilities. The FAO species catalogues published in the *FAO Fisheries Synopsis* series are most useful but a large number of families remain to be covered. The *FAO Species Identification Sheets for Fishery Purposes* are also helpful but often lack keys to species level.

Unfortunately, to be reasonably sure of identifying most of the species in a tropical estuary in, for example, the Indo–West Pacific, the field worker needs to carry around a large number of books, papers and keys. Even then, the identification of some groups, such as gobies, is a job for the specialist, and fish need to be sent to the relevant expert. The consequence of these problems, together with the difficulties of carrying out fieldwork in remote tropical areas with few scientific facilities, is that many species caught in tropical estuaries are incorrectly identified or are identified only to genus or family. This impression may then render comparisons between areas or estuaries difficult.

## 1.1  RESEARCH CHALLENGES

The fishes of tropical estuaries have been studied rather less than those of other tropical waters, partly because freshwater fish biologists tend to regard them as primarily marine, while their marine counterparts have concentrated more on other habitats such as coral reefs. The volume of research directed at coral reef fishes has not been related primarily to their fisheries importance, but rather to their attractiveness, both scientific and aesthetic, and the relative ease with which detailed ecological and behavioural studies can be carried out in warm and clear, shallow waters. This is not to detract from the fundamental importance of many of the findings of such studies, but rather to contrast the situation with that in tropical estuaries. Here the water is usually muddy with little or no underwater visibility, often with strong tidal currents, muddy substrata, with large and dangerous (unseen) predators such as crocodiles, and surrounded by frequently impenetrable mangroves with large numbers of mosquitoes and sandflies. For these reasons, fish research in

**Table 1.1** Range of fish yields from different tropical habitats. Modified after Marten and Polovina (1982)

| Habitat type | Fish yield (tonnes km$^{-2}$ year$^{-1}$) | |
|---|---|---|
| | Minimum | Maximum |
| Tropical lakes | 0.1 | 23.0 |
| Tropical reservoirs | 0.1 | 35.0 |
| Tropical rivers | 0.02 | 78.0 |
| Tropical estuaries | 1.7 | 250.0 |
| Tropical demersal shelf | 0.2 | 21.9 |
| Tropical pelagic | 0.2 | 17.5 |
| Coral reefs | 0.09 | 18.0 |

tropical estuaries is somewhat challenging, requiring different sampling techniques and a certain amount of fortitude. Much of the work has been confined to species of economic importance because the fisheries productivity of subtropical and tropical estuaries is relatively high (Table 1.1). The economic importance of tropical estuaries outweighs that of most other tropical aquatic habitats, including coral reef areas, except perhaps in relation to tourism.

## 1.2  DEFINITIONS

Almost all books written about estuaries begin by defining what is meant by the word estuary. The most widely accepted definitions are those of Pritchard (1967), Dyer (1973) and Day (1981). These all provide very sound sets of characteristics that help determine whether a particular body of water is an estuary. For example, Pritchard (1967) defined an estuary as follows: *"An estuary is a semi-enclosed coastal body of water which has a free connection with the open sea and within which sea water is measurably diluted with freshwater derived from land drainage."* However, this definition ruled out blind estuaries and those that are hypersaline, and Day (1981) modified this as follows: *"An estuary is a partially enclosed coastal body of water which is either permanently or periodically open to the sea and within which there is a measurable variation of salinity due to the mixture of sea water with freshwater derived from land drainage."* In this book I have not set out to change these definitions in any way, but from a tropical fish point of view, have had to broaden the definition to include shallow coastal waters that are contiguous with estuaries and have similar reduced salinities. This is because the fish faunas of many tropical estuarine areas extend onto the shallow shelf and there is only a gradual change to marine conditions. There may be no clearly defined boundary between estuary and sea. Essentially estuarine conditions with an estuarine fish fauna may extend many kilometres off shore and cover huge areas in, for example, parts of South East Asia, South America and West Africa. In this way, in relation to fishes, there is actually more estuarine area outside the confines of traditionally defined estuaries than within them. This 'estuarization' of the shallow continental shelf as it has been defined by Longhurst and Pauly (1987) is discussed further in Chapter 2.

## 1.3 GEOGRAPHICAL EXTENT

Moving away from the Equator, tropical estuaries grade into subtropical systems beyond the tropics of Cancer and Capricorn. Seasonal water temperature differences become more marked further from the tropics and it is these differences between summer and winter conditions that separates tropical from subtropical estuaries. A winter water temperature low of about 12 °C is used in this book to mark the southern and northern limits of subtropical estuaries (Whitfield, 1994a; Ayvazian and Hyndes, 1995). The latitudinal extent of subtropical estuaries depends primarily on sea temperatures and the influence of currents. For example, in South East Africa and eastern Australia, the warm Moçambique and East Australian currents maintain water temperatures above 12 °C as far south as about 32 °S – hence a tropical to subtropical fish fauna extends further south than is otherwise the case in, for example, western South America.

The coverage of estuaries in this book approximately follows the distribution of mangroves, which are the dominant intertidal vegetation in subtropical and tropical estuaries (Chapman, 1976) (Fig. 1.1). Hutchings and Saenger (1987) reviewed the data concerning the worldwide distribution of mangroves in relation to temperature and concluded that the presence of mangroves correlates with areas where the water temperature of the warmest month exceeds 24 °C; also that their northern and southern limits correlate reasonably well with the 16 °C isotherm for the air temperature of the coldest month.

The situation regarding the estuaries of the southern and south-eastern USA, where a great deal of fish research has taken place is important. While many such estuaries have been described as 'tropical' or 'subtropical' (McPherson and Miller, 1987; Powell *et al.*, 1991), this is not really the case in a world context, with the possible exception of south Florida mangrove-lined systems. Winter water temperatures in many US

Fig. 1.1 The distribution of mangroves (hatched shading) in relation to the 24 °C isotherm (black lines) (modified from Chapman, 1976 and Hutchings and Saenger, 1987).

Gulf of Mexico estuaries fall as low as 5 °C (Deegan and Thompson, 1985). Many of the paradigms developed for fishes of these estuaries (e.g. estuarine dependence among juveniles – Chapter 7) are most applicable to warm temperate, temperate and even cold water estuaries, but cannot be extrapolated without many qualifications to tropical estuaries. Similarly, generalizations about estuarine fishes, particularly ecological interactions, based on data from temperate estuaries must be treated with caution in relation to the tropics, where environmental conditions and habitats are different and species diversity is usually much higher.

## 1.4 TROPICAL ESTUARIES AND HUMANS

Estuaries have long been the focal point for much human activity. As the meeting place of sea and river, they have provided quiet and sheltered waters for harbours and historically gave the easiest or only access to the interior, for trade or settlement. For peoples living in the interior they have acted as a route to the sea for trade, fishing and migration.

Hence most large tropical or subtropical estuaries have port cities associated with them. Port cities brought industries, which in many areas pour their effluents into the estuary; together with discharges from shipping, this has resulted in the lower reaches of many estuaries becoming badly polluted. Industries bring population increases and growth of cities which discharge their sewage into the estuary. Despite these problems, most tropical estuaries are still vitally important as a source of fish to the local people and fishing plays an important role in the economy.

Fish and fisheries must be considered in relation to all these uses and abuses (Chapters 8 and 9). Conflicts of usage and whether industry or fisheries are more important, whether fisheries or waste disposal are more important; and whether estuaries can be 'cleaned up' so that the various users can coexist, provide a challenge to planners in many developing as well as developed countries. Within the fisheries sector, resource allocation between commercial, artisanal, subsistence and recreational interests can be a frequent source of conflict. How these issues are resolved – whether on economic, aesthetic or cultural values – requires decision making based at least partly on knowledge of the fishes. Unfortunately, even where the requisite knowledge is available, decisions are frequently made on political grounds and a balance between users may not be achieved. In developing countries, political instability, together with lack of infrastructure, has often made rational fisheries-management planning difficult and its implementation impossible. Nevertheless, despite all these problems, there is an increasing realization throughout the tropical world that the conservation and wise exploitation of estuarine fishes is vital, for feeding people, for recreation and for their heritage value. The main difficulty remains in translating thoughts and documents into practical actions.

# Chapter two

# Tropical estuaries and their habitats

## 2.1 INTRODUCTION

Tropical estuarine environments range in size from tiny seasonally flowing systems of 1–2 km to the estuaries of some of the world's largest rivers such as the Amazon and the Zaire. They also encompass extensive coastal lakes in Africa and Asia, and the reduced-salinity estuarine waters extending along the coast in parts of South East Asia, South America and Africa. There is thus a lack of uniformity among tropical estuaries in terms of size, depth, habitats and physical regimes. These factors and many others, such as the nature of the adjacent marine and freshwater habitats, can have a profound influence on the fish faunas and fisheries of tropical estuaries.

In this chapter the tropical estuarine environment has been divided into four broad categories to facilitate the description of those factors most relevant to their fish and fisheries. However, it must be emphasized that this is a division of convenience, and intermediates are found between the categories of open estuary, estuarine coastal waters, blind estuary and coastal lake (Fig. 2.1). Within each of these broad divisions, examples have been drawn from a diversity of types covering a wide geographical range.

## 2.2 OPEN ESTUARIES

The 'open estuaries' category includes all the larger, well-known estuaries of most of the medium and larger rivers of the tropics, and comes closest to the classical European and North American concepts of an estuary. These estuaries are never isolated from the sea and are subject to tidal influence with all its physical consequences of regular salinity, turbidity

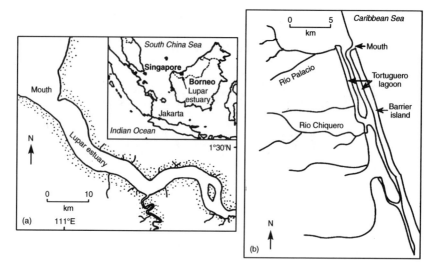

**Fig. 2.1** Representative examples of subtropical and tropical estuaries: (a) Lupar estuary, Borneo, a large open estuary (redrawn from Blaber *et al.*, 1996); (b) Tortuguero estuary, Costa Rica, a small open estuary; (c) Mhlanga estuary, KwaZulu-Natal, South Africa, a blind estuary (redrawn from Harrison and Whitfield, 1995); (d) St Lucia, KwaZulu-Natal, South Africa, a coastal lake system (redrawn from Blaber, 1976); (e) Terminós Lagoon, Mexico, a coastal lake (redrawn from Yáñez-Arancibia *et al.*, 1985).

and current flow changes, tidal prisms, haloclines and intertidal habitats. Extensive deltas occur at the mouths of most of the larger estuaries, such as those of the Zambezi and Niger in Africa, the Orinoco in South America, the Mekong in Asia and the largest of them all, the Ganges (Meghna) in India and Bangladesh. The size and depth of this sort of estuary has made them a focus of human activity, including ports and harbours with their associated industrial and city developments and consequent effluent problems. Fishery activities have increased in many of these estuaries, often developing from a subsistence base, through an artisanal phase to fully fledged commercial operations, to supply the demand from burgeoning cities. This development increases the pressure on fish resources and may lead to changes in the fish fauna and fish community structure.

Seventeen open estuaries have been chosen to illustrate the variety of size and form, as well as features they have in common, throughout the tropics and subtropics (Fig. 2.2). Their major features, particularly those of relevance to fish and fisheries, are summarized below.

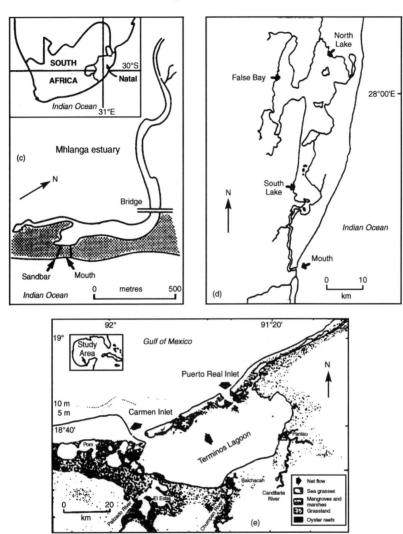

**Fig. 2.1** (*continued*)

## East Africa

The continental shelf along the East African coast varies considerably in width. Along much of the Zululand coast of South Africa it is less than 25 km wide, but broadens to more than 200 km in Moçambique before narrowing again to a width of about 50 km along much of the Tanzanian and Kenyan coasts. Similarly, the coastal plain gradually increases in width from Zululand northwards before narrowing again in northern

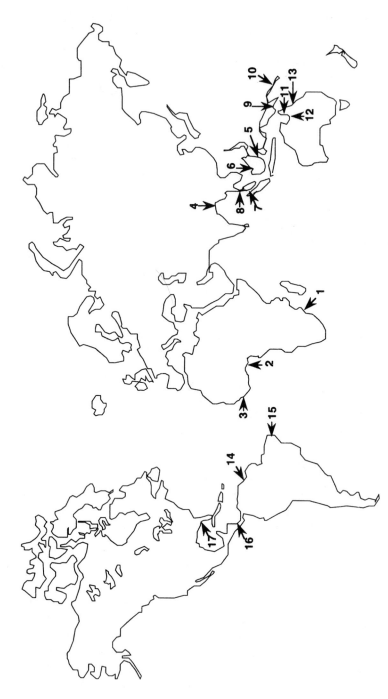

**Fig. 2.2** Location of open estuaries described in the text: (1) Morrumbene, (2) Niger, (3) Sierra Leone, (4) Ganges, (5) Mahakam, (6) Lupar, (7) Matang, (8) Ranong, (9) Fly, (10) Solomon Islands, (11) Embley, (12) Norman, (13) Alligator, (14) Orinoco, (15) Amazon, (16) Tortuguero, (17) Apalachicola.

Moçambique. The extent of the coastal plain and continental shelf play an important role in shaping the form of the estuaries and the composition of their fish faunas.

## Morrumbene estuary, Moçambique

This 20 km estuary opens into Inhambane Bay at 23°S on the coast of Moçambique. A comprehensive study of the physical nature and flora and fauna of this relatively quiet, mangrove-fringed estuary was undertaken by Day (1974). It is fed by five main tributary streams with a total drainage area of about 650 km$^2$. There is sufficient flow in the rivers, even in the dry season (April–November), to maintain a tidal salinity gradient in the estuary from 30–35‰ at the mouth to 0.5‰ 20 km from the mouth with a pronounced salt wedge. The maximum tidal range is 3.2 m and tidal currents in the mangroves reach $0.28 \, \text{m s}^{-1}$, and $1 \, \text{m s}^{-1}$ at the mouth on the ebbing tide. Turbidity varies greatly with the tide and substratum type. This latter varies along the length of the estuary, from poorly sorted sandy mud in the upper reaches to well-sorted sand at the mouth. The percentage of silt and organic matter declines towards the mouth.

The habitats are varied and range from an open channel 2–3 km wide in the lower reaches, which forms an extensive lagoon at high tide. This channel narrows to 50 m in the mangroves of the upper reaches where there is a maze of small creeks in the extensive mangrove forests (Fig. 2.3) There are large areas of intertidal sandflats in the lower reaches and mudflats adjacent to the mangroves.

## West Africa

The West African coast is rich in all categories of estuaries. Open systems range from the large Niger and Volta to medium-size estuaries such as the Sierra Leone and large numbers of smaller estuaries, particularly along the Liberia, Guinea and Senegal coasts. West of about 12° longitude the continental shelf increases in width to over 100 km.

## Niger delta

The Niger delta (5°N, 7°E) extends over more than 5000 km$^2$ and is the largest in West Africa. More than half of this area originally consisted of tidally inundated mangrove swamps but has now been reduced or modified by humans. There are definite wet and dry seasons which give rise to changes in river flow and salinity regimes. During the wet season (May–October), salinities fall to almost zero throughout the delta. River

**Fig. 2.3** A small mangrove creek in the Morrumbene estuary, Moçambique.

flow in the dry season (November–March) is still sufficient to keep the maximum salinity at the mouth to 28‰. Detailed studies of the Bonny estuary part of the delta have been carried out (Dublin-Green, 1990) and results showed that salinity, conductivity, pH, dissolved oxygen and alkalinity exhibited spatial and temporal variations. Minimum salinities of 10–24‰ and maxima of 19–31‰ were recorded in the late wet season and late dry season respectively, in both the upper and lower reaches of the estuary. On the basis of salinity, the Bonny estuary could be classified into three zones: upper reaches (mesohaline at all seasons except in the late dry season, salinity < 18‰), middle reaches (polyhaline at all seasons, salinity 18–27‰) and lower reaches (polyhaline at all seasons, salinity above 27‰).

### Rokel estuary, Sierra Leone

This estuary in Sierra Leone (9°35′N, 13°W) covers an area of about $160 \, km^2$ and has a tidal length of about 70 km. It has been the subject of a number of detailed studies (Watts, 1958; Bainbridge, 1960; Longhurst, 1983). The following summary is taken largely from Findlay (1978). It is a young drowned river valley and shallow except for a deep southern channel (30–40 m) close to the Freetown shoreline. The upper reaches merge into a network of creeks and channels fringed by large areas of mangrove swamps. It is tidal with oceanic water entering on a diurnal cycle. The climate of Sierra Leone is marked by a very distinct change between a very wet rainy season and a dry season. The tidal range of the estuary is low (spring 3.03 m; neap 2.28 m) and the harbour can be used all the time. The volume of fresh water entering the estuary is large during the rainy season and greatly reduced during the dry season. Consequently there is a marked fall in salinity during the rainy season (mean value 15‰ in August–October) and higher salinities (mean value of 33‰ in April–May) due to the marine influence prevailing during the dry season. Freshwater discharge in the wet season affects salinities for at least 100 km offshore (Longhurst, 1963; pp.26–27 below).

## Asia

The Asian region has some of the most extensive estuarine areas in the world, including the Ganges delta in the Bay of Bengal, the Irrawaddy delta of Burma, the Indus of Pakistan, the Mekong of Vietnam and the large estuaries of Borneo and Sumatra. All play vital roles as shipping highways leading far into the hinterland, as well as supporting productive fisheries. There is also a large number of smaller estuaries, particularly in India, Malaysia and Indonesia, that likewise support diverse fish faunas and productive fisheries.

### Ganges (Meghna)–Brahmaputra delta

The extensive delta of the Ganges–Brahmaputra rivers in India and Bangladesh forms the largest estuarine system in the Indian Ocean, extending over at least 6000 km$^2$ at the northern end of the Bay of Bengal. It consists of a series of branching and interconnected channels. In some areas, such as the Sundarbans, there are extensive mangrove forests, whereas along others, such as the Meghna in Bangladesh, there are vast flat areas of seasonally inundated mudflats and grazing land. In the main Meghna estuary in Bangladesh, the river is tidal at least as far as Dhaka, and estuarine salinities reach about 160 km from the mouth. However, the influence of salinity varies according to season, with flood tides and higher salinities reaching further in the dry season (May and June), but being overcome in the south-west monsoon (July to October) by huge amounts of freshwater – during this time the surface salinity is close to 0‰ at the mouth and in the sea adjacent to the whole delta. Turbidities follow a similar pattern, and are lower in the dry season and greatest during the monsoon, when both low salinities and high turbidities extend more than 50 km from the shore. Tidal currents are strong, generally in excess of $1\,\mathrm{m\,s^{-1}}$, but vary according to season and the relative strengths of marine and freshwater flows.

### Borneo estuaries

The extensive lowlands of Borneo contain a number of large estuaries such as the Kapuas and Mahakam in Kalimantan and the Lupar in Sarawak. They are characterized by high flood volumes, rapid tidal currents, high tidal ranges and extreme turbidities. Two contrasting examples are described.

The drainage basin of the Mahakam river covers almost one-third of Borneo and its delta (0°21'S to 1°10'S) on the east coast was classified by Dutrieux (1991) as a 'physically controlled environment'. The vast array of mangrove channels making up the delta constitute the estuary. This is influenced primarily by two water masses: a marine high-salinity (30‰), low-turbidity influx and a freshwater high-turbidity flow from the land. (Rainfall occurs throughout the year, averaging between 6 and 8 m per annum.) These water masses meet about 25 km from the sea, the distance varying with the tidal cycle (maximum tidal range 2.5 m). In the more marine areas of the delta, the channel edge is dominated by mangroves (mainly *Avicennia*, *Brugiera* and *Aegiceras*) behind which are dense stands of *Nypa*.

The Lupar river of Sarawak (Fig. 2.1), which drains northwards into the South China Sea, is more than 150 km long with an estuary of about 60 km (Blaber *et al.*, 1996). Like the Mahakam it is dominated by physical processes, but unlike the Mahakam it has not formed a delta. Its mouth gradually widens to more than 20 km as it enters the sea (Fig. 2.4). The maximum tidal range exceeds 5 m and tidal currents $> 1\,\mathrm{m\,s^{-1}}$. The high tidal exchange prevents the formation of a salt wedge and leads to the formation of a tidal bore (section 2.6). The high volumes of fresh water entering the whole system ensure that salinities seldom exceed 20‰, even at the mouth. Turbidities reach values of well over 1000 NTU* in the upper reaches of the estuary. The habitat is a very uniform, deep channel (up to 30 m) with steep banks and a muddy substratum. The banks are narrowly fringed by mangroves (mainly *Avicennia* ) and *Nypa* (Fig. 2.5) which broaden to form dense stands only at the confluences of tributaries.

---

*Nephelometric turbidity units. This standardized unit measures transmission of light through water; it cannot be directly translated to Secchi disc readings, which vary according to what is causing the turbidity.

**Fig. 2.4** The mouth of the Lupar estuary in Sarawak, Borneo.

### Smaller estuaries

Most of the smaller estuary systems of South East Asia have (or had) extensive and diverse mangrove forests and, like the larger estuaries, grade gradually into a shallow, low-salinity, turbid sea. Good examples include the Matang system of West Malaysia (4°48'N, 100°30'E) and the Ranong system (9°50'N, 98°35'E) between Thailand and Burma. Both these systems have been studied in detail, the former by Chong *et al.* (1990) and Sasekumar *et al.* (1994a,b) and the latter by a UNDP/UNESCO team (UNDP/UNESCO, 1991).

The Matang estuary system forms part of the Matang Mangrove Forest Reserve of 40 700 ha. It consists of a network of small interconnected estuaries, most about 12–15 km long, with mangrove islands and mudflats. Annual rainfall is about 4 m, most of it falling in the monsoon periods from October to December and from April to May. Turbidities are generally high and salinities vary according to season and tide. The tides range between 0.2 and 2.7 m and hence induce fairly strong tidal currents. A salt wedge occurs during neap tides but not spring tides when greater turbulence mixes the water column.

The Ranong system is smaller (11.5 km$^2$ of mangroves) and consists of

**Fig. 2.5** The mangrove palm, *Nypa*, growing along a creek of the Sarawak estuary, Borneo.

a series of interconnected waterways, one of the largest of which is the tidal mangrove estuary of the Kra Buri river. The area receives 4 to 5 m of rain a year, much of it falling in the south-west monsoon which lasts from May to October. There is a maximum tidal range of 4 m and tides have a strong influence on the estuary; even in the wet season, salinities 8 km from the mouth do not fall below 14‰ and there is a strong salt wedge. During the dry season, salinities throughout the estuary remain above 30‰. The distribution of the more than 12 species of mangroves differs according to topography, soil type and tidal influence.

## Australasia

### New Guinea estuaries

The island of New Guinea, comprising Papua New Guinea and Irian Jaya (the easternmost province of Indonesia), is rich in open estuaries. The central highlands of New Guinea receive some of the highest rainfall totals in the world with an average of 10 m a year, hence high freshwater river flows are typical. The largest systems are the Sepik and Mamberamo, which flow northwards, and the Fly, which discharges south into the Gulf of Papua.

The Fly River rises in the Star Mountains where there is heavy, year-round rainfall (13 m a year). It receives the flow from hundreds of large and small tributaries on its way to the sea 1000 km away. In terms of volume of water the Fly is only exceeded by the Amazon and Zaire (Roberts, 1978). The estuarine portion of the river is confined mainly to the delta, which is 90 km wide at the mouth, and salt water intrudes only about 100 km. Even at the mouth of the delta, surface salinities do not exceed 15‰. Mangroves extend up the delta as far as the salt intrusion, but there are numerous mangrove-covered islands in the delta (Robertson *et al.*, 1991). Tidal bores affect the lowermost 200 km and tidally induced fluctuations in water level are an important factor affecting fish distribution. Turbidities are very high and there is frequently strong wave action.

### Solomon Island estuaries

In contrast to the estuaries of New Guinea, those of the Solomon Islands to the east are mainly small and isolated from one another by fringing coral reef lagoons. They are very quiet, have relatively clear water and are slow flowing. Thirteen were studied by Blaber and Milton (1990); all are lined by extensive intertidal stands of mangroves and open into coral reef lagoons. They vary in length from 1 to 14 km and in depth from 1 to 5 m and are tidal, but the tidal range is small and does not exceed 1 m. Salinity varies from near-fresh water to full sea water depending on the tide, distance from the mouth and the freshwater inflow. Temperatures are fairly constant from 27 to 32 °C. In general, turbidities are relatively low (1–18 NTU) and the water clear, but in all cases turbidities are higher than in adjacent coral lagoons ( < 0.1–0.5 NTU). Current speeds are low and do not usually exceed $0.08 \, \mathrm{m \, s^{-1}}$.

Habitats consist of substrata with soft muds and an intertidal fringe of tall (up to 20 m) mangroves comprising mainly *Rhizophora apiculata*, or harder sands with a mangrove of mainly *Brugiera* species. Estuaries with hard sand substrata are invariably clogged by large numbers of mangrove logs, branches and fallen trunks, which provide a highly structured habitat.

### Australian estuaries

#### Embley estuary
The 50 km long Embley estuary in the eastern Gulf of Carpentaria was studied from 1986 to 1991 by Blaber *et al.* (1989, 1990a), Brewer *et al.* (1989, 1991) and Salini *et al.* (1990). It is almost entirely mangrove lined, mainly by *Rhizophora stylosa, Brugiera gymnorhiza, Ceriops tagal* and

*Avicennia marina.* Extensive intertidal mudflats and areas of seagrass (mainly *Enhalus acoroides*) in shallow water are found on the south side of the lower reaches. Intertidal sandy mud beaches occur on the north side of the lower reaches and on both sides at the mouth. Open-water channels and small mangrove creeks occur throughout the estuary, but are the only habitats in the middle and upper reaches. There is a maximum tidal range of about 2.6 m. Surface water temperatures in the lower reaches vary from 25 °C in August to 32 °C in February. In the middle and upper reaches the lowest temperature recorded was 27°C. Surface salinity is at about seawater level (35‰) throughout the estuary during the dry season (August–November) but falls markedly in the wet season, particularly in the middle and upper reaches. Turbidity is generally inversely related to salinity, with maximum values in the middle and upper reaches during the wet season. Turbidity is low over much of the estuary during the dry season and in the lower reaches all year. It should be noted, however, that the lowest values in the estuary (2 to 5 NTU) are higher than the levels recorded in the adjacent Albatross Bay (usually < 2 NTU).

### Norman estuary

In contrast to the Embley, this 60 km long estuary in the south-eastern Gulf of Carpentaria has less diverse habitats. Open-water channels predominate – the banks are steep and hence the intertidal mud areas are small and are largely restricted to the lower reaches. The extensive fringe of mangroves (mainly *Avicennia marina*) is consistently narrow ( < 20 m wide) (Fig. 2.6). The maximum tidal range is about 4 m. The Norman river has strong seasonal fluctuations in discharge because the only rains fall in the monsoon from December to April, and hence although tidal currents are similar throughout the year, in the dry season salinities are high (29–34‰) and turbidities relatively low (18–44 NTU) and in the wet season the inverse is found (salinity 13–19‰; turbidity 85–2656 NTU). Temperatures in wet and dry seasons are similar, ranging from 27 to 34 °C (Staples, 1980; Blaber *et al.*, 1994a).

### Alligator Creek

Alligator Creek (19°21′S, 146°57′E) is a short ( ~ 4 km) mangrove-lined estuary 10 km east of Townsville in north-east Queensland. This creek was studied in detail by Robertson and Duke (1987, 1990) from 1985 to 1987, and is typical of those in this region of the dry tropics. The dominant fish in Alligator Creek are also abundant in other estuaries along the north-east coast of Queensland (Blaber, 1980; Robertson and Duke, 1987, 1990). The mangroves are dominated by *Rhizophora stylosa*, *Ceriops tagal* and *Avicennia marina*. The annual range of mean water temperatures in the main channel of the estuary is 21–31 °C. Salinities range from 30‰ in the wet

**Fig. 2.6** A meander of the Norman estuary in the south-east Gulf of Carpentaria, Australia, showing the narrow mangrove fringes along the main channel and tributaries. Seasonally flooded salt pans lie behind the mangroves.

season to 38‰ in the late dry season. Rainfall is very seasonal with most falling from December to May, although there is a marked interannual variation. The mean annual rainfall for Townsville is 1215 mm. Tides are semi-diurnal and the maximum tidal range is 3.5 m. More details of the hydrology of estuaries in this region are given by Wolanski (1986) and Wolanski and Ridd (1986).

Three main habitat types were identified in this estuary: the main channel with its shallow marginal mudbanks; small tributary creeks (depth at low tide ~0.5 m); and the mangrove forests that are flooded at high tide.

## South America

### Orinoco estuary

The Orinoco river rises in the highlands of Venezuela and Colombia and flows over 1500 km before forming an estuarine delta on the north-east coast of South America (between 8°30′N and 10°N) (Cervigón, 1985). This delta occupying an area of some 22 500 km², with its apex about 150 km from the sea, is made up of a complex of interconnected large and small channels. Estuarine waters stretch for a maximum of about 80 km from the sand bars at the mouth. The region experiences marked wet (April–October) and dry (November–March) seasons, with freshwater inflow reaching a peak in July and August. Consequently salinities are least in the wet season, when estuarine waters penetrate only about 60 km up river. The tidal range is 1–2 m and tidal effects (not salinity) are felt up to 200 km from the sea. Unusually, the salinity may be affected from the seaward end because during April, northward-flowing waters from the Amazon mixing zone pass within 50 km of the mouth of the Orinoco when its current speeds are from 15 to 33 cm s$^{-1}$. However, during autumn, the increased discharge of the Orinoco displaces the fresher Amazon waters seaward (Moore and Todd, 1993). The interrelationships with waters of Amazon origin lead to a complex stratification in the dry season with three layers: a layer of fluid mud near the bottom, a layer of low salinity and low turbidity at the surface, and a salt wedge of high salinity and low turbidity in between. In addition, about twice as much sediment is supplied by the Amazon to the Orinoco delta as by the Orinoco river (Eisma *et al.*, 1978).

The habitats of the Orinoco consist largely of open-water channels and tributaries with muddy substrata, and intertidal mudflats, all fringed by extensive intertidal mangrove forests dominated by *Rhizophora mangle*.

### Itamaraca estuary

This much smaller system lies on the north-east coast of Brazil (7°45′S, 38°50′W) and consists of a main channel of 22 km (the Santa Cruz) into which four rivers discharge (Paranagua and Eskinazi-Leça, 1985). The Santa Cruz flows into the sea through two sand bar openings separated by an island. Tidal flows from the two openings meet in the middle of the channel and in this dynamic environment there are no haloclines or thermoclines. Salinities vary from 5‰ to 35‰ depending on season but salinities at the mouths never fall below 29‰ even in the wet season (March to August). This system has a variety of habitats: in the high-current areas near the sea, the channels and intertidal areas have a sandy substratum, in the middle and upper reaches muddy substrata predominate, while mangroves occupy the channel banks.

Other well-studied open estuaries of tropical South America include the Sinnamary, Maroni and Mahury of French Guiana (Boujard and Rojas-Beltran, 1988). They are medium-size systems (e.g. the Sinnamary,

5°30'N, 53°00'W, has a drainage basin of 6 565 km) with changes in flow and salinity related to the wet and dry seasons.

## Central America

One of the best-studied estuaries of this region is the Tortuguero, which is located on the Caribbean coast of Costa Rica (10°37'N, 83°33'W) (Caldwell *et al.*, 1959; Gilbert and Kelso, 1971; Nordlie and Kelso, 1975; Winemiller and Leslie, 1992). It is formed from the junction of two rivers and consists of a 20 km long lagoon-like system parallel to the coast that opens to the sea at its northward end (Fig. 2.1). The lagoon averages 7.5 m in depth. The region is very wet, with an annual average rainfall of 5 m, much of it falling in two wet seasons, July–August and November–January. However, no month receives less than 50 mm of rainfall. Differences between high and low tide rarely exceed 0.25 m but as much of the lagoon is below sea level, tidal influences are strong throughout the lagoon area. A salt wedge (8.7–10.5‰) has been recorded deeper than 5 m throughout the lagoon in both the wet and dry seasons (Nordlie and Kelso, 1975). High-tide surface salinities in the wet season are low (0.1‰) but increase to only about 1‰ in the dry season. Temperatures vary little throughout the year (27–30 °C) and turbidities are relatively low, increasing only during periods of heavy run-off. The environment is not very diverse, much of the substratum is sandy and there is little intertidal area. However, rooted macrophytes such as *Najas* and *Potamogeton* grow on the muddier areas in the lower reaches and rafts of floating macrophytes, particularly *Eichornia crassipes*, sometimes cover a large surface area in the dry season, thus providing additional habitats.

## Gulf of Mexico

Most of the tropical estuaries of Mexico and the subtropical systems of the United States coast of the Gulf of Mexico fall into the category of estuarine coastal waters or coastal lakes (sections 2.3 and 2.5). However, in Florida there are a number of open estuaries. One of the best-studied is the Apalachicola (29°50'N, 85°00'W) in the north-east Gulf (Livingston, 1982). It has a high sediment load, much of which is deposited in the Apalachicola delta and estuary, including Apalachicola Bay. The estuarine waters are well mixed and there is a salinity gradient between sea and river. Just east of Apalachicola is the St Marks River estuary, which was the subject of a detailed study by Subrahmanyam (1985): it is a small, shallow sandy estuary in which saltwater influence at high tide penetrates 6 km. The overall salinity range is from 1‰ to 35‰ according to distance from the mouth and season. Water temperatures also fluctuate considerably with values as low as 10 °C in January – indicating that this system might better be considered as warm temperate, rather than subtropical, as is also the case with many other southern US estuaries.

## 2.3   ESTUARINE COASTAL WATERS

Nowhere is the problem of definitions and boundaries of estuaries and estuarine fish faunas better illustrated than in tropical coastal waters. Many of the larger open estuaries, particularly those of the Indian subcontinent and South East Asia, grade almost imperceptibly into the adjacent sea. The effects of the discharge from the Amazon are felt up to 400 km from the mouth and those of the Orinoco up to 50 km away from the mouth. The shallow nature of such tropical coastal waters and their physical conditions of lowered salinity and high turbidity make them at least partly estuarine in character, particularly as regards their fish faunas. The cut-off point between these waters and true marine conditions may best be defined by the fauna (Pauly, 1985) rather than by physical factors, but estuarine conditions seldom extend beyond depths of about 15–20 m, often related to the thermocline depth. For example, off Guyana these waters extend 40 km from shore (Lowe-McConnell, 1987), in the Gulf of Carpentaria about 10 km from shore (Blaber *et al.*, 1990b), in the South China Sea 20–50 km from the Borneo coast and in the Bay of Bengal > 100 km from the mouth of the Ganges. For more detailed descriptions and analyses of the shallower tropical continental shelves, readers are urged to consult Longhurst and Pauly (1987) or Pauly and Murphy (1982).

### Gulf of Carpentaria, Australia

The Gulf of Carpentaria in northern Australia is a shallow (maximum depth 60 m) bowl-shaped tropical sea. Many large estuaries, such as the Embley and Norman (section 2.2), flow into it and large areas have depths of less than 20 m. The fishes and associated physical conditions of the shallow estuarine coastal waters of part of the north-east coast were studied by Blaber *et al.* (1995a). In this region, salinity varies little throughout the year, ranging from 38‰ in the dry season to 29.4‰ in the wet season. Temperature also varies little: the lowest temperature, 26.2 °C, was recorded in the dry season (August) and the highest temperature, 30.8 °C, in the wet season (January). The lowest recorded turbidity was 0.6 NTU and the highest 26.0 NTU. Turbidity generally varies both seasonally and according to the wind direction and strength, but is almost always lower than that in the adjacent Embley estuary. It is also lowest in the dry season when the coast is sheltered from the prevailing south-east winds. Wind is seldom a significant factor in smaller estuaries, but is important in such coastal waters as well as in larger coastal lakes and very large estuaries. When the wind is from the south-east quarter, the east coast of the Gulf of Carpentaria is completely sheltered and the water calm. Wind has the most influence, even at low speeds, when it is from a westerly direction when the fetch across the Gulf is 500 km: waters are then turbulent with waves up to 1 m in height. The lowest recorded wind speed is $16 \, \text{km h}^{-1}$ and the highest $47 \, \text{km n}^{-1}$ (disregarding cyclones), regardless of direction. The greatest tidal range of 2.5 m occurs during spring tides and the smallest tidal range of 0.1 m during neap tides. Hence conditions in these coastal waters may be much more turbulent than are normally experienced in estuaries.

## Bay of Bengal

Much of the northern part of the Bay of Bengal is estuarine and salinities of less than 20‰ prevail at the end of the south-west monsoon in September and October, caused mainly by massive outflows of fresh water from the Ganges delta. This outflow also brings much sediment and the shallow-water substrata are predominantly muddy and the water highly turbid. As the monsoon changes to the north-east, the current circulation pattern in the bay changes and the reduced-salinity water moves eastwards and penetrates down the coasts of Bangladesh and Burma. These patterns have seasonal effects on the important fisheries in this area.

## San Miguel Bay, Philippines

This bay in the northern Philippines has been studied in detail by Pauly (1985). It is about 20 km wide and 30 km from north to south and although wide open to the sea, is essentially estuarine in character. It is lined by mangroves. It receives a freshwater inflow from the Bicol River and there is a salinity gradient from nearshore to the open sea. Shallow-water substrata are mainly muddy but these change to sand in the deeper waters towards the sea. In these deeper waters there are also rocky outcrops and coral reefs.

## Gulf of Mexico

Many of the shallower waters of the Gulf of Mexico have reduced salinities and relatively high turbidities and hence qualify as estuarine coastal waters. Most are the sites of highly productive commercial fisheries. The tropical waters of Campeche Sound (18°50′N, 91°50′W) outside of

Terminós Lagoon (Fig. 2.1) in Mexico have been extensively surveyed by Yáñez-Arancibia *et al.* (1985) who found that there is a seasonal pattern of temperature, turbidity and salinity which is correlated with climatic conditions. The shelf in this area can be divided into western and eastern sections based on the influence of outflows from Terminós Lagoon. The west, which receives the lagoonal discharge, has high turbidities, lime-clay sediments and high organic content, whereas the east is more marine in character with clearer water, calcareous sediments and seagrasses growing on the bottom. In both west and east zones, temperatures range from 22 °C to 29 °C and salinities from 32‰ to 38‰. Rainfall in this area of Mexico is high (3–4.5 m per annum) compared with areas to the north, especially the more arid subtropics of the United States.

The estuarine systems of the subtropical United States coast of the Gulf of Mexico are very extensive and are well summarized by Gunter (1967). There are 33 bay systems or sounds of various shapes covering over 800 km$^2$. Mobile Bay (30°30′N, 88°00′W) is stated by Gunter (1967) to be the best general example of a large bay on the north coast. It is oblong, about 40 km long and separated from the open waters of the gulf by a relatively narrow opening. There is a salinity gradient from the Mobile River mouth through the bay to the sea with full seawater conditions usually prevailing around the mouth, but these decrease rapidly to near freshwater at the mouth of the Mobile River. During dry periods, saline waters penetrate the river and conversely during times of floods, the whole bay can become more or less fresh. Turbidities are usually high, partly through river discharge and partly due to wind-induced turbulence which keeps fine sediment in circulation and stirs up the muddy substratum. The coastal waters west of Mobile Bay are affected by the outflow of the Mississippi River, which apart from reducing salinities, dumps between one and two million tonnes of sediment a day. Along the west coast of Florida, there are a number of estuarine bays of which Tampa Bay (27°40′N, 84°50′W) is the largest and one of the best studied (Killam *et al.*, 1992).

## West Africa

Reduced salinities exist along much of the Gulf of Guinea coast during the wet season (Longhurst, 1962, 1963; Longhurst and Pauly, 1987) and the turbidity and muddy substrata of coastal waters qualifies them as essentially estuarine in character. The greatest reduction in salinities occurs in the Bight of Benin in connection with the outflows from the Niger delta and further west off Liberia and Sierra Leone. Longhurst (1963) recorded a salinity gradient off Cape Sierra Leone going from 25‰ 16 km offshore to 33‰ at the shelf edge at 120 km. The coastal estuarine waters are separated from deeper, cooler water by a thermocline, usually at about 20 m (Longhurst and Pauly, 1987).

## 2.4 BLIND ESTUARIES

The most useful definition of blind or closed estuaries is that of Day (1981) who described them in the following terms "These are estuaries which are temporarily closed by a sand bar across the sea mouth." (Fig. 2.7) "At such times there is no tidal range and thus no tidal currents. Freshwater enters from the river and the circulation is dependent on the residual river current and the stress of the wind on the water surface. According to the ratio between evaporation and seepage through the bar on the one hand, and freshwater inflow plus precipitation on the other, the salinity will vary."

Estuaries in this category are usually relatively small, both in length

**Fig. 2.7** The sandbar of the blind Mdloti estuary in KwaZulu-Natal, South Africa, showing contrasts between the turbulence of the sea and estuary and the relative differences in water level.

and catchment, and are characteristic of areas of low or highly seasonal rainfall. Although best developed along the warm temperate and Mediterranean climate coasts of southern Australia and South Africa (Potter *et al.*, 1990), they do extend into the subtropics and tropics of both Africa and India. In addition, equatorial examples occur in South East Asia.

Although small, blind estuaries often have very diverse fish faunas and may play an important ecological role in the life history of many marine fishes, these estuaries have been less subject to port and industrial development than open estuaries because they are relatively small and often cut off from the sea. In addition, their fishery potential is usually limited and most fishing activities are of a subsistence nature in developing countries, or recreational in developed countries. Two forms of development are, however, having an impact on such systems. Firstly, in areas such as the Tamil Nadu coast of India, they are being converted to intensive aquaculture; and secondly, along the Queensland, KwaZulu-Natal and Florida coasts, many have been used as the basis for residential canal estates (section 9.6).

### East Africa

#### *Mhlanga estuary, KwaZulu-Natal*

This estuary, which lies 17 km north of Durban on the subtropical KwaZulu-Natal coast (29°42′S, 31°05′E) (Fig. 2.1), is a good example of the large numbers of small blind estuaries in this region. It is cut off from the sea for much of the year by a sand bar, opening only when flooding after rains breaches the barrier and causes the level to drop suddenly as the water flows into the sea. The following details have been taken from the extensive studies of Mhlanga by Whitfield (1980a,b,c) and Harrison and Whitfield (1995). The river is 28 km long and the catchment about 118 km$^2$. The banks are lined by *Phragmites* reeds (backed by coastal forest). The mouth of the estuary opens frequently during the summer (wet season, September–March) and the periods that the mouth is open vary from 2 to 25 days. When the rate of outflow drops, the mouth soon closes due to longshore drift of sand. Depths range from less than 0.5 m to 2.2 m with greater depths occurring during the closed period in the dry season. Water temperatures range from 14 °C in July to 30 °C in February. Salinities range between 0‰ and 34‰ but are generally less than 10‰. During periods when the mouth is open, higher salinities occur and tidal influences cause some vertical stratification. Turbidity is linked to river flow and estuary mouth condition; during closed periods the water is clear with values of <1 NTU. After heavy rains, suspended sediment increases and there is a turbid outflow through the mouth – at this time values of up to 62 NTU have been recorded over periods of a few days. An important feature of blind estuaries is that their bottom waters may become slowly deoxygenated during the closed phase due to lack of circulation and the decomposition of organic matter. Values as low as $2.9\,\text{mg}\,l^{-1}$ have been recorded in the bottom waters of Mhlanga estuary, and more polluted systems may become almost anoxic.

## West Africa

The hydrology of a number of blind estuaries that occur along the coast of Ghana has been described by Kwei (1977). They are small systems and most are kept open artificially by concrete culverts to prevent flooding of the main coastal road. One that is not kept open is the Mukwe ($0°04'$N, $5°38'$W). It is 2 km long and is usually closed by a sand bar for ten months of the year. When the bar is breached in the rainy season (February–June), it becomes tidal and is extremely shallow at low tide. When closed the maximum depth is only 1.5 m. Temperatures vary between 26 and 30 °C. Salinities reach their lowest (about 3‰) when the mouth is open, but after closure rise steeply due to evaporation, and the system is hypersaline for much of the year. In the middle of the dry season (January), salinities may exceed 70‰ and at this time dissolved oxygen levels fall as low as $2\,\text{mg}\,l^{-1}$.

## India

Blind estuaries are common along both east and west coasts of India. Most are small, and good examples include the Manakkudy estuary, about 8 km north of Cape Comorin in Tamil Nadu ($8°10'$N, $77°40'$E) and the Talapady lagoon, a blind estuary, near Mangalore ($12°55'$N, $74°55'$E). In both these systems the mouth opens only during the rainy season, and at that time a normal estuarine salinity gradient is established. When the bar is closed, the salinity varies between 7‰ and 13‰. Substrata are mainly sandy or sandy silt with high concentrations of organic carbon, possibly due to decay of mangrove foliage during the closed periods following the monsoon.

## South East Asia

In Sarawak, for a distance of 160 km along the north coast of Borneo between Igan and Bintulu, there are about twenty small blind estuaries, two examples of which are the Penipah and the Penat. They drain the peat swamp forest lowlands, have small catchments and most are less than 10 km long. They are closed during the dry season (April–

**Fig. 2.8** The breaching of the sandbar of a small blind estuary in Sarawak, Borneo, during the wet season.

November) but open during the monsoon when strong outflows break the sand bar at the mouth (Fig. 2.8). Their waters are highly acidic and stained a deep 'tea' colour, and during the dry season are mainly fresh with only residual salinities in the lower reaches. Wave action and currents along the exposed shoreline build up sand bars across the mouths as outflow decreases at the end of the wet season. The tidal range along this part of the coast is only about 2 m. The development of sand bars further west along the coast is prevented by the greater tidal range – up to 6 m. Vegetation along these estuaries consists mainly of *Nypa* and swamp forest.

## 2.5 COASTAL LAKES

Although lacustrine in appearance, many of these large bodies of water behind tropical shorelines on most continents are estuarine in character. Their faunas are mainly marine or estuarine, and most coastal lakes have some form of connection to the sea. The form and constancy of this connection largely determines the prevailing salinities and the nature of the fish fauna. The range of habitats is also determined by this connection as well as by the bathymetry, which is a result of local geological history and processes. They are sometimes referred to as coastal lagoons (e.g. Barnes, 1980) but the term 'lake' is used here to avoid confusion with many of the small lagoons found at the mouths of blind estuaries. What makes these systems unique is their relatively large surface area. Many have been extensively modified by humans and most have important fisheries, recreational or conservation value. Important chains of coastal lakes occur along the south-east and West African coasts, in India, in the Gulf of Mexico and in northern South America.

The main features of the fifteen examples (Fig. 2.9) drawn from throughout the tropics are described below. They are very different from one another. Their physical characteristics vary from freshwater to hyper-saline, from highly turbid to clear, from depths of about 1 m to over 40 m, and from high to low habitat diversity. What they all have in common is an estuarine fish fauna derived primarily from the sea.

### East Africa

A chain of coastal lakes stretches from Lagoa Poelela in Moçambique to St Lucia and Richard's Bay in KwaZulu-Natal (Fig. 2.10). St Lucia, which has an area of 300 km$^2$, is the largest estuarine system in East Africa. To quote Barnes (1980) they "... grade into other coastal habitat types: into semi-enclosed marine bays, into freshwater lakes, and into estuaries: and some of these intergradations may represent stages in an evolutionary sequence." The south-east African coastal lakes show all stages in this evolutionary sequence and their present states are a result of their geological history, degree of isolation from the sea, the number of inflowing rivers, or if they are totally isolated, the length of time they have been cut off. Their salinity varies from fresh water to values greater than sea water, and is controlled mainly by the nature of the marine connection, the amount of fresh water and their surface-to-volume ratio. The main physical features of the four largest lakes and their fish faunas are shown in Table 2.1. Some such as Lake Nhlange (1–4‰) and Lagoa Poelela (4–7‰) have relatively stable long-term salinities, whereas others, notably St Lucia, undergo extreme salinity fluctuations related to 5–7-year climatic cycles (Blaber, 1988). Salinities during wet cycles are usually between 10‰ and 25‰, but in drought periods may reach 102‰ (Day, 1981). The St Lucia system, which consists of three interconnected lakes draining to the sea through a 21-km-long, narrow estuary, is at sea level and receives the inflow of

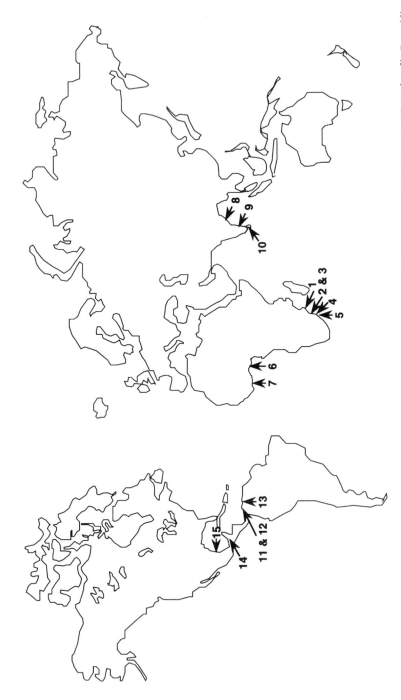

**Fig. 2.9** Location of coastal lakes described in the text: (1) Poelela, (2) Nhlange, (3) Mpungwini, (4) St Lucia, (5) Richard's Bay, (6) Lagos Lagoon, (7) Ébrié Lagoon, (8) Lake Chilka, (9) Lake Pulicat, (10) Negombo Lagoon, (11) Bahía de Cartagena, (12) Ciénaga de Tesca, (13) Ciénaga Grande de Santa Marta, (14) Terminós Lagoon, (15) Laguna Madre.

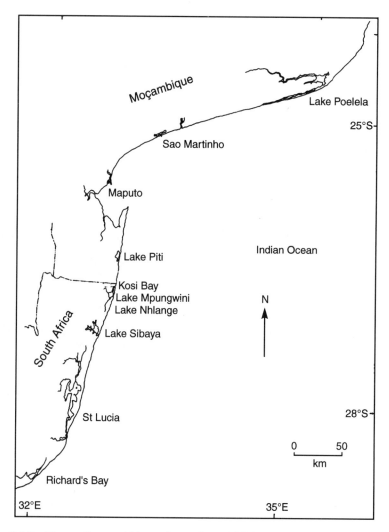

**Fig. 2.10** The south-east African chain of coastal lakes.

four rivers with a catchment of 9000 km² (Fig. 2.1). Most rain falls in the summer in the coastal area and the inland region is dry in winter. St Lucia's surface area fluctuates between 420 and 215 km² and has a mean depth of less than 1 m. Tidal effects penetrate 14 km up the narrows. The St Lucia system has changed substantially since the 1930s due to bad farming techniques within its catchment. Large amounts of sediment have been deposited in the estuary mouth, causing it to close

*Tropical estuaries and their habitats*

**Table 2.1** The main physical features (after Allanson and van Wyk, 1969; Hill *et al.*, 1975; Millard and Broekhuysen, 1970) and dominant fish faunas of four of the largest south-east African coastal lakes

| | Lake | | | |
|---|---|---|---|---|
| Characteristic | Poelela | Nhlange | Sibaya | St Lucia |
| Isolation | 75 km tenous link with sea | Estuary linked | Isolated | Estuary linked |
| Depth (m), max and (mean) | 24 (13.7) | 31 (7.2) | 40 (13) | 2 (<1) |
| Area (km²) | 65 | 31 | 65 | 300 |
| Salinity (‰) | 6.5–8.0 | 0–5.0 | 0 | 20–110 |
| Temperature (°C) | 22–23* | 19–32 | 19–32 | 14–38 |
| Turbidity (Secchi, m) | 7 | 1–2 | 3 | <0.5 |
| Dominant fish group | Euryhaline freshwater | Euryhaline marine | Freshwater and estuarine relict | Euryhaline marine |
| Dominant iliophagous species | *Oreochromis mossambicus* | *Mugilidae and Chanos chanos* | *Oreochromis mossambicus* | Mugilidae |
| Dominant filter-feeding species | None recorded | *Gilchristella aestuarius* | *G. aestuarius and Hepsetia breviceps* | *Thryssa, Hilsa, Gilchristela* |
| Dominant invertebrate-feeding species | *Glossogobius, Croilia, Pseudocrenilabrus* | Gerreidae, Pomadasys, Rhabdosargus | *Glossogobius, Croilia, Pseudocrenilabrus* | Wide variety |
| Dominant piscivores | *Tylosurus leiurus* | Sphyraenidae, Carangidae | *Clarias gariepinus* | *Argyrosomus and Elops* |
| No. species of fish recorded | 12† | 37‡ | 18§ | 108¶ |

* Only cool season temperatures available.
† From Hill *et al.* (1975).
‡ From Blaber and Cyrus (1981).
§ From Bruton (1980).
¶ From Whitfield (1980a).

during low-flow periods, although today it is kept open artificially by dredging. Poelela, Nhlange and Sibaya have deep basins and mainly sandy substrata, and hence relatively low turbidities. In contrast, St Lucia is very shallow over most of its area and has predominantly muddy substrata and high turbidities: values in excess of 1000 NTU have been recorded at the northern end. Considerable wave action may develop on these coastal lakes and wave-washed sandy beaches are a feature of Poelela, Nhlange and Sibaya. The marginal vegetation consists mainly of extensive *Phragmites* reed beds, although in St Lucia and Richard's Bay there are large areas of mangroves where salinities are higher.

## West Africa

There are two important chains of coastal lakes along the West African coast. The first stretches from the Niger delta westwards as far as the Volta estuary, and the second is on the Ivory Coast.

The largest system extends 200 km from Cotonou in Benin past Lagos to the western edge of the Niger delta. It opens to the sea only at Cotonou and Lagos. Where large rivers enter the system it is wide and shallow, but where there are few rivers the system is narrow and elongate. In the west there are large sheets of shallow water, especially in the wet season when salinities fall to almost zero. Where brackish conditions prevail for most of the year there are stands of mangroves such as *Rhizophora* and *Avicennia*.

The hydrological features of Lagos Lagoon were studied by Hill and Webb (1958): it shows both diurnal salinity fluctuations due to tidal effects and much greater seasonal changes caused by the influx of fresh water in the rainy months. During the dry season, the salinity is high (25–35‰) and brackish conditions spread inland for about 35 km. In the rainy season (April–October) the salinity falls and almost freshwater conditions prevail (0–10‰). Sediments vary, with shallow shoal sands that are well sorted and contain little or no silt in the north and north-eastern rims of the lagoon, and muds in central areas, while the western industrialized parts are highly modified by humans and receive a complex mixture of domestic and industrial wastes (Dufour, 1987). A less polluted coastal lake is Epé Lagoon (6°29′N, 3°30′W). This lake is fed by the Oshun River and lies between the Lagos and Lekki Lagoons. The Lagos, Epé and Lekki Lagoons open into the Gulf of Guinea via the Lagos Harbour. Epé Lagoon has a surface area of about 225 km$^2$ with a maximum depth of 6 m. However, a large area of the lagoon is relatively shallow with minimum depth of about 1 m (Balogun, 1987).

The main coastal lake on the Ivory Coast with Abidjan at its centre extends for over 100 km and is fed by several large rivers. The fauna and physical characteristics of Ébrié Lagoon have been extensively studied by ORSTOM in Abidjan (Albaret and Écoutin, 1989, 1990; Charles-Dominique, 1982; Durand and Chantraine, 1982).

## India

Two of the best examples of coastal lakes in India are Lakes Chilka (19°50′N, 85°30′E) and Pulicat (13°30′N, 80°10′E) on the east coast. Both support important fisheries. The former is the largest brackish-water lagoon in India, being 65 × 16 km with a surface area of more than 1000 km$^2$. It receives a number of rivers, the largest of which, the

Mahanadi, enters at the north-east end. It is shallow, only about 2 m over most of its area, although this can increase to 3–4 m in the wet season. In this respect it is quite similar to the St Lucia system of south-east Africa. One zigzag outer channel 35 km long connects the main body of the lake with the Bay of Bengal. The reduction in the area of Lake Chilka from 1914 to 1989 can be attributed mainly to heavy siltation through the Daya, Bhargavi and Nuna rivers, distributaries of the Mahanadi River, and obstructed movement of sedimented water through the constricted outer channel, combined with littoral transport of sediments in the coastal region. There is also frequent shifting of the inlet due to longshore drift and localized coastal erosion (Kumar *et al.*, 1989). During the September monsoon, most of the lake becomes almost fresh, but with the advent of the dry season, salinities reach seawater values. There is a distinctive and pronounced salinity gradient along the north–south axis of the lagoon. The southern zone, in general, shows a lower salinity range during the year and seems to be least affected by seawater and freshwater incursions. The opening of the Palur Canal to the sea in 1964, coupled with lesser river discharge in 1965, appears to have led to comparatively higher salinities in the southern zone. The river discharge into the lake is influenced mainly by hinterland rainfall rather than by coastal rainfall (Sarkar, 1979).

Lake Pulicat of 280 km$^2$ is also only about 2 m deep, is connected to the sea by a single channel and has a similar salinity regime. There are a series of coastal lakes on the west coast of India, often referred to as 'backwaters' – examples include the Cochin backwater (Vembanad) (9°30′N, 76°20′E) and Parur backwater. They are likewise shallow, with salinities that fluctuate between sea water and almost fresh water according to rainfall.

Sri Lanka has a number of coastal lakes, of which Negombo Lagoon (7°10′N, 79°50′E) is the best studied (Wijeyaratne and Costa, 1987a,b, 1988; Samarakoon, 1991), and one of the most important for fisheries. It is about 12 km long and 3–4 km wide and opens to the sea at the north end through a narrow channel. Tidal ranges are small and the narrow channel chokes the tidal amplitude, which is only 7 cm compared with approximately 25 cm in the sea. The tidal retention time in the system is influenced both by freshwater supply and by tides, but is long and averages one week (Wickbom, 1992). Salinity fluctuations are regular and are determined by rainfall in the monsoons, with almost fresh conditions prevailing in November–December and in May. Values approach sea water at the peak of the dry seasons in February–March and in August–September (Silva and De Silva, 1981). The habitats consist of muddy and sandy substrata and some fringing mangroves, although the lake is much modified by human activity.

## South America

Three good examples of coastal lake systems in tropical South America are the Bahia de Cartegena (82 km$^2$, 10°20'S, 75°30'W), Cienaga de Tesca (22.5 km$^2$, 10°25'S, 75°30'W) and Cienaga Grande de Santa Marta (480 km$^2$, 11°00'S, 74°30'W) on the Caribbean coast of Colombia. A bibliography of the large number of studies on these systems is provided by León and Racedo (1985), who also summarize the main features of each system. Cienaga Grande is shallow with an average depth of 2.3 m whereas Bahia de Cartegena has depths of up to 15 m. Water temperatures in all three fluctuate between 27 and 32 °C and salinities between 0‰ and 35‰. The variations in salinity are controlled primarily by freshwater inflow. Bahia de Cartegena and Cienaga de Tesca lie in an area that receives an average of 920 mm of rain a year with a wet season from July to December. Cienaga Grande is in a more arid area that has only between 400 and 800 mm of rain a year, but it receives freshwater input from a number of rivers, including the Magdelena, which rise in the 5000 m high mountains of the Sierra Nevada de Santa Marta. All these systems have mangrove forests of *Rhizophora* and *Avicennia* although the Bahia de Cartegena is much modified and polluted by human activity.

## Gulf of Mexico

The best-known tropical coastal lake in the Gulf of Mexico is Terminós Lagoon, which has been intensively studied by Yáñez-Arancibia and his colleagues since 1977. Comprehensive reviews are contained in Yáñez-Arancibia *et al.* (1980, 1985), from which the following account is drawn.

Terminós Lagoon is in the southern Gulf of Mexico adjacent to Campeche Sound and is large (2500 km$^2$) and shallow (mean depth 3.5 m) with eastern and western connections to the sea (Fig. 2.1). Three rivers enter Terminós and the period of high river discharge is from August to November. The freshwater flows and coastal currents cause a strong net inflow into the eastern inlet and an outflow from the western. The tidal range is small, from 0.3 to 0.7 m, temperatures are between 22 and 31 °C and salinities in the central basin vary between 12‰ and 30‰. Higher salinities and lower turbidities prevail on the eastern side of the system because the main river discharge flows through the western part, bringing turbid, lower-salinity conditions. Most of the shoreline is lined by dense *Rhizophora* mangroves and there are extensive seagrass (*Thalassia*) meadows in the north-east. In the south-east there are reefs of the oyster *Crassostrea virginica* near the river mouths.

Just north of the Mexican border between latitudes 26°00'N and 27°45'N lies the Laguna Madre of Texas, a 1000 km$^2$ hypersaline coastal lake. It is a 200 km long series of coastal lagoons separated from the Gulf of Mexico by a narrow strip of land, Padre Island. The region is semi-arid and receives only 600 mm of rain a year whilst evaporation reaches 500 mm a year. No large rivers flow into the system and until recently it was closed off from the sea except across a sill at Port Aransas and through a narrow opening in the south. The main features of Laguna Madre have been reviewed by Hedgpeth (1967). Salinities often exceed 80‰ and values above 100‰ are not uncommon. Most rainfall comes in short, heavy cloudbursts and these result in temporary surface salinities as low as 2‰. The surface area varies greatly because the average depth is only about 1 m and sections of the system dry out. A shipping channel (The Intracoastal Waterway) was constructed along the whole length of Laguna Madre in 1949 and this has ameliorated salinities somewhat, although extremes still occur in peripheral areas. Nevertheless, in spite of its physical extremes, the Laguna Madre hosts one of the most important fisheries in Texas. There is little shoreline vegetation and the substratum is mainly sand or mixed silt and sand. In the north and south there are large meadows of seagrass, particularly *Thalassia testudinum,* and *Ruppia maritima* is common along the waterway.

## 2.6  PHYSICAL FACTORS INFLUENCING THE NATURE OF TROPICAL ESTUARIES

The size, shape and physical regime of a tropical estuary are determined largely by an interplay of three major factors, which then in turn exert a strong influence on the composition of its fish fauna.

The first major factor is the geomorphology of the catchment and coastal area together with its geological structure. The extensive coastal areas of sedimentary rocks found in parts of India, much of South East Asia and the Gulf region of northern Australia have large meandering river floodplains and open estuaries. Most of the estuaries of Borneo, Sumatra and Indo–China fall into this category, as well as those flowing into the Bay of Bengal. All have their origins in distant highland areas. In contrast, many of the estuaries of the east Australian and south-east African coast have smaller catchments and the coastal plain is relatively narrow and the estuaries less meandering. Some of the most extreme examples of this type occur on the northern Transkei coast of South Africa, where the estuaries have been cut through ancient igneous rocks by falls in sea level and are fjordlike (Fig. 2.11).

Second, the amount and seasonal distribution of rainfall has a profound influence on the salinity and flow patterns of estuaries. In equatorial regions, significant amounts of rainfall occur throughout the year, so

**Fig. 2.11** Msikaba estauary, a fjordlike open estuary on the Transkei coast of South Africa.

river flow is relatively constant. For example, the estuaries of Borneo and Sumatra originate in catchments that receive a yearly average of about 6 m of rainfall, that of the Fly River of Papua New Guinea, 13 m, and the Orinoco in South America, 2 m. In contrast, many regions experience definite wet and dry seasons that exert a strong influence on estuaries. This is well illustrated by the estuaries of the Gulf of Carpentaria in northern Australia where almost all the rainfall (about 1 m a year) falls between November and March. For example, in the Embley estuary in the north-east Gulf of Carpentaria, salinity is at about seawater level (35‰) throughout the estuary during the dry season, but falls markedly in the wet season, particularly in the middle and upper reaches. Turbidity is generally inversely related to salinity, with maximum values in the middle and upper reaches during the wet season. Turbidity is low over much of the estuary during the dry season and in the lower reaches all year. It should be noted, however, that the lowest values in the estuary (2–5 NTU) are still higher than the levels recorded in the adjacent Albatross Bay (usually < 2 NTU).

Blind estuaries usually occur in small catchments where rainfall is highly seasonal and may be relatively low even in the wet season, such as in parts of Africa and Australia. The closure of the mouth is usually a result of a lack of flow during the dry season coupled with a combination of longshore drift of coastal sands and a small tidal range (Day, 1981).

Third, the nature of the adjacent sea, particularly its depth, currents and tides, helps to shape the physical characteristics of an estuary, as well as having a major influence on the origins, species composition and behaviour of the fish fauna.

*Where the continental shelf is wide and coastal waters shallow, tropical estuaries fall into two broad types depending on the amount of river discharge (i.e. rainfall), tidal range and local and oceanic current patterns*

'Type 1' estuaries (Fig. 2.12) have high river discharge, large tidal range and weak coastal currents. Vast quantities of fresh water and a high sediment load from these estuaries dominate the coastal waters and there is no clear demarcation between the turbid, low-salinity estuarine and marine waters. Many of the large estuaries of South East Asia and Bay of Bengal fall into this type. Tidal ranges in excess of 5 m in, for example, the large estuaries of north Borneo, lead to tidal currents in excess of $1 \, \mathrm{m \, s^{-1}}$ and in some cases tidal bores. Such a tidal bore occurs in the Lupar estuary in Sarawak. This was described in the following terms by Beccari (1904) "... woe betide those who are unfortunate enough to be caught by it on the river. The bore-wave which is about six feet high, advances with a foaming crest across the entire width of the river with a velocity of several miles per hour. It is felt about 10 miles inside the mouth of the river, and penetrates also the Lingga, which is the first affluent of the Batang–Lupar, continuing up the main stream for about thirty miles, a loud roar announcing its advent."

Similar tidal ranges occur in northern Australia in the Gulf of Carpentaria, where the Norman estuary has a tidal range of about 4 m and tidal current speeds of up to $0.7 \, \mathrm{m \, s^{-1}}$; further west, tidal ranges reach a massive 10 m at the Fitzroy River estuary in north-western Australia.

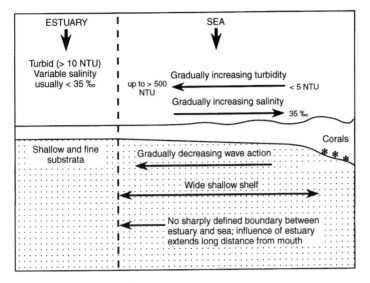

**Fig. 2.12**   Profile of a type 1 (see text) open estuary.

'Type 2' estuaries (Fig. 2.13) have low and/or strongly seasonal river discharge, small tidal range and well-developed marine currents. The low sediment loads and small amount of fresh water entering the sea usually have little influence on the marine waters adjacent to such estuaries, particularly where longshore and offshore currents are strong. It is in these shallow marine waters that coral reefs have usually developed – the best-known example is the Great Barrier Reef of eastern Australia, but similar reefs occur along some of the East African coast from Moçambique to Kenya. In these areas there is a clear demarcation between the very clear reef waters and the more turbid, lower-salinity estuarine waters. It is here that during times of high rainfall, turbid plumes of low-salinity water enter the sea, standing out in stark contrast to the coral reef waters. Such plumes deposit sediment very rapidly, sometimes smothering and killing corals. The outflows of river water also contain increased concentrations of low-relative molecular-mass (< 10 000) fulvic acids that become incorporated as fluorescent bands in coral skeletons. The discovery by the Australian Institute of Marine Science of these yellow-green fluorescent bands in the skeletons of massive corals from inshore waters of the Great Barrier Reef (Boto and Isdale, 1985), and the temporal and spatial correlations of these bands with adjacent river flow, together with the great age of massive corals, means that extremely long, highly resolvable records of coastal run-off exist throughout tropical nearshore seas. Such records

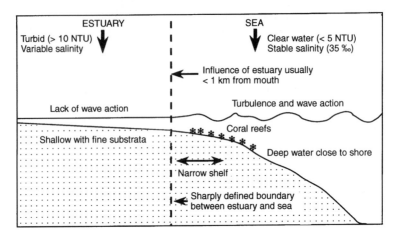

**Fig. 2.13**  Profile of a type 2 (see text) open estuary.

have been discovered for freshwater flow from the Everglades to Florida Bay (Smith *et al.*, 1989), and in Papua New Guinea and Indonesia (Scoffin *et al.*, 1989), and for seasonal rainfall in the Sinai Desert in corals of the Red Sea (Klein *et al.*, 1990).

Lack of freshwater inflow during the dry season, a small tidal range and longshore currents will combine to close seasonally the mouths of many smaller estuaries – 'blind estuaries'. The timing of the closing and opening of such systems has a marked effect on the composition of their fish faunas.

*Where the continental shelf is narrow, and deep oceanic waters come close to the shore, strong marine influences, both abiotic and biotic, may extend considerable distances from the mouth of the estuary.*

Such estuaries occur on the KwaZulu-Natal coast of South Africa, the west coast of South America, and on many of the larger islands of the South Pacific, such as the Solomons and Fiji. At the Kosi estuary on the borders of South Africa and Moçambique, the continental shelf is very narrow and clear oceanic water penetrates several kilometres up the estuary on the high tide. Clear tropical seawater conditions prevail in the lower reaches, permitting the maintenance of a small rocky reef and its associated coral fauna in one corner of the lagoon (Blaber, 1978).

## 2.7  CATASTROPHIC EVENTS

Tropical estuaries occur in parts of the world where cyclones (hurricanes, typhoons) and volcanic activity take place. Such catastrophic or episodic events can cause long-lasting changes to the morphology and habitats of estuaries and hence markedly influence their fish faunas.

Tropical cyclones occur in a band between about 8° and 20° north and south of the Equator. Areas close to the Equator, such as Borneo and Sumatra, do not suffer cyclones but Bangladesh, the Philippines, New Guinea, northern Australia, the South Pacific, the Caribbean and the Gulf of Mexico are battered rather unpredictably by cyclones every wet season.

Tropical cyclones affect estuaries in two ways: firstly, rises in sea level at high tide brought about by winds in excess of $100 \, \text{km} \, \text{h}^{-1}$ can physically change the configuration of the mouth; and secondly, vast amounts of rainfall dumped by a cyclone as it crosses the coast cause flooding, with associated reductions in salinity, high flow rates and destruction of aquatic vegetation and mangroves. Although the effects of a cyclone are usually restricted in area, the changes caused may be profound and long lasting. Good examples of the long-term effects are provided by the changes wrought to most of the estuaries of KwaZulu-Natal, South Africa, by a cyclone in 1987, documented in detail by Perry (1989). This cyclone brought very heavy rain which resulted in massive flooding of most rivers. Sand spits or bars at the mouths of estuaries were washed away, resulting in wide open mouth conditions, or in some cases a new mouth was breached in the coastal dunes. Silt and turbidity plumes stretched for more than a kilometre offshore, extending southwards due to the prevailing current. There was considerable destruction of natural vegetation such as *Phragmites* reed beds and mangroves. Earlier cyclones in 1984 (Demoina followed 2 weeks later by Imboa) caused extensive flooding of the St Lucia coastal lake system, reducing salinities from a maximum of 60‰ to 5‰ over a period of 2 weeks. The flooding had relatively long-term effects on the distribution and numbers of some estuarine resident and marine migrant species which utilize the estuary as a nursery ground (Cyrus, 1988a; Forbes and Cyrus, 1992; Martin *et al.*, 1992). The rise in lake levels and the floodwaters also inflicted damage on mangrove stands (Steinke and Ward, 1989; Martin *et al.*, 1992; Wright and Mason, 1993).

The Texas coast of the Gulf of Mexico suffers regular hurricanes which may have profound effects on the salinities of coastal lakes. Salinities in Laguna Madre were 60‰ before Hurricane Beulah in September 1967, but fell to 30‰ after its passage and 760 mm of rain in 4 days. Such hurricanes also scour the mouths of estuarine systems due to flooding and higher-than-usual tides. More sediment is eroded, transported and deposited during a hurricane than in many years of normal activity. Water is piled against the coast by wind-induced tides that are superimposed on normal tides, and the ensuing storm surge can raise estuary or lagoon water levels 3–6 m above normal levels (Hayes, 1967; Fisher *et al.*, 1972). As the hurricane passes, these high levels combined with river flood water, are flushed out through the estuary mouth. The fisheries production of Texas lagoons, particularly of sciaenids, may be positively correlated with the occurrence of hurricanes due to transport of higher numbers of larvae to nutrient-rich bays and higher survival rates in lowered salinities (Matlock, 1987).

Volcanic activity occurs in a band through the South Pacific, South East and East Asia – 'The Fiery Ring of the Pacific' – and eruptions of, for example, Mount Pinatubo in the Philippines in 1991 and the Rabaul volcanoes in New Guinea in 1994, caused significant changes to coastal waters. The eruption of Mount Pinatubo was accompanied by a typhoon and heavy rainfall which added to the devastation over an area of 2500 ha. The major effects were caused by ash fall from the volcano and mudflow from the heavy rain and ash combined. High-turbidity waters extended 10 km from the coast, and estuaries and rivers suffered heavy siltation. The Daan–Bapor estuary which used to be 6 m deep, is now only 0.5 m deep, and 26 000 ha of brackish-water fish ponds were silted up or cut off from tidal influence. The effects of volcanic ash on seagrasses and mangroves were significant, but only ongoing studies will determine the recovery dynamics of such habitats (Infante-Guevara, 1992).

# Chapter three

# The fishes of tropical estuaries

## 3.1 INTRODUCTION

The variations in the size, depth and physical characteristics of estuaries of the subtropics and tropics are reflected in the great variety and diversity of their fish faunas. The composition of these faunas depends upon the interplay of a whole range of factors, among which the most important are:

- estuary size, depth and physical regimes, particularly salinity and turbidity, as well as the types of habitats;
- the nature and depth of adjacent marine waters and, to a much lesser extent, fresh waters;
- the geographical location of the estuary in relation to marine features such as ocean currents, canyons and reefs.

More details of these and other influences are described in Chapter 4.

The classifications of estuarine fishes have been almost as numerous as the classifications of estuaries (Day, 1951; Gunter, 1967; McHugh, 1967; Green, 1968; Perkins, 1974), with taxonomic, physiological and ecological groupings based on attributes such as salinity tolerance, breeding, feeding and migratory habits. Most research on estuarine fishes has been in temperate and warm temperate regions of Europe and North America, where salinity has long been regarded as a key factor regulating the composition of estuarine fish faunas. Hence it has been possible to define species in relation to their salinity tolerances as euryhaline or stenohaline (Green, 1968) with various subdivisions. The distributions of estuarine fishes can then be described in relation to their various tolerances to reduced salinities. Gunter (1967) considered the lower salinity of estuarine waters to be the outstanding difference between these waters and sea water, with the large reduction in numbers of species between sea and estuary an indication of the overriding influence of salinity. While this phenomenon has been amply demonstrated for warm temperate and temperate areas, it is not at all clear in the subtropics and tropics. Here the number of fish species in estuaries is usually an order of magnitude greater. Where individual temperate estuaries may have about 20 species (Potter *et al.*, 1986; Elliott and Taylor, 1989; Pomfret *et al.*, 1991; Elliott and Dewailly, 1995) and warm temperate ones about 50 (Darnell, 1961; Lenanton and Hodgkin, 1985), most medium-to-large subtropical and tropical estuarine areas have at least 100, with some reaching over 200. Additionally, the number of species (often exceeding 300 in adjacent marine and estuarine areas) may be similar or differences can be attributed to the presence of specific habitats such as coral reefs or seagrass beds. Indeed, where such habitats occur within estuaries, such as the reef in the lower reaches of the Kosi estuary in northern KwaZulu-Natal, South Africa, or the seagrass beds in the Embley estuary in the Gulf of Carpentaria, Australia, they are inhabited by species found in those habitats in the sea. Salinity is only one of an array of factors that determine which species are found in any one subtropical or tropical estuary. The relative importance of each of these factors is discussed in Chapter 4. Undoubtedly tolerance of salinity plays a significant role in the distribution of fishes in tropical estuaries, but as these systems may undergo very great fluctuations in salinity (often from almost 0‰ to 35‰), both diurnally and seasonally, a considerable degree of euryhalinity is a precondition and prerequisite for their inhabitants, as well as for many species in marine areas that suffer seasonal reductions or fluctuations in salinity (section 2.3). For this reason, a grouping of tropical estuarine fishes based mainly on their salinity tolerances is not very useful or practical, especially as there has been little experimental work conducted on the salinity tolerances of most of the species. Based on their occurrence in low-salinity waters, almost all fit into a similar, very euryhaline category.

Day *et al.* (1981) went some way towards developing a meaningful grouping of estuarine fishes for temperate as well as tropical areas and this has been modified below. It includes the following five groups.

1. Marine migrants from the sea. This is usually the largest group in subtropical and tropical estuaries. They may occur in estuaries both as adults and juveniles (e.g. *Lates calcarifer* of South East Asia and Australia; the circumtropical *Mugil cephalus*, as well as most Leiognathidae and Pomadasyidae), or only as juveniles (e.g. some species of Mugilidae, Carangidae and Polynemidae) or only as adults (e.g. some species of Ariidae).

2. Anadromous species that breed in fresh water and spend time in estuaries on their way to and from their spawning grounds. This group is important in temperate estuaries and includes salmon (*Salmo* and *Oncorhynchus*), shad (*Alosa* spp.) and lampreys (Petromyzontidae). It is a rare category in subtropical and tropical estuaries but includes the commercially important tropical shads *Tenualosa ilisha* and *T. toli* of South Asia and Sarawak respectively.

3. Catadromous species that ascend to fresh waters as juveniles but return to the sea to breed. This grouping includes the anguillid eels, some species of which occur in the tropics, but it is a rare category in tropical systems.

4. Estuarine species that complete their entire life-cycle in the estuary. They represent a small but significant part of the tropical estuarine fish fauna and include a number of clupeids (e.g. *Gilchristella aestuaria* in Africa), gobies (hundreds of species), engraulids (e.g. several species of *Coilia* in South East Asia), ambassids, atherinids and syngnathids.

5. Freshwater migrants that move varying distances down estuaries but usually return to fresh water to breed. In the tropics this group is best represented in South and Central American estuaries where a number of pimelodid catfish, poeciliids and characids penetrate estuarine waters (Cervigón, 1985; Winemiller and Leslie, 1992). The freshwater cichlid *Oreochromis mossambicus* is a well-known African example and has even been found in hypersaline waters of up to 110‰ in St Lucia, KwaZulu-Natal (Whitfield *et al.*, 1981); also other tilapias occur in the hypersaline waters of the Casamance River in Senegal, West Africa (Albaret, 1987). The archerfishes (Toxotidae) are a well-known South East Asian and Australasian example of freshwater species that are abundant in mangrove-lined estuaries.

## 3.2 DIVERSITY AND PATTERN OF LIFE CYCLES

Among the large numbers of fishes in tropical estuaries, there is much variety in the form of their life histories. They can, however, be grouped according to the categories listed above; representative examples of four of these are illustrated below.

### Marine migrants

Four examples are shown in Fig. 3.1. They have somewhat similar life histories in that they spend time in both the sea and estuaries, but each has distinctive features.

The circumtropical mullet species, *Mugil cephalus*, has been the subject

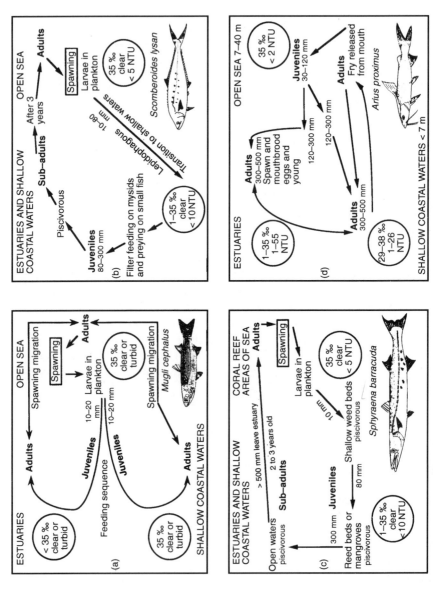

**Fig. 3.1** The life cycles of four marine migrant species from Indo-West Pacific subtropical and tropical estuaries.

of intensive study (Thomson, 1966; Odum, 1970). Like most Mugilidae they are iliophagous and feed on the surface layer of the substratum, ingesting inorganic and organic material and passing out the indigestible sand grains. Adults occur widely in estuaries and coastal waters and do not appear to have very specific habitat requirements. They spawn only in the sea and undertake what are often spectacular spawning migrations, where many thousands of fish gather at estuary mouths before moving to the open sea (Wallace, 1975a). It is thought that the postlarvae must seek estuaries and shallow waters at a small size (10–20 mm) in order to be able to go through a complex sequence of changes in food and feeding, leading to iliophagy. More is described about this phenomenon in Chapter 5.

The queenfish or leatherjacket, *Scomberoides lysan*, occurs throughout the Indo–West Pacific but only in relatively clear estuaries and coastal waters. Its congener *S. commersonianus* is abundant in more turbid areas. Adults live in clear waters of the sea where they spawn. Juveniles seek shallow coastal waters of estuaries at lengths of 10–80 mm. To start with they are lepidophagous, that is, they feed by pulling off the scales of other fishes (Major, 1973; Blaber and Cyrus, 1983). In the shallow clear waters of suitable estuaries they change to filter feeding on mysids and take small fish. The proportion of fish in the diet increases with size until they are entirely piscivorous at a size of about 300 mm (approximately 3 years old), after which they leave the estuaries and do not return.

*Sphyraena barracuda* is another circumtropical species found in all warm seas (De Sylva, 1963, 1973). A notable feature of this species is that it is a voracious piscivore at all sizes, from about 10 mm upwards (Blaber, 1982). Adults of up to 2 m in length live around the edges of coral reefs where they spawn. Postlarvae seek shallow weedy areas on the margins of clear-water estuaries or quiet coastal waters. At a length of about 80 mm they move to the deeper waters of adjacent reed beds or mangroves; from a length of about 300 mm they move to open waters. Once above a size of about 500 mm they move out of estuaries to the vicinity of coral reefs.

*Arius proximus* is one of a large number of species of catfish of the family Ariidae that live in the fresh waters, estuaries and seas of the tropics. The family exhibits a number of life-cycle strategies (Kailola, 1990) of which *Arius proximus* is an Australo–Papuan example. Adults are found in turbid estuaries and both shallow and deep coastal waters. As with most ariids they are mouthbrooders and there is no pelagic larval phase. Adults carrying eggs in the mouth are common in estuaries and shallow coastal waters, but the juvenile catfish are not released except in deeper waters where they live until they move into estuaries at a length of about 120 mm (Blaber *et al.*, 1995a). This species thus differs from most marine migrants in that the juvenile stage of the life history is marine.

## Anadromous species

This category is rare in tropical estuaries but does include the abundant and commercially important *Tenualosa ilisha* of the Bay of Bengal and its affluents. Details of its life cycle can be found in the extensive bibliography of Jafri and Melvin (1988). Its close relative *Tenualosa toli* of Sarawak estuaries and coastal waters has a more complex life history (Figs 3.2 and 8.1) because it changes sex and only lives for between two and three years (Blaber *et al.*, 1996; Chapters 6 and 7). Males spawn towards the end of their first year in the middle reaches of large, low-salinity, turbid estuaries and spawn as females in their second year. The postlarvae move to the upper reaches but do not, unlike *T. ilisha*, enter fresh water. As they grow they spread throughout the estuary. After spawning, the males move to the lower reaches and to contiguous middle-salinity, turbid coastal waters, where they change sex to female. These females then return to the middle reaches of the estuaries to spawn the following season.

## Estuarine species

*Gilchristella aestuaria* is a small south-east African clupeid that is truly estuarine and completes its entire life-cycle in estuaries (Fig. 3.2) (Whitfield, 1994b). It is a zooplankton feeder throughout its short 1–2 year life and is tolerant of a wide range of salinities and turbidities. Growth is rapid and the fish can spawn at the end of their first year when they deposit sticky eggs on the substratum or on weeds. The completion of the life cycle in 1 year may be an adaptation that has evolved to allow the species to survive in the relatively unstable African estuarine environment.

## Freshwater migrants

The euryhaline, iliophagous freshwater cichlid *Oreochromis mossambicus* is a common component of many estuaries of East Africa, South and South East Asia and Australasia. The factors affecting its distribution and numbers in estuaries have been reviewed by Whitfield and Blaber (1979a). A complex of six interrelated factors appear to control its distribution (Fig. 3.2; Chapter 4). When five or six of these are favourable then the species is abundant. As the suitability of the habitat declines, the species becomes more and more restricted to the upper reaches and fresh-

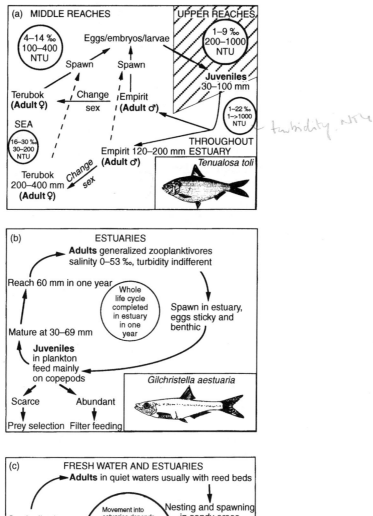

(a)  MIDDLE REACHES                                                  UPPER REACHES

4–14 ‰ 100–400 NTU          Eggs/embryos/larvae          1–9 ‰ 200–1000 NTU

Spawn          Spawn

Juveniles 30–100 mm

Terubok (Adult ♀)          Change sex          Empirit (Adult ♂)

SEA          1–22 ‰ 1–>1000 NTU

16–30 ‰ 30–200 NTU                    THROUGHOUT
                                       ESTUARY
Terubok 200–400 mm          Change sex          Empirit 120–200 mm (Adult ♂)

(Adult ♀)          Tenualosa toli

turbidity. NTU

(b)                    ESTUARIES

**Adults** generalized zooplanktivores
salinity 0–53 ‰, turbidity indifferent

Reach 60 mm in one year

Whole life cycle completed in estuary in one year

Spawn in estuary, eggs sticky and benthic

Mature at 30–69 mm

**Juveniles** in plankton feed mainly on copepods

Gilchristella aestuaria

Scarce          Abundant

Prey selection   Filter feeding

(c)          FRESH WATER AND ESTUARIES

**Adults** in quiet waters usually with reed beds

Movement into estuaries depends on:
1. Salinity stability
2. Current strength
3. Breeding sites
4. Marginal vegetation
5. Marine competitors
6. Marine predators

Nesting and spawning in sandy areas

Gradually change to iliophagous diet

**Fry** in marginal reed or weed beds

**Juveniles** carnivorous mainly on Crustacea in and near aquatic vegetation

Oreochromis mossambicus

**Fig. 3.2** The life cycles of an anadromous species (*Tenualosa toli*), estuarine species (*Gilchristella aestuaria*) and a freshwater migrant species (*Oreochromis mossambicus*) from subtropical and tropical Indo-West Pacific estuaries.

water zones. In general, *O. mossambicus* occurs most commonly in blind estuaries and coastal lakes.

## 3.3   SPECIES ASSEMBLAGES

Tropical estuarine fishes fall into the overall zoogeographic realms of the Indo–West Pacific, tropical East Atlantic, tropical West Atlantic and East Pacific (Lowe-McConnell, 1987) and these divisions are followed here in describing examples of fish communities from a representative range of tropical estuaries.

When comparing the fish faunas of tropical estuaries it must be remembered that not all workers have sampled using the same types of nets or gear, and hence many species lists are incomplete. Ideally, to obtain a comprehensive description of the fish community, various different nets, such as seines, gill nets, cast nets and fyke nets should be used to collect fishes from different habitats. In addition, in limited areas that cannot be netted, such as among mangroves or thick seagrass, a piscicide such as rotenone can be used. Data from subsistence, artisanal or commercial fishers may also be a valuable source of information.

### Indo–West Pacific

This huge area stretching from the east coast of Africa through South and South East Asia to Australia and the central Pacific has the highest diversity of fishes in the world (at least 600 species recorded in estuaries). Many species occur throughout the region while others are restricted to particular regions. The ubiquitous species come from a wide variety of families and often only one species of a genus has an Indo–West Pacific distribution. Examples of species that occur in most estuaries of the region include the sly bream, *Acanthopagrus berda*, the glassfish, *Ambassis gymnocephalus*, the trevally, *Caranx sexfasciatus*, the Zambezi or bull shark, *Carcharhinus leucas*, the wolf herring, *Chirocentrus dorab*, the tenpounder, *Elops machnata*, the pursemouth, *Gerres filamentosus*, the ponyfish, *Leiognathus equulus*, the mangrove jack, *Lutjanus argentimaculatus*, the flathead, *Platycephalus indicus*, the flounder, *Pseudorhombus arsius*, the whiting, *Sillago sihama*, and the thornfish, *Terapon jarbua*. The mullets (Mugilidae) serve to illustrate the point that whereas some members of a family are very widespread, some are regional whilst yet others are endemic to one or two estuaries (Fig. 3.3). They are among the most abundant fish in shallow coastal waters and estuaries of the subtropics and tropics. The family consists of some 95 species (Nelson, 1984), of which more than 40 are Indo–West Pacific. Many species are closely related and ecologically

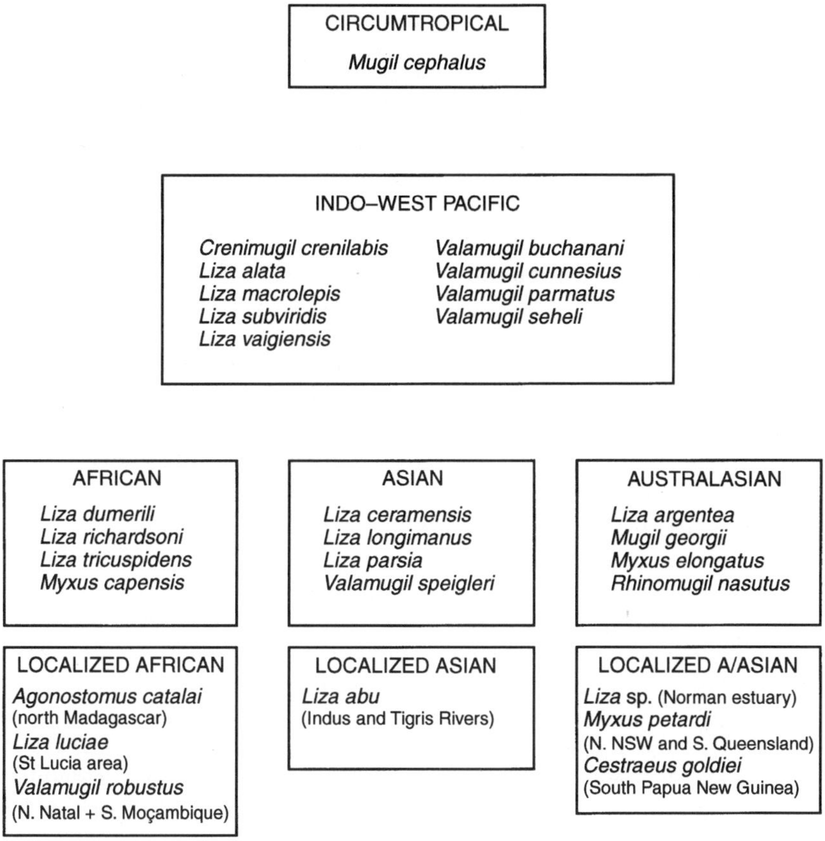

```
┌─────────────────────────────────┐
│        CIRCUMTROPICAL           │
│                                 │
│         Mugil cephalus          │
└─────────────────────────────────┘
```

```
┌───────────────────────────────────────────────────────┐
│                 INDO–WEST PACIFIC                      │
│                                                        │
│   Crenimugil crenilabis      Valamugil buchanani       │
│   Liza alata                 Valamugil cunnesius        │
│   Liza macrolepis            Valamugil parmatus         │
│   Liza subviridis            Valamugil seheli           │
│   Liza vaigiensis                                       │
└───────────────────────────────────────────────────────┘
```

| AFRICAN | ASIAN | AUSTRALASIAN |
|---------|-------|--------------|
| *Liza dumerili* | *Liza ceramensis* | *Liza argentea* |
| *Liza richardsoni* | *Liza longimanus* | *Mugil georgii* |
| *Liza tricuspidens* | *Liza parsia* | *Myxus elongatus* |
| *Myxus capensis* | *Valamugil speigleri* | *Rhinomugil nasutus* |

| LOCALIZED AFRICAN | LOCALIZED ASIAN | LOCALIZED A/ASIAN |
|-------------------|-----------------|-------------------|
| *Agonostomus catalai* (north Madagascar) | *Liza abu* (Indus and Tigris Rivers) | *Liza* sp. (Norman estuary) *Myxus petardi* (N. NSW and S. Queensland) *Cestraeus goldiei* (South Papua New Guinea) |
| *Liza luciae* (St Lucia area) | | |
| *Valamugil robustus* (N. Natal + S. Moçambique) | | |

**Fig. 3.3** Widespread, regional and endemic examples of species of Mugilidae (mullet) from the Indo-West Pacific region.

similar and their relatively uniform morphology has led to taxonomic confusion. The most ubiquitous species, *Mugil cephalus*, is circumtropical, inhabiting estuaries, the open sea and even hypersaline lagoons. In contrast, some species are highly endemic: for example, *Valamugil robustus* and *Liza luciae* occur in and adjacent to only a few estuaries of KwaZulu-Natal and Moçambique.

## East Africa

### Morrumbene estuary

The fish fauna of this open estuary in central Moçambique (p.13) was surveyed in some detail by Day (1974), who recorded 114 species from the several different habitats. Unfortunately, only seine netting was employed so larger fish in the channels and smaller cryptic or burrowing species were probably overlooked. Nevertheless, the survey is one of the few detailed studies of a tropical East African estuary.

The shallow sandy areas of the lower reaches are home to large numbers of small species such as *Ambassis natalensis, Atherinomorus lacunosus, Sillago sihama, Stolephorus indicus* and *Terapon jarbua*, together with juvenile carangids, lethrinids and the sparid *Acanthopagrus berda*. The seagrass beds near the mouth are dominated by the at least partially herbivorous species, *Siganus rivulatus, Crenidens crenidens* and *Pelates quadrilineatus*, while three species of *Hippocampus* and three species of pipefish live concealed in the vegetation. In all, 46 species were collected from the lower reaches.

The middle reaches, which consist of a main open-water channel, inter-tidal mudbanks, a few rocks and are lined by mangroves, have a more diverse fauna with 98 species recorded. Many of the species found in the lower reaches also occur here, together with large numbers of the small clupeid *Harengula ovalis* and the half-beak *Hemiramphus far*. Mugilidae are a dominant component of this area with five species, *Liza dumerili, L. macrolepis, Valamugil buchanani, V. robustus* and *V. seheli*, occurring in the channel and over the intertidal mudflats at high tide. Other species common in this area include three species of grunter, *Pomadasys kaakan, P. maculatus* and *P. multimaculatus*; three gerreids, *Gerres acinaces, G. filamentosus* and *G. oblongus*; four snappers, *Lutjanus argentimaculatus, L. fulviflamma, L. sanguineus* and *L. vaigiensis*; juveniles of five carangids, *Caranx ignobilis, C. melampygus, C. sexfasciatus, Scomberoides lysan* and *S. tol*; and two barracuda, *Sphyraena jello* and *S. putnamiae*.

Only 26 species were collected from the upper reaches of thick mangrove forest, small creeks and muddy channels. A number of middle-reaches fish occur in this area, including mugilids, *Lutjanus argentimaculatus* and *L. fulviflamma*. Of particular note in the mangroves are the large numbers of gobies, mostly *Glossogobius giuris* and the mudskippers *Periophthalmus* spp. Other common residents of the mangroves are the flathead *Platycephalus indicus* and the flatfish *Pseudorhombus arsius* and *Solea bleekeri*.

*Kosi estuary system*
This subtropical system on the border of South Africa and Moçambique (Fig. 2.1) consists of a chain of four interconnected coastal lakes of gradually increasing salinity that drain to the sea through lagoon-like lower reaches. The characteristics of the largest of the lakes, Nhlange, have been described in section 2.5 and further details of the physical characteristics of the system can be found in Allanson and van Wyk

(1969), Hill (1969), Day (1981) and Kyle (1988). A detailed account of the fish fauna is given by Blaber (1978) and Blaber and Cyrus (1981) and a full range of sampling methods were used. A total of 163 species have been recorded from the system, of which 75% are restricted to the lower reaches and the reef within. The majority of species are marine migrants which penetrate the system to varying extents, with the remainder consisting of estuarine resident species such as Ambassidae, the clupeid *Gilchristella aestuaria* and Hemiramphidae, and freshwater species such as *Oreochromis mossambicus*. Further details of the distribution of species in relation to physical factors such as salinity are discussed in Chapter 4.

*Species that occur throughout the system.* The common species that penetrate beyond the lower reaches as far as Lake Nhlange are listed in terms of their abundances in summer (wet season, October–March) and winter (dry season, April–September) in Chapter 4 (Table 4.2). Brief accounts of the dominant species are given below.

*Chanos chanos:* the milkfish is one of the most abundant large fish (often exceeding 1 m in length) in Lakes Nhlange, Mpungwini and Makhawu-lani, but is seldom encountered in the lower reaches. Large schools of this fast-swimming, detritus-feeding species enter the system and move to the lakes which provide a large source of detritus.

Carangidae: the queenfish, *Scomberoides lysan*, is the most commonly occurring carangid in Kosi. Juveniles and adults occur throughout the system at all times of year but are absent from Lake Nhlange in winter. Adults of the kingfishes or trevallies, *Caranx ignobilis*, *C. papuensis* and *C. sexfasciatus*, are found in all the lakes and lower reaches in the summer and juveniles occur throughout the year. *Caranx melampygus* is common only in the lower reaches.

Gerreidae: five closely related species of *Gerres* have been recorded but only three are common: *Gerres acinaces*, which is common throughout, *G. filamentosus*, which is plentiful at the estuary but is progressively more scarce further from the mouth, and *G. rappi*, which is abundant in the lakes but less so in the lower reaches. All occur over shallow sandy areas, often in mixed-species schools. They form an important component of the subsistence fish traps of the local people. Their biology in Kosi was studied in detail by Cyrus and Blaber (1982, 1983, 1984a,b,c).

*Pomadasys commersonni:* the spotted grunter occurs commonly throughout the system in both summer and winter. Juveniles, subadults and adults have been recorded at all times of the year. This species is also an important part of the subsistence fishery (Kyle, 1988).

*Acanthopagrus berda:* juveniles and adults of the river or sly bream are present throughout the system, but the species is very wary and seldom netted. However, adults and subadults are caught in large quantities in the subsistence fish traps throughout the year.

*Rhabdosargus sarba*: the Natal stumpnose or tarwhine is abundant in the lower reaches and common in the lakes in summer. Adults and juveniles were captured in winter but were less numerous than in the wet season. Juveniles occur over shallow sandy areas, often in association with Gerreidae.

Mugilidae: the complex distribution of the 11 species of grey mullet at Kosi has been described by Blaber (1977) and Blaber and Whitfield (1977a). All species are common in the lower reaches with the exception of *Liza alata*. *Mugil cephalus* is the only species occurring in any numbers in Lake Nhlange; large schools move through the reed channel connecting Lake Nhlange to Lake Mpungwini in both wet and dry seasons.

Sphyraenidae: juveniles and subadults of four species of barracuda occur throughout the Kosi system at all seasons: *Sphyraena barracuda*, *S. jello*, *S. putnamiae* and *S. qenie*.

*Estuarine resident species*. These form a relatively insignificant proportion of the fish fauna and all are small species. The most abundant are *Ambassis productus*, *A. natalensis*, *Gilchristella aestuaria*, *Hyporhamphus capensis* and the gobies *Croilia mossambica*, *Glossogobius giuris* and *Periophthalmus* sp.

*Freshwater species*. Of the ten or so species found in the system, only the cichlid *Oreochromis mossambicus* is of significance, but it seldom penetrates past Lake Mpungwini. It is most common in Lake Nhlange, where it is largely restricted to the extensive fringing *Phragmites* reed bed areas. It is an extremely euryhaline species and its distribution in the system is not restricted by salinity (Whitfield and Blaber, 1979a; Chapter 4).

*Species occurring only in the lower reaches*. These can be divided into two groups, firstly the species associated with the reef, and secondly, those distributed throughout the lower reaches.

More than 20% of the species in the system are found only around the reef about 300 m inside the estuary mouth. They represent an extension of a marine reef fauna into the relatively quiet water of the estuary. The unusual rocky substrata and clear water of this estuary makes this an uncommon estuarine community. Salinities fall to at least 25‰, the water is clear and of oceanic origin except during periods of heavy flooding, and even then it is clear on the flowing tide; and the reef is too large and high to become inundated by sand or mud. Its fishes are typical of western Indian Ocean reefs and it is numerically dominated by Acanthuridae, Pomacentridae, Chaetodontidae and Labridae. Moray eels

(Muraenidae), stonefish, *Synanceja verrucosa*, and scorpionfish (Scorpaenidae) occur in crevices and caves on the reef. Larger teleosts typical of reefs frequent the edges of the reef in the main channel, for example: *Sarpa salpa, Diplodus sargus*, Lutjanidae, *Neoscorpis lithophilus* and *Monodactylus falciformis*. Predators such as Carangidae and large carcharhinid sharks, as well as large schools of Mugilidae, are always present in the vicinity of the reef.

The non-reef species that are confined to the lower reaches include clupeids such as *Sardinella melanura* and *Spratelloides delicatulus*, juveniles of the emperor *Lethrinus nebulosus* and adults of the garfish *Tylosurus crocodilus*. The atherinid *Atherinomorus lacunosus* is extremely abundant over the sandbanks in summer but absent in winter.

*The St Lucia system of northern KwaZulu-Natal*
This largest coastal lake system in East Africa has been the subject of intensive study using all types of gear and the fish fauna is well documented (various summaries and checklists are published in Day *et al.*, 1954; Millard and Broekhuysen, 1970; Wallace, 1975a,b; Wallace and van der Elst, 1975; Whitfield, 1980d; Blaber, 1988). Aspects of responses to salinity fluctuations and the trophic relationships and spawning behaviour of St Lucia fishes are dealt with in Chapters 4, 5 and 6.

About 110 species have been recorded from the system, although their distribution and occurrence changes according to the large cyclical salinity changes (from virtually fresh to over 100‰ in some areas). For much of the time the open-water areas are dominated by subadults and adults of the following taxa: *Argyrosomus hololepidotus, Caranx ignobilis, C. sexfasciatus, Chanos chanos, Elops machnata, Lichia amia, Muraenesox bagio, Otolithes ruber* and *Pomadasys commersonni*. Eleven species of Mugilidae are also a numerically dominant part of the fauna with *Liza macrolepis, L. dumerili, Mugil cephalus, Valamugil buchanani* and *V. cunnesius* the most abundant.

Pelagic filter feeders are also abundant in the open waters, with *Gilchristella aestuaria, Hilsa kelee* and *Thryssa vitrirostris* the most common. A wide variety of small-to-medium benthic and benthopelagic species may occur throughout the system, including *Acanthopagrus berda, Johnius dussumieri, Leiognathus equulus, Polydactylus sextarius, Pseudorhombus arsius, Rhabdosargus holubi* and *R. sarba*. Most of the fishes can be categorized as marine migrants, either juveniles or adults or both. About 15% are small estuarine residents such as *Gilchristella aestuaria, Glossogobius giuris, Solea bleekeri* and *Syngnathus djarong*. The only freshwater species that penetrates any distance into the system is the cichlid *Oreochromis mossambicus*. During hypersaline conditions it was the only species that survived in areas where the salinity was over 100‰.

The strong connection with the sea and diversity of habitats leads to vast numbers of juvenile marine fish migrating into the lake and spreading along the lengths of its shores (Wallace and van der Elst, 1975). Wallace (1975a,b) also showed that a greater proportion of adult fish occur in St Lucia than in most other estuaries, some of them such as *Mugil cephalus* and *Pomadasys commersonni* undergoing annual 'spawning runs' to the sea (Chapter 6).

*Mhlanga estuary, KwaZulu-Natal*
The fish fauna of Mhlanga (pp.28-29) is among the best studied of any small, blind subtropical estuary and a full range of sampling methods have been used (Whitfield, 1980a,b,c; Harrison and Whitfield, 1995). The following is taken largely from the latest account by Harrison and Whitfield (1995). Forty-seven species of 19 families have been collected from Mhlanga with *Gilchristella aestuaria* (46% by numbers), *Oreochromis mossambicus* (18%) and *Valamugil cunnesius* (9%) the most abundant species. In terms of biomass, *Oreochromis mossambicus* accounted for 47% of the total, followed by *Valamugil cunnesius* (20%), *Liza alata* (10%), *Myxus capensis* (8%) and *Mugil cephalus* (8%). The community was numerically dominated by estuarine resident species, which made up 48% of the total. Marine migrants accounted for 34% of the total catch and freshwater species accounted for 18%. In terms of biomass the situation is different, reflecting the small size of estuarine residents compared with other categories: marine migrants made up 52% of the catch, freshwater species 47% and estuarine residents 1% (Fig. 3.4).

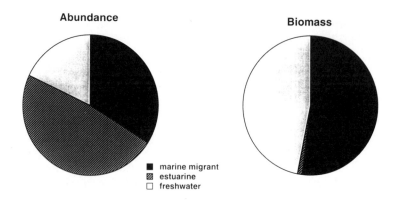

**Fig. 3.4** Relative abundances and biomasses of marine migrant, estuarine and freshwater species in the blind Mhlanga estuary in KwaZulu-Natal, South Africa (data from Harrison and Whitfield, 1995).

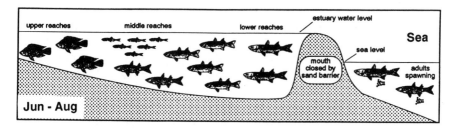

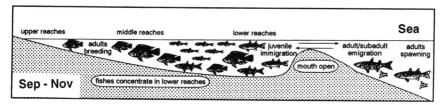

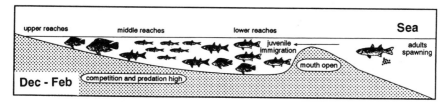

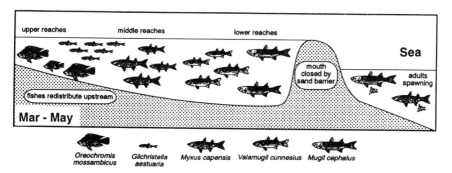

**Fig. 3.5** Seasonal changes in fish community structure in the blind Mhlanga estuary in KwaZulu-Natal, South Africa (redrawn from Harrison and Whitfield, 1995).

Mhlanga estuary is dominated by different categories of fishes according to the season and state of the mouth. The situation is summarized in Fig. 3.5. In winter (dry season), when the mouth is closed, freshwater species predominate in the upper reaches while marine migrants inhabit the

middle and lower reaches; when the system opens with the spring/
summer rains, adult and subadult marine migrants emigrate to the sea
and juvenile marine migrants enter the estuary; freshwater species move
towards the lower reaches. During this open phase, food resources and
habitat are reduced and this reduction, coupled with the influx of juvenile
marine migrants, results in a decrease in the proportion of estuarine and
freshwater species during summer. In late summer and autumn the
mouth closes and the estuary fills, providing more habitat and food, and
the freshwater species are displaced further up the estuary as the marine
migrants spread upstream.

### Taitinga River estuary, Comoro Islands

This tiny, clear-water estuary provides a marked contrast to other
estuaries in the region. It is on the island of Anjouan in the Comoro
group, north of Madagascar. It is only 500 m long and discharges into
deep oceanic waters. In this respect it is comparable with similar small
estuaries in the South Pacific. It is isolated from other similar systems and
in the Taitinga fishes cannot penetrate the river beyond 500 m due to a
large waterfall. The fauna was studied by Balon and Bruton (1994), who
found only seven species in the estuary: the gobies *Sicyopterus lagocephalus*
and *Stenogobius genivittatus*; the eleotrids *Eleotris fusca* and *E. mauritianus*;
the eel *Anguilla marmorata*; the flagtail *Kuhlia rupestris*; and juveniles of
the reef-dwelling lutjanid *Lutjanus monostigma*. Only two of these, *S. lagocephalus* and *A. marmorata*, occur above the waterfall, and no other species
were found in the fresh water. All seven species are euryhaline and of
marine origin, but the extent to which the stocks are maintained from
other rivers via the sea is at present unknown.

### South Asia

### Vellar–Coloroon (Pichavaram) system, India

This medium-size open estuary system is situated on the east coast of
Tamil Nadu (11°29′N, 79°49′E) and is shallow (mainly less than 1 m in
depth) and thickly vegetated with luxuriant mangroves. The fishes have
been documented in a series of studies by Chandrasekaran and Natarajan
(1992), Jeyaseelan and Krishnamurthy (1980) and Krishnamurthy and
Jeyaseelan (1981). They have recorded 195 species from 68 families with
the following 17 families being the most numerous: Clupeidae (6 species),
Engraulididae (6), Ariidae (3), Plotosidae (2), Mugilidae (7), Centropomidae (3) (including Ambassidae), Teraponidae (3), Carangidae (11),
Sillaginidae (3), Leiognathidae (6), Gerreidae (4), Lutjanidae (4), Haemulidae (2), Sciaenidae (4), Cichlidae (3), Siganidae (2), and Gobiidae ( > 20
species).

As in other Indo–West Pacific estuaries, this system is dominated by juveniles and subadults with relatively few large fish. One of the exceptions to this is the commercially important *Lates calcarifer* (barramundi, cock-up or seabass), where adults of over 1 m in length, as well as juveniles, are present. Other common species, mainly of small size or juveniles, include *Polynemus sextarius, Sillago sihama, Gerres oblongus, G. setifer, Terapon jarbua, Sphyraena jello, Strongylura strongylura, Arothron hispidus* and *Solea ovata* (Thangaraja, 1984; Shameem, 1992). Most can be classed as marine migrants and there are about 10 freshwater species (5%) including the three cichlids *Etroplus maculatus, E. suratensis* and *Oreochromis mossambicus*, and the three bagrid catfishes *Mystus gulio, M. keletius* and *M. vittatus*. About 15% of species, mostly of small size and mainly Gobiidae, are estuarine residents.

## Lake Chilka, India

This large coastal lake on the east coast of Orissa, India (pp.35–36) has a fish fauna of about 152 species (Jones and Sujansingani, 1951; Pillay, 1967a; Mohanty, 1975; Kowtal, 1976; Jhingran, 1991). The species composition varies seasonally and with the changing depth (< 1 to over 4 m according to the rains). The dominant families are Mugilidae (15–29% by numbers), Centropomidae (predominantly *Lates calcarifer*, 2–15%), Polynemidae (mainly *Eleutheronema tetradactylum*, 4–11%), Sciaenidae (4–9%), Cichlidae (1–6%) and Clupeidae (6–11%).

The most numerous mullets are *Mugil cephalus* and *Liza macrolepis* but *Liza subviridis, Valamugil speigleri* and *Rhinomugil corsula* are also abundant. The dominant clupeids are *Tenualosa ilisha* and *Nematalosa nasus* while several species of *Thryssa* and *Stolephorus* make up the bulk of the engraulids (anchovies). Other abundant species are *Megalops cyprinoides, Elops machnata, Harpadon nehereus, Chanos chanos, Gerres setifer, Pseudosciaena coibor* and the cichlid *Etroplus suratensis*. There are at least five species of ariid catfishes. In the lower-salinity areas of the lake, several genera of freshwater catfish and cyprinids are present and include *Mystus cavasius, M. vittatus, Heteropneustes fossilis, Clarias batrachus, Puntius* spp. and *Chela* spp. In the more marine parts there are numerous euryhaline marine species and in the occasional rocky area, reef species such as *Abudefduf bengalensis* are found.

## South East Asia

### Matang system, Peninsular Malaysia

The fish of the Matang estuary system with its series of creeks, channels and mangroves (pp.108–9) on the west coast of Malaysia has been documented by Khoo (1989), Chong *et al.* (1990) and Sasekumar *et al.*

(1994a). A total of 117 species have been recorded with the most abundant families being the Sciaenidae (28% by numbers), Engraulididae (17%), Ambassidae (10%), Clupeidae (9%), Ariidae (8%), Leiognathidae (4%) and Scatophagidae (4%). Approximately 15% of the species could be considered estuarine residents with the remainder being in the marine migrant category (the majority being juveniles or subadults), although it should be emphasized once again that there is a very gradual gradation between the estuarine and marine conditions in this area, hence rigid categorization is difficult. Only two freshwater species are listed, both catfish.

There are considerable differences between the abundances of the various families in the different habitats (Table 3.1). The most abundant species in the mangrove channels are *Ambassis gymnocephalus, Anodontostoma chacunda, Arius venosus, A. maculatus, A. caelatus, Johnius carouna, J. weberi, Nematalosa nasus, Scatophagus argus, Stolephorus tri* and *Thryssa kamalensis*. Over the mudflats, clupeids and engraulids were more abundant, particularly *Ilisha megaloptera, I. melastoma* and *Hilsa* sp. and *Stolephorus tri, Thryssa kamalensis* and *Thryssa* sp. In the estuarine nearshore waters *Johnius carouna, Ilisha melastoma* and *Stolephorus tri* were the most abundant species.

The mean biomass of fish was highest in the mangrove channels (3.1 g $m^{-2}$), followed by the mudflats (1.1 g $m^{-2}$) and the inshore waters (0.25 g $m^{-2}$).

*The Ranong estuary system, western Thailand*
This area on the Thai–Burma border was the subject of an intensive study by UNESCO/UNDP in association with the National Research Council of

**Table 3.1** Abundance (%) of the seven most common fish families in the Matang estuary system of west Malaysia (after Sasekumar *et al.*, 1994a)

| Family | Habitats | | | Mean abundance |
| --- | --- | --- | --- | --- |
| | Mangrove channels | Mudflats | Inshore water | |
| Ambassidae | 12.1 | 5.0 | 0.2 | 10.0 |
| Ariidae | 10.7 | 1.5 | 1.8 | 8.3 |
| Clupeidae | 7.1 | 11.3 | 26.9 | 9.0 |
| Engraulididae | 8.8 | 36.8 | 43.4 | 16.9 |
| Leiognathidae | 3.3 | 2.9 | 0.1 | 3.9 |
| Scatophagidae | 4.9 | 0.5 | 0.2 | 3.7 |
| Sciaenidae | 33.3 | 15.8 | 7.6 | 28.3 |

Thailand (1991). Most information came from monitoring subsistence fish traps and by using a commercial Thai 'push-net'. The latter is a 'V' shaped net pushed along the bottom ahead of a motorized boat (Fig. 8.9).

A total of 198 species of fish of 55 families have been recorded and the most abundant families are Leiognathidae, Clupeidae, Engraulididae, Ambassidae, Mugilidae and Carangidae. There was considerable seasonal variation in the relative proportions of the main families. For example Leiognathidae made up 48% of numbers in March 1988 but only 29% in August 1988. The most abundant species were *Ambassis gymnocephalus*, *A. kopsii*, *Clupea fimbriata*, *Leiognathus fasciatus*, *L. brevirostris*, *L. lineolatus*, *Liza* spp. (probably mainly *L. subviridis*), *Lutjanus russelli*, *Secutor ruconius*, *S. insidiator*, *Stolephorus indicus* and *S. tri*. The majority of the fishes are small species or juveniles, but a significant number of adults of three species of ariid catfish as well as of *Eleutheronema tetradactylum*, *Pomadasys kaakan* and Sphyraenidae also enter the system. The majority of species can be considered as marine migrants, but as with the Matang area, rigid categorization is difficult. Approximately 20% of the fishes are estuarine; the majority are gobies (7.6%), eleotrids (2.5%), hemiramphids (3.5%), engraulids (2%), ambassids (2%) and clupeids (2%). No freshwater species have been recorded.

*The Lupar estuary, Sarawak, east Malaysia*
This large open estuary (p.16; Figs 2.1, 2.4) opens gradually into estuarine coastal waters with which much of its fauna is contiguous. No published species lists exist, but the author has recorded over 100 species of 35 families from the system, based upon gill netting and market sampling. Undoubtedly the full species list will be much greater. There is relatively little habitat diversity and most of the fish occur in the main channel and in the narrow fringe of *Nypa* (Fig. 2.5) and mangroves. The dominant families are the Ariidae (5 species), Clupeidae (8 species), Engraulididae (10 species), and Sciaenidae (11 species), and the most abundant taxa are *Arius utik*, *Coilia coomansi*, *C. macrognathos*, *C. rebentischii*, *Harpadon nehereus*, *Ilisha* spp., *Otolithoides* sp., *Polynemus borneensis*, *Setipinna taty*, *Tenualosa toli*, *Thryssa mystax* and *Zenarchopterus* spp.

Sharks are well represented in the lower reaches, with *Eusphyrna blochii*, *Rhizoprionodon* sp. and *Scoliodon laticaudis*, as well as *Carcharhinus leucas*, being taken by fishermen. A number of other large marine species also penetrate well into the lower reaches, including the scombrids *Scomberomorus guttatus* and *Euthynnus affinis*. In contrast, in the upper and middle reaches a variety of freshwater species occur, including the catfishes *Mystus planiceps*, *Pangasius niewenhuisi* and *P. tubbi*, and the ambassids *Parambassis macrolepis* and *P. wolffi*. The freshwater stingray *Himantura oxyrhynchus* is also found in the upper reaches of the estuary.

Approximately 55% of species fall into the marine migrant category, 14% are freshwater species and the remaining 31% occur only in estuaries and estuarine coastal waters where they are known to complete their entire life cycle. These latter include among others, *Coilia* spp., *Ilisha* spp., *Tenualosa toli* and various eleotrids and gobies. The proportion in the marine migrant category would be likely to increase with additional and more varied sampling.

## Australasia

In this area we examine the fishes of a small estuary in north-east Queensland, a large estuary in the north-east Gulf of Carpentaria and a group of small estuaries in the Solomon Islands. In all cases a comprehensive suite of sampling gears was used.

### Alligator Creek

Robertson and Duke (1987, 1990) recorded 150 species of 44 families from Alligator Creek (pp.108–9) and the following account has been taken mainly from the review by Robertson and Blaber (1992). The fish community is dominated numerically by the families Engraulididae, Ambassidae, Leiognathidae, Clupeidae and Atherinidae. Two species, *Ambassis gymnocephalus* and *Stolephorus carpentariae*, make up 52% of the total catch and together with 14 other species (Table 3.2) make up > 95% of total numbers.

A large number of small species enter the intertidal forests at high tide and leave on the ebbing tide. *Ambassis gymnocephalus*, *Stolephorus carpentariae*, *S. nelsoni*, *Encrasicholina devisi*, *Leiognathus equulus*, *L. splendens* and *Toxotes chatareus* are all important components (by numbers and biomass) of the fish community using intertidal forests, although their abundances showed definite seasonal patterns. Greater densities of fish occur in the wet season than the dry season. The wet season is the period of greatest recruitment of juvenile fish to the community (Robertson, 1988) and is also the time when zooplankton abundance is highest in the mangrove forests (Robertson *et al.*, 1988). *Stolephorus* anchovies are most abundant in the wet season while other anchovies, *Thryssa brevicauda* and *T. hamiltoni*, and the clupeid *Escualosa thoracata* are more common later in the year. Postlarvae and juveniles of engraulids and clupeids move into the intertidal forest areas in vast numbers on flood tides and feed on the abundant zooplankton during the wet season, and move out of the estuary altogether in the dry season. In contrast, large numbers of juvenile and subadult *Ambassis* and *Leiognathus* use the forest habitat throughout the year. Larger species such as *Lates calcarifer*, *Lutjanus argentimaculatus* and *Acanthopagrus berda* also enter the forests at high tide, in

**Table 3.2** Number of individuals of 16 species that together made up 95% of total fish catch at Alligator Creek (after Robertson and Duke, 1990)

| Species | Total numbers |
| --- | --- |
| *Ambassis gymnocephalus* | 32 966 |
| *Stolephorus carpentariae* | 25 673 |
| *Leiognathus equulus* | 15 504 |
| *Encrasicholina devisi* | 7 699 |
| *Stolephorus nelsoni* | 7 580 |
| *Leiognathus splendens* | 4 307 |
| *Escualosa thoracata* | 3 955 |
| *Atherinomorus endrachtensis* | 2 193 |
| *Sardinella albella* | 1 495 |
| *Leiognathus decorus* | 1 439 |
| *Pseudomugil signifer* | 885 |
| *Herklotsichthys castelnaui* | 791 |
| *Thryssa brevicauda* | 774 |
| *Chelonodon patoca* | 759 |
| *Ambassis nalua* | 721 |
| *Thryssa hamiltoni* | 666 |

search of food. The overall mean biomass calculated for the mangrove forest fish was $10.9 \, \mathrm{g \, m^{-2}}$ (Robertson and Duke, 1990).

The two dominant fish in terms of both numbers and biomass in small creeks were *Ambassis gymnocephalus* and *Leiognathus equulus*, which together form more than 50% of the community. Other seasonally abundant species in this habitat are *Leiognathus splendens*, *Thryssa hamiltoni*, *Drombus ocyurus* and *Acanthopagrus berda*. Many of the fish which use the forest at high tide move into shallow creeks on the ebb tide and hence such creeks may support very high standing stocks of fish at low tide. Mean biomass for these small creeks was $29.0 \, \mathrm{g \, m^{-2}}$ (Robertson and Duke, 1990). Pelagic schooling species such as the *Stolephorus* species and the clupeids *Escualosa thoracata*, *Sardinella albella* and *Herklotsichthys castelnaui* move into the main stream of the estuary at low tide.

The main channel of the estuary is dominated by a suite of small species including the benthic clupeid *Nematalosa come*, the pelagic clupeid *Herklotsichthys castelnaui* as well as *Ambassis gymnocephalus*. Larger species common in the main stream are *Lates calcarifer*, *Scomberoides* species, *Lutjanus argentimaculatus* and *Toxotes chatareus*.

*Embley estuary*
The fishes of this large estuary and the neighbouring bay (Fig. 3.6) were studied from 1986 to 1991 by Blaber *et al.* (1989, 1990a,b), Brewer *et al.*

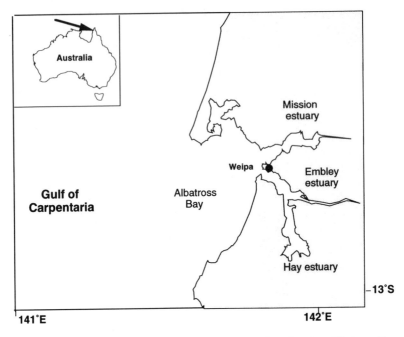

**Fig. 3.6**   The north-east Gulf of Carpentaria in Australia, showing Albatross Bay, a shallow tropical marine embayment, and its estuaries, including the Embley.

(1989, 1991) and Salini *et al.* (1990), from which the following summary is drawn.

A total of 197 species have been recorded in the Embley estuary, most of which (92%) are found in the lower reaches, whereas only 69 species (35%) and 40 species (20%) live in the middle and upper reaches respectively. Of the 243 species recorded in Albatross Bay, 91 also occur in the Embley estuary. However, 25 of these are incidentals (fewer than three occurrences). One hundred and six species from the Embley estuary were not recorded in Albatross Bay. A complete list of species recorded in each area is given by Blaber *et al.* (1990a). Below is a summary of the species composition in different estuarine habitats of the Embley estuary.

*Open-water channels*
A total of 127 species have been recorded and of these, 80 species (63%) also occur offshore in Albatross Bay. The 37 species that make up 97% of the biomass are listed in Table 3.3 in terms of overall catch rate and individual catch rates for lower, middle and upper reaches. The catch rate of most species declines from the lower to the upper reaches. The excep-

tions are *Lates calcarifer, Liza subviridis, Lutjanus argentimaculatus, Megalops cyprinoides, Negaprion acutidens* and *Pristis pristis*, which are most abundant in the middle reaches; and *Arius argyropleuron, Chanos chanos, Epinephelus malabaricus* and *Nematalosa erebi*, which are most abundant in the upper reaches. *Arius argyropleuron* and *Nematalosa erebi* are usually more abundant in the middle and upper reaches and are relatively uncommon in the lower reaches. *Lates calcarifer, Polydactylus sheridani* and *Pomadasys kaakan* show no definite seasonal patterns, but catch rates in the upper reaches are generally lower. Carcharhinid sharks are mainly confined to the lower and middle reaches, but most species extend to the upper reaches by the end of the dry season (November).

The mean biomasses of fishes from the upper reaches were estimated at $16.066\,g\,m^{-2}$ and from the middle reaches at $7.098\,g\,m^{-2}$. The higher overall biomass in the upper reaches was largely a result of larger numbers of carcharhinid sharks and the primarily freshwater species *Nematalosa erebi*. The dominant species by weight in both regions is *Scomberoides commersonianus*.

### Sandy mud beaches in the lower reaches

Fish of 72 species with a mean biomass estimate of $5.03\,g\,m^{-2}$ have been recorded. The dominant species (75% of the biomass) are *Acanthopagrus berda, Arrhamphus sclerolepis, Himantura uarnak, Lates calcarifer, Liza vaigiensis* and *Scomberoides commersonianus*. The 31 species that also occurred offshore in Albatross Bay represent only 15.9% of the total biomass.

### Seagrass areas in the lower reaches

Fish of 53 species with a mean biomass of $0.484\,g\,m^{-2}$ have been recorded from beam trawl sampling in this habitat. Most are small ($< 10\,cm$ standard length, SL); *Epinephelus suillus, Pelates quadrilineatus, Siganus canaliculatus, Apogon ruppelli, Lutjanus russelli* and *Monacanthus chinensis* contribute most to biomass (49%). Of the 53 species caught, 28 (65% of the biomass) also occur offshore in Albatross Bay. Rotenone sampling in the seagrass, which yielded only 108 g of fish of 14 species, gave a higher biomass estimate ($1.841\,g\,m^{-2}$) than that from the beam trawls. Three of the species had not been caught in the beam trawl.

### Intertidal mudflats adjacent to mangrove forests in the lower reaches

In July 1988, a stake net which isolated $9167\,m^2$, trapped 647.3 kg of fish of 39 species giving a biomass estimate of $70.6\,g\,m^{-2}$ (Table 3.4). These fish had moved on to the mudflats at high tide. The catch was dominated by four teleosts and two species of Dasyatidae (stingrays), which made up 76% of overall weight. Numerically the most abundant

**Table 3.3** Catch rates (grams per metre of net per hour) for the 37 most abundant species (listed alphabetically) captured in gill nets in the Embley estuary, shown in overall terms and separately for lower, middle and upper reaches ($\overline{X}$, mean catch rate; SE, standard error; –, absent; %, per cent of total catch rate)

| Species | Overall | | | Lower reach | | | Middle reach | | | Upper reach | | |
|---|---|---|---|---|---|---|---|---|---|---|---|---|
| | $\overline{X}$ | SE | % | $\overline{X}$ | SE | % | $\overline{X}$ | SE | % | $\overline{X}$ | SE | % |
| *Arius leptaspis* | 0.227 | 0.001 | 0.60 | 0.121 | 0.001 | 0.23 | 0.445 | 0.001 | 1.48 | 0.086 | 0.001 | 0.43 |
| *Arius macrocephalus* | 0.667 | 0.001 | 1.76 | 0.227 | 0.001 | 0.44 | 0.980 | 0.002 | 3.25 | 1.102 | 0.004 | 5.54 |
| *Arius mastersi* | 0.218 | 0.001 | 0.57 | 0.446 | 0.003 | 0.86 | 0.011 | <0.001 | 0.04 | 0.070 | <0.001 | 0.35 |
| *Arius* sp. 2 (of Kailola) | 0.218 | 0.001 | 0.58 | 0.005 | <0.001 | 0.01 | 0.513 | 0.002 | 1.71 | 0.183 | 0.001 | 0.92 |
| *Arius proximus* | 5.119 | 0.009 | 13.52 | 8.878 | 0.020 | 17.10 | 3.020 | 0.009 | 9.99 | 0.496 | 0.002 | 2.49 |
| *Carcharhinus amblyrhynchoides* | 0.291 | 0.001 | 0.77 | 0.604 | 0.003 | 1.16 | – | – | – | 0.105 | 0.001 | 0.53 |
| *Carcharhinus amblyrhynchos* | 0.990 | 0.002 | 2.61 | 1.637 | 0.005 | 3.15 | 0.375 | 0.002 | 1.25 | 0.626 | 0.004 | 3.15 |
| *Carcharhinus cautus* | 1.205 | 0.004 | 3.18 | 2.475 | 0.008 | 4.77 | 0.289 | 0.001 | 0.96 | – | – | – |
| *Carcharhinus dussumieri* | 0.362 | 0.001 | 0.96 | 0.781 | 0.003 | 1.51 | – | – | – | 0.064 | 0.001 | 0.32 |
| *Carcharhinus leucas* | 0.404 | 0.002 | 1.07 | – | – | – | 0.773 | 0.004 | 2.58 | 0.656 | 0.004 | 3.30 |
| *Carcharhinus limbatus* | 1.180 | 0.005 | 3.12 | 2.480 | 0.011 | 4.78 | – | – | – | 0.354 | 0.003 | 1.78 |
| *Carcharhinus sorrah* | 0.261 | 0.002 | 0.69 | 0.586 | 0.005 | 1.13 | – | – | – | – | – | – |
| *Chanos chanos* | 0.352 | 0.002 | 0.93 | 0.273 | 0.002 | 0.53 | 0.003 | <0.001 | 0.01 | 1.120 | 0.009 | 5.62 |
| *Drepane punctata* | 0.155 | <0.001 | 0.41 | 0.257 | 0.001 | 0.50 | 0.101 | <0.001 | 0.34 | 0.024 | <0.001 | 0.12 |
| *Eleutheronema tetradactylum* | 0.144 | <0.001 | 0.38 | 0.293 | 0.001 | 0.56 | 0.038 | <0.001 | 0.13 | – | – | – |
| *Epinephelus malabaricus* | 0.079 | <0.001 | 0.20 | – | – | – | – | – | – | 0.193 | 0.002 | 0.97 |
| *Gnathanodon speciosus* | 0.405 | 0.001 | 1.07 | 0.651 | 0.002 | 1.25 | 0.213 | 0.001 | 0.71 | 0.196 | 0.001 | 0.99 |

| Species | | | | | | | | | | | | |
|---|---|---|---|---|---|---|---|---|---|---|---|---|
| *Lates calcarifer* | 4.873 | 0.006 | 12.87 | 4.990 | 0.010 | 9.61 | 6.344 | 0.010 | 21.13 | 1.780 | 0.006 | 8.95 |
| *Liza subviridis* | 0.184 | <0.001 | 0.49 | 0.199 | 0.001 | 0.38 | 0.252 | 0.001 | 0.84 | 0.035 | <0.001 | 0.17 |
| *Liza vaigiensis* | 0.257 | 0.001 | 0.68 | 0.549 | 0.003 | 1.06 | 0.036 | <0.001 | 0.12 | – | – | – |
| *Lutjanus argentimaculatus* | 0.098 | <0.001 | 0.26 | 0.025 | <0.001 | 0.05 | 0.224 | 0.001 | 0.75 | 0.043 | <0.001 | 0.22 |
| *Megalops cyprinoides* | 0.137 | 0.001 | 0.36 | 0.007 | <0.001 | 0.01 | 0.253 | 0.001 | 0.84 | 0.222 | 0.001 | 1.12 |
| *Negaprion acutidens* | 0.772 | 0.002 | 2.04 | 0.639 | 0.002 | 1.23 | 1.352 | 0.003 | 4.51 | 0.071 | 0.001 | 0.36 |
| *Nematalosa erebi* | 2.537 | 0.006 | 6.70 | 0.031 | <0.001 | 0.06 | 2.059 | 0.006 | 6.86 | 8.565 | 0.027 | 43.08 |
| *Polydactylus sheridani* | 1.428 | 0.003 | 3.77 | 2.003 | 0.006 | 3.86 | 1.295 | 0.003 | 4.31 | 0.400 | 0.003 | 2.01 |
| *Pomadasys argenteus* | 0.115 | <0.001 | 0.03 | 0.072 | <0.001 | 0.14 | 0.224 | 0.001 | 0.75 | 0.021 | <0.001 | 0.11 |
| *Pomadasys kaakan* | 0.897 | 0.001 | 2.37 | 1.095 | 0.003 | 2.11 | 1.043 | 0.002 | 3.48 | 0.216 | 0.001 | 1.09 |
| *Pristis pectinata* | 0.212 | 0.001 | 0.56 | 0.169 | 0.001 | 0.33 | 0.390 | 10.001 | 1.30 | – | – | – |
| *Pristis pristis* | 1.045 | 0.008 | 2.75 | – | – | – | 2.993 | 0.020 | 9.97 | – | – | – |
| *Rachycentron canadus* | 0.087 | <0.001 | 0.22 | 0.194 | 0.001 | 0.37 | – | – | – | 0.025 | <0.001 | 0.12 |
| *Rhizoprionodon acutus* | 1.043 | 0.004 | 2.75 | 2.326 | 0.010 | 4.48 | – | – | – | – | – | – |
| *Rhynchobatus djiddensis* | 0.246 | 0.001 | 0.65 | 0.494 | 0.003 | 0.95 | 0.073 | <0.001 | 0.24 | – | – | – |
| *Scomberoides commersonianus* | 8.662 | 0.010 | 22.87 | 13.986 | 0.022 | 26.94 | 5.607 | 0.000 | 18.68 | 2.266 | 0.008 | 11.40 |
| *Scomberomorus semifasciatus* | 0.259 | 0.001 | 0.68 | 0.474 | 0.001 | 0.91 | – | – | – | 0.232 | 0.001 | 1.17 |
| *Sphyrna lewini* | 0.801 | 0.004 | 2.11 | 1.727 | 0.009 | 3.33 | – | – | – | 0.147 | 0.001 | 0.74 |
| *Sphyrna mokarran* | 0.348 | 0.003 | 0.92 | 0.780 | 0.006 | 1.50 | – | – | – | – | – | – |
| *Valamugil buchanani* | 0.391 | 0.001 | 1.03 | 0.453 | 0.001 | 0.87 | 0.456 | 0.001 | 1.52 | 0.145 | 0.001 | 0.73 |
| Total catch rates (including minor species not shown here) | 37.87 | | | 51.91 | | | 30.02 | | | 19.88 | | |

**Table 3.4** Biomasses of fishes (listed alphabetically) from intertidal mudflats in the lower reaches of the Embley estuary obtained from stake netting an area of $9167.3 \ m^2$ (*n*, number of individuals)

| Species | Total weight (g) | Biomass (g m$^{-2}$) | % Biomass | *n* |
|---|---|---|---|---|
| *Acanthopagrus berda* | 3 605 | 0.393 | 0.56 | 6 |
| *Aetobatus narinari* | 2 650 | 0.289 | 0.41 | 2 |
| *Arius proximus* | 73 210 | 7.986 | 11.31 | 166 |
| *Chelonodon patoca* | 150 | 0.016 | 0.02 | 3 |
| *Dasyatis annotatus* | 7 300 | 0.796 | 1.13 | 4 |
| *Dasyatis leylandi* | 8 180 | 0.892 | 1.26 | 7 |
| *Dasyatis sephen* | 128 030 | 13.965 | 19.78 | 37 |
| *Drepane punctata* | 113 087 | 12.336 | 17.47 | 271 |
| *Echeneis naucrates* | 20 | 0.002 | <0.01 | 1 |
| *Gazza minuta* | 27 | 0.003 | <0.01 | 1 |
| *Gerres abbreviatus* | 43 292 | 4.722 | 6.69 | 295 |
| *Himantura granulata* | 23 000 | 2.509 | 3.55 | 2 |
| *Himantura uarnak* | 87 820 | 9.580 | 13.57 | 55 |
| *Lactarius lactarius* | 440 | 0.048 | 0.07 | 2 |
| *Lates calcarifer* | 24 395 | 2.661 | 3.77 | 12 |
| *Leiognathus equulus* | 38 | 0.004 | 0.01 | 1 |
| *Liza subviridis* | 2 665 | 0.291 | 0.41 | 11 |
| *Liza vaigiensis* | 8 904 | 0.971 | 1.38 | 29 |
| *Lutjanus argentimaculatus* | 2 750 | 0.300 | 0.42 | 2 |
| *Lutjanus russelli* | 170 | 0.019 | 0.03 | 2 |
| *Negaprion acutidens* | 5 100 | 0.556 | 0.79 | 2 |
| *Nematalosa come* | 164 | 0.018 | 0.03 | 3 |
| *Platycephalus endrachtensis* | 140 | 0.015 | 0.02 | 1 |
| *Platycephalus indicus* | 1 860 | 0.203 | 0.29 | 5 |
| *Polydactylus sheridani* | 8 150 | 0.889 | 1.26 | 2 |
| *Pomadasys argenteus* | 840 | 0.092 | 0.13 | 2 |
| *Pomadasys kaakan* | 8 420 | 0.918 | 1.30 | 11 |
| *Pseudorhombus elevatus* | 242 | 0.026 | 0.04 | 2 |
| *Rhinobatos* sp. 1 | 15 240 | 1.662 | 2.35 | 8 |
| Sciaenidae (unidentified) | 100 | 0.011 | 0.02 | 1 |
| *Scomberoides commersonianus* | 13 115 | 1.431 | 2.03 | 20 |
| *Selenotoca multifasciata* | 574 | 0.063 | 0.09 | 2 |
| *Sphyraena putnamiae* | 5 950 | 0.649 | 0.92 | 2 |
| *Strongylura strongylura* | 87 | 0.009 | 0.01 | 2 |
| *Taeniura lymna* | 10 520 | 1.148 | 1.63 | 10 |
| *Thryssa hamiltoni* | 58 | 0.006 | 0.01 | 1 |
| *Tylosurus crocodilus* | 550 | 0.060 | 0.08 | 1 |
| *Valamugil buchanani* | 46 340 | 5.055 | 7.16 | 64 |
| *Valamugil cunnesius* | 100 | 0.011 | 0.02 | 1 |
| Totals | 647 283 | 70.608 | | 1049 |

species were *Arius proximus*, *Drepane punctata* and *Gerres abbreviatus*. Twenty-nine species (76% of the biomass) were also recorded offshore in Albatross Bay.

*Small mangrove creeks and inlets (all reaches)*
Fish of 66 species with a mean biomass estimate of $8.2\,g\,m^{-2}$ have been recorded in this habitat. Most are less than 10 cm SL. The dominant species are *Tetraodon erythrotaenia*, *Liza subviridis*, *Anodontostoma chacunda* and *Toxotes chatareus* (46% of biomass). Twenty-three species (19% of biomass) also occur offshore.

Data from offshore sampling (Blaber *et al.*, 1990a,b) show that of the 197 species in the Embley estuary, 106 are not found in the bay; 59 of these latter are, however, known to spawn or occur in shallow marine areas. These 59 species were probably not captured in Albatross Bay because depths < 7 m could not be trawled. *Negaprion acutidens*, *Rhynchobatus djiddensis*, *Scatophagus argus* and *Scomberomorus semifasciatus*, which were trawled in low numbers offshore (> 7 m), are probably essentially estuarine or inshore marine species.

The 91 species common to both the estuary and bay contribute most to the estuarine biomass in the open-water channels, intertidal mudflats and seagrass areas, but are poorly represented in sandy mud beach areas and mangrove creeks. Of these species, 25 are incidental (< 3 occurrences) in the estuary whereas 16 are incidental offshore. Another four species (*Arius* sp. 4, *Echeneis naucrates*, *Euristhmus nudiceps* and *Pomadasys argenteus*) are uncommon in both areas. The estuarine/shallow marine component can be categorized on the basis of their known distributions from the general literature. The length ranges of the remaining 46 species common to both areas are shown in Table 3.5. Based upon these data and information drawn from standard texts (Munro, 1967; Smith and Heemstra, 1986), we split these species into six groups:

1. juveniles found only in the estuary (14 species): *Anodontostoma chacunda*, *Chelonodon patoca*, *Gerres filamentosus*, *Gerres oyena*, *Leiognathus equulus*, *Lutjanus argentimaculatus*, *Monacanthus chinensis*, *Pelates quadrilineatus*, *Sardinella albella*, *Sillago lutea*, *Stolephorus indicus*, *Terapon jarbua*, *T. puta*, *Thryssa hamiltoni*;
2. juveniles found only offshore (3 species): *Arius argyropleuron*, *Arius proximus*, *Gnathanodon speciosus*;
3. juveniles found both in the estuary and offshore (24 species): *Carcharhinus amblyrhynchos*, *C. cautus*, *C. dussumieri*, *C. limbatus*, *C. sorrah*, *Drepane punctata*, *Gerres subfasciatus*, *Leiognathus decorus*, *L. leuciscus*, *L. splendens*, *Lethrinus lentjan*, *Lutjanus russelli*, *Platycephalus indicus*, *Pomadasys kaakan*, *Pseudorhombus arsius*, *P. elevatus*, *Rhizoprionodon*

*The fishes of tropical estuaries*

**Table 3.5** Minimum and maximum lengths of fish species (listed alphabetically) common to the Embley estuary and Albatross Bay (min, minimum length in mm; max, maximum length in mm; $n$, number of individuals measured)

| Species | Embley estuary | | | Albatross Bay | | |
|---|---|---|---|---|---|---|
| | Min | Max | $n$ | Min | Max | $n$ |
| *Anodontostoma chacunda* | 20 | 140 | 721 | 93 | 140 | 29 360 |
| *Arius macrocephalus* | 160 | 400 | 182 | 70 | 390 | 1 421 |
| *Arius proximus* | 120 | 350 | 1481 | 30 | 280 | 1 191 |
| *Carcharhinus amblyrhynchos* | 430 | 1350 | 47 | 325 | 992 | 39 |
| *Carcharhinus cautus* | 510 | 1230 | 41 | 384 | 790 | 19 |
| *Carcharhinus dussumieri* | 405 | 1350 | 33 | 507 | 830 | 48 |
| *Carcharhinus limbatus* | 495 | 1320 | 57 | 638 | 1017 | 18 |
| *Carcharhinus sorrah* | 610 | 1100 | 18 | 750 | 1150 | 12 |
| *Chelonodon patoca* | 11 | 56 | 39 | 59 | 194 | 250 |
| *Dasyatis annotatus* | 360 | 510 | 7 | 211 | 268 | 5 |
| *Dasyatis sephen* | 280 | 650 | 37 | 940 | 1460 | 9 |
| *Dasyatis* sp. cf. *leylandi* | 230 | 480 | 4 | 155 | 235 | 7 |
| *Drepane punctata* | 100 | 400 | 66 | 51 | 310 | 1 773 |
| *Gerres filamentosus* | 9 | 185 | 283 | 72 | 198 | 14 760 |
| *Gerres oyena* | 30 | 66 | 61 | 69 | 141 | 91 |
| *Gerres subfasciatus* | 31 | 82 | 15 | 52 | 143 | 5 952 |
| *Gnathanodon speciosus* | 225 | 655 | 45 | 36 | 560 | 29 |
| *Herklotsichthys lippa* | 31 | 46 | 34 | 79 | 145 | 77 |
| *Leiognathus decorus* | 14 | 53 | 114 | 27 | 114 | 9 910 |
| *Leiognathus equulus* | 10 | 58 | 557 | 50 | 159 | 62 104 |
| *Leiognathus leuciscus* | 48 | 66 | 9 | 50 | 162 | 9 103 |
| *Leiognathus splendens* | 20 | 38 | 69 | 28 | 125 | 109 425 |
| *Lethrinus lentjan* | 67 | 116 | 17 | 82 | 220 | 197 |
| *Lutjanus argentimaculatus* | 59 | 405 | 12 | 584 | 584 | 4 |
| *Lutjanus russelli* | 35 | 124 | 37 | 85 | 405 | 86 |
| *Monacanthus chinensis* | 21 | 64 | 71 | 60 | 60 | 4 |
| *Pelates quadrilineatus* | 19 | 70 | 375 | 84 | 143 | 6 546 |
| *Platycephalus indicus* | 193 | 385 | 20 | 157 | 383 | 8 |
| *Pomadasys kaakan* | 53 | 510 | 108 | 80 | 555 | 4 419 |
| *Pseudorhombus arsius* | 45 | 171 | 14 | 92 | 203 | 51 |
| *Pseudorhombus elevatus* | 35 | 210 | 19 | 84 | 174 | 65 |
| *Rhizoprionodon acutus* | 320 | 720 | 225 | 225 | 880 | 70 |
| *Sardinella albella* | 18 | 100 | 145 | 66 | 120 | 10 308 |
| *Scomberoides commersonianus* | 30 | 950 | 1086 | 114 | 806 | 94 |
| *Secutor insidiator* | 31 | 40 | 38 | 35 | 98 | 27 672 |
| *Secutor ruconius* | 10 | 27 | 142 | 22 | 84 | 25 349 |
| *Selenotoca multifasciata* | 140 | 250 | 11 | 66 | 105 | 16 |
| *Siganus canaliculatus* | 33 | 102 | 181 | 64 | 150 | 228 |
| *Sillago lutea* | 56 | 113 | 5 | 112 | 207 | 602 |
| *Sphyraena putnamiae* | 245 | 435 | 13 | 249 | 320 | 155 |
| *Sphyrna lewini* | 1200 | 1790 | 9 | 960 | 960 | 3 |
| *Stolephorus indicus* | 31 | 86 | 4 | 55 | 144 | 298 |
| *Terapon jarbua* | 8 | 76 | 45 | 63 | 152 | 376 |
| *Terapon puta* | 17 | 64 | 151 | 64 | 160 | 2 666 |
| *Thryssa hamiltoni* | 45 | 195 | 23 | 123 | 195 | 902 |
| *Trixiphichthys weberi* | 22 | 168 | 11 | 8 | 322 | 911 |

*acutus, Scomberoides commersonianus, Secutor insidiator, S. ruconius, Selenotoca multifasciata, Siganus canaliculatus, Sphyraena putnamiae, Trixiphichthys weberi;*

4. adults found only in the estuary: *Selenotoca multifasciata;*

5. adults found only offshore (23 species): *Chelonodon patoca, Gerres filamentosus, Gerres oyena, Gerres subfasciatus, Herklotsichthys lippa, Leiognathus decorus, L. equulus, L. leuciscus, L. splendens, Lethrinus lentjan, Lutjanus russelli, Monacanthus chinensis, Pelates quadrilineatus, Pseudorhombus arsius, P. elevatus, Secutor insidiator, S. ruconius, Siganus canaliculatus, Sillago lutea, Stolephorus indicus, Terapon jarbua, T. puta, Thryssa hamiltoni;*

6. adults found both in the estuary and offshore (22 species): *Anodontostoma chacunda, Arius argyropleuron, A. proximus, Carcharhinus amblyrhynchos, C. cautus, C. dussumieri, C. limbatus, C. sorrah, Drepane punctata, Gnathanodon speciosus, Lutjanus argentimaculatus, Platycephalus indicus, Pomadasys kaakan, Rhizoprionodon acutus, Sardinella albella, Scomberoides commersonianus, Sphyraena putnamiae, Sphyrna lewini, Trixiphichthys weberi, Dasyatis annotatus, D. sp. cf. leylandi, D. sephen.*

Except for *Carcharhinus leucas,* which was found only in the estuary, carcharhinid sharks are equally well represented in the estuary and offshore.


*Species recorded only in the estuary*

The estuarine/shallow marine component, consisting particularly of Ambassidae, Belonidae, Hemiramphidae, Mugilidae and Sillaginidae, dominates sandy beaches in the lower reaches (79% of biomass) and forms approximately one-third of the biomass in other habitats. The juveniles of 20 species of the group are found only in the estuary.

Twenty-four species from the Embley can be considered truly estuarine as they complete their entire life history in the estuary (Table 3.6). Most are small species, living mainly among mangroves and in adjacent creeks. In terms of species the group is dominated by Gobiidae (15 species), Eleotridae (3 species) and Hemiramphidae (2 species), but the biomass is largely composed of three species: *Tetraodon erythrotaenia* (19.7%), *Toxotes chatareus* (5%) and *Zenarchopterus buffonis* (10.1%). Three additional species included in Table 3.6 – *Arius leptaspis, Nematalosa erebi* and *Pseudomugil gertrudae* – are primarily freshwater (Taylor, 1964; Allen and Cross, 1982) but are commonly found in estuaries.

Of the remaining 17 species recorded only in the estuary, 15 are known to occur also in offshore waters (Gloerfelt-Tarp and Kailola, 1984; Sainsbury *et al.*, 1985) but were not captured in Albatross Bay. No information is available on the other two species, *Callionymus* sp. and *Trachinotus* sp. cf. *mookalee*.

**Table 3.6**　Estuarine fish species recorded in the Embley estuary that complete their entire life cycle in the estuary (listed alphabetically)

| Species | References |
|---|---|
| *Acentrogobius caninus* | Hoese – unpublished keys |
| *Acentrogobius gracilis* | Hoese – unpublished keys |
| *Acentrogobius viridipunctatus* | Hoese – unpublished keys |
| *Amoya* sp. | Hoese (1986) |
| *Apogon hyalosoma* | Allen – unpublished key |
| *Arius leptaspis** | Rimmer and Merrick (1982) |
| *Butis butis* | Hoese *et al.* (1980) |
| *Cryptocentrus* sp. | Hoese (1986) |
| *Drombus globiceps* | Hoese – unpublished keys |
| *Drombus ocyurus* | Hoese – unpublished keys |
| *Drombus triangularis* | Hoese – unpublished keys |
| *Glossogobius biocellatus* | Hoese (1986) |
| *Glossogobius celebius* | Munro (1967), Hoese (1986) |
| *Glossogobius circumspectus* | Hoese (1986) |
| *Monodactylus argenteus* | McDowall (1980) |
| *Nematalosa erebi** | Taylor (1964) |
| *Ophieleotris aporos* | Munro (1967) |
| *Ophiocara porocephala* | Hoese (1986) |
| *Ophichthus* sp. | Castle pers. comm. |
| *Pandaka rouxi* | Hoese (1986) |
| *Pseudogobius* (3 undescribed spp.) | Hoese unpublished |
| *Pseudomugil gertrudae** | Allen and Cross (1982) |
| *Siganus vermiculatus* | Woodland (1984) |
| *Siphamia roseigaster* | Allen unpublished (1987) |
| *Tetraodon erythrotaenia* | Hardy (1981) |
| *Toxotes chatareus* | Munro (1967) |
| *Zenarchopterus buffonis* | Collette (1974) |
| *Zenarchopterus dispar* | Collette (1974) |

*Primarily freshwater.

*Solomon Islands estuaries*

Estuaries in the Solomon Islands are mainly small and isolated from one another by extensive fringing reef lagoons. A total of 136 species of fish were recorded by Blaber and Milton (1990) from 13 estuaries in four island groups (Kolombangara, New Georgia, Rendova and Florida) (Table 3.7), but none contained more than 50 species.

*Open-water areas*

These were limited and consisted only of the main channels in the lower reaches. Forty-four species were caught of which the most numerous were *Lactarius lactarius*, *Scomberoides commersonianus*, *Valamugil seheli*, *Anodontos-*

**Table 3.7** Physical characteristics and numbers of species in Solomon Island estuaries (–, no data). The sampling methods used in each estuary are shown

| Island and estuary | Length (km) | Depth max. (m) | Salinity (‰) | Turbidity (NTU) | Mangrove genus | Substratum | Number of fish spp. | Fishing method* |
|---|---|---|---|---|---|---|---|---|
| Kolombangara | | | | | | | | |
| Dulo | 3 | 1 | 0–2 | 11.0 | *Brugiera* | Hard sand/logs | 30 | R |
| Sambira | 14 | 2 | 5–33 | 12.0 | *Brugiera* | Hard sand/logs | 44 | R, S |
| Niupilesi | 1 | 1 | 3 | 4.0 | *Brugiera* | Hard sand/logs | 23 | R |
| Lady Lever | 7 | 2 | 3–10 | 1.2–6.7 | *Rhizophora* | Soft mud | 31 | G, R |
| Ringgi Cove 1 | 1 | 2 | – | – | *Rhizophora* | Soft mud | | G, R |
| Ringgi Cove 2 | 1 | 2 | 6–24 | 4.5–14.0 | *Rhizophora* | Soft mud | 93† | G, R |
| New Georgia | | | | | | | | |
| Lembu | 6 | 3 | 6–33 | 0.9–7.7 | *Rhizophora* | Soft mud | 8 | G |
| Ondongo | 3 | 5 | – | 3.1–33.0 | *Rhizophora* | Sandy mud/ soft mud | 24 | G, S |
| Mbareke | 3 | 2 | 1–35 | 1.0–9.5 | *Brugiera* | Sand/logs | 25 | G, S |
| Rendova | | | | | | | | |
| Tivu | 2 | 1 | 30 | – | *Brugiera* | Hard sand/logs | 16 | R |
| Poko | 5 | 1 | 8–35 | 1.2–18.0 | *Rhizophora* | Soft mud | 38 | R |
| Kenela | 3 | 1 | 1–35 | – | *Brugiera* | Hard sand/rock/ logs | 15 | R |
| Florida | | | | | | | | |
| Ha'a | 5 | 2 | – | – | *Brugiera* | Sand/rock/logs | 38 | G, R |

*Methods: G, gill net; R, rotenone; S, seine net.
†Number of species for Ringgi Cove 1 + 2.

*toma chacunda, Carcharhinus melanopterus* and *Rastrelliger kanagurta* as well as juveniles of Gerreidae (four species), Clupeidae (three species), Carangidae (three species) and Mullidae (three species). Six species of carcharhinid sharks were a notable feature of New Georgia estuaries.

### Mangrove creeks and intertidal areas

Eighty-five species of 25 families were recorded of which almost one-quarter were Gobiidae. The next largest families were Apogonidae (9 species), Eleotridae (7), Lutjanidae (6), Serranidae (5) and Pomacentridae (4). Twenty-one species, mainly of the families Carangidae, Leiognathidae, Lutjanidae and Serranidae, were present only as juveniles. The remaining species were small estuarine resident species such as gobies, eleotrids and hemiramphids. Cluster analysis revealed two patterns of species composition in the estuaries. The first group, in which Gobiidae were the most numerous taxon, inhabited the soft, muddy-bottom estuaries; the second, dominated by Pomacentridae, lived in harder-bottom, mangrove-log-strewn estuaries.

### Estuarine coastal waters of the Indo–West Pacific

In Chapter 2 (section 2.3) the similarity of many of the shallow coastal waters of the region was indicated, together with the difficulties in separating their faunas from those of adjacent estuaries. Although a detailed account of their species composition is beyond the scope of this book, a brief account to place these areas in context is necessary. Such waters reach their greatest extent in the Bay of Bengal, South East Asia, northern Australia and West Africa. They are insignificant off the East African coast. They form the basis of extremely important fisheries and so have been well studied (Pauly, 1988). Their faunas are similar and are dominated by such families as Carangidae, Clupeidae, Dasyatidae, Gerreidae, Haemulidae, Leiognathidae, Mullidae, Nemipteridae, Polynemidae, Priacanthidae and Sciaenidae. The biomass and patterns of abundance of these demersal fish communities are influenced by a complexity of factors including depth, substratum, seasons, rainfall and turbidity (Blaber *et al.*, 1990b) and in many areas, such as the Gulf of Thailand and South China Sea, by very heavy fishing pressure (Chapter 8).

The Gulf of Carpentaria in northern Australia is one such region that does not yet suffer heavy fishing pressure. In a comprehensive trawl survey during 1990 (Blaber *et al.*, 1994c), over 300 species from 85 families were caught by trawling a systematic grid of 107 stations (Fig. 3.7). The absolute mean biomass was $124.8\,\mathrm{kg\,ha^{-1}}$ (SE = 44.1) for day trawls and $53.7\,\mathrm{kg\,ha^{-1}}$ (SE = 6.0) for night trawls. The overall mean catch rates were $421.3\,\mathrm{kg\,h^{-1}}$ (SE = 128.5) for day trawls and 198.6 kg

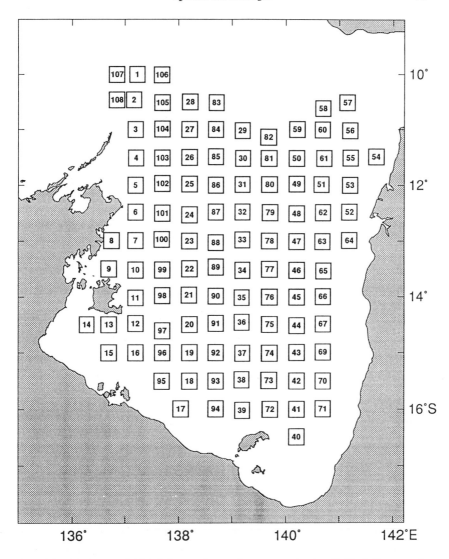

**Fig. 3.7** Positions of 107 fish trawl sample stations in the Gulf of Carpentaria surveyed during 1990 (station 68, inshore of station 69, was not worked). Reproduced with permission from Blaber *et al.* (1994c).

$h^{-1}$ (SE = 21.5) for night trawls. Biomasses were twice as high in the prawn-trawling grounds of Albatross Bay, the south-east Gulf and around Groote Eylandt. Twenty-five species made up 75% of the biomass; the dominant families were Haemulidae, Carangidae, Leiognathidae and

Nemipteridae. Community structure and distribution patterns were analysed by numerical classification techniques and principal coordinates analysis. These indicated 15 fish community groups (Fig. 3.8) based on fish species occurrences and biomasses. To illustrate the diversity and complexity of the system these are outlined below.

Group 1 contained mostly reef-associated species such as *Apogon albimaculosus*, *Coradion chrysozonus*, *Dipterygonotus balteatus*, *Pterocaesio diagramma*, *Sargocentron rubrum* and *Cycliththys jaculiferus*.

Group 2 consisted of 69 species that were widely distributed throughout the Gulf but with higher biomasses in northern and central areas. Abundant species included in this group were *Carangoides chrysophrys*, *Terapon jarbua*, *Saurida undosquamis* and *Lutjanus sebae*.

Group 3 comprised 22 species clustered mainly at stations in the eastern Gulf. It consisted mainly of pelagic and benthopelagic species such as *Dussumieria acuta* and *D. elopsoides* as well as the benthic rays *Dasyatis thetidus* and *D. leylandi* and the sharks *Carcharhinus sorrah* and *Sphyrna lewini*.

Group 4 comprised only four, usually reef-associated species, *Carangoides fulvoguttatus*, *Epinephelus malabaricus*, *Leiognathus fasciatus* and *Fistularia commersonni*, caught mainly at one station in the north-west.

Group 5 comprised only three species, *Asterhombus intermedius*, *Lethrinus genivittatus* and *Inimicus sinensis*, that occurred only at one station in the eastern Gulf.

Group 6 comprised 24 species that occurred mainly in the northern Gulf and particularly at a group of sites in the north and north-east Gulf. It included mainly larger species, some of commercial significance, such as *Lethrinus laticaudis*, *Lutjanus argentimaculatus*, *L. erythropterus*, *L. lemniscatus*, *Diagramma pictum* and *Pristipomoides multidens*, as well as the carangids *Caranx caeruleopinnatus* and *Caranx ignobilis*, and the tuskfishes *Choerodon monostigma* and *C. schoenleinii*.

Group 7 contained 15 species that were caught almost entirely at three stations in the south-west Gulf. They were mainly reef-associated species such as *Cephalopholis boenack*, *Chelmon mulleri*, *Chaetodontoplus duboulayi*, *Lutjanus vittus*, *Epinephelus areolatus* and *Plectropomus maculatus*.

Group 8 contained only four benthic, probably reef-attached species, *Congrogadoides amplimaculatus*, *Cottapistus cottoides*, *Tetrabrachium ocellatum* and *Pseudochromis quinquedentatus*, that were caught at one station in the southern Gulf.

Group 9 comprised 35 species that were widely distributed in shallow waters. Abundant species in this group were *Caranx bucculentus*, *Lethrinus lentjan*, *Leiognathus decorus*, *Secutor insidiator*, *Pomadasys maculatus*, *P. kaakan* and *Johnieops vogleri*.

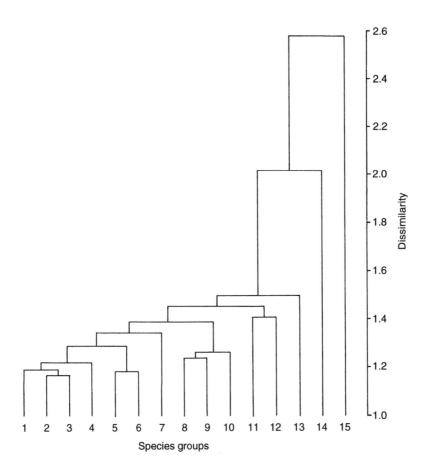

**Fig. 3.8** Relationships between 286 fish species from the Gulf of Carpentaria, based on their distribution and abundance (biomass), Fifteen species groups are indicated. Reproduced with permission from Blaber *et al.* (1994c).

Group 10 consisted of four predominantly pelagic species, including *Rastrelliger faughni*, *Scomberoides tala* and *S. commersonianus*, that were caught only immediately north of Groote Eylandt in the western Gulf.

Group 11 was made up of eight species that occurred primarily at station 95 in the southern Gulf. It consisted mainly of elasmobranchs including *Pristis cuspidatus*, *Rhizoprionodon taylori*, *Himantura* sp. and *Gymnura australis*.

Group 12 consisted of 44 widespread and abundant species that occurred at almost all stations. It included *Himantura toshi, Trixiphichthys weberi, Apogon quadrifasciatus, Fistularia petimba, Carcharhinus dussumieri, Leiognathus bindus, Nemipterus hexodon, N. nematopus, Pentaprion longimanus, Priacanthus tayenus, Saurida micropectoralis, Saurida* sp. 2, *Upeneus sulphureus, Carangoides talamporoides* and *Lutjanus malabaricus.*

Group 13 comprised 21 species that occurred mainly in the same stations as group 9 but were also present at other shallow-water stations in the southern Gulf. It consisted mainly of small demersal species including sillaginids, monacanthids and pleuronectids.

Group 14 comprised 12 primarily benthopelagic or pelagic species and included *Sardinella albella, S. gibbosa, Pelates quadrilineatus* and *Pomadasys trifasciatus.*

Group 15 consisted of nine species that occurred almost entirely at station 39 in the southern Gulf. The species were mainly reef-associated and included *Acanthurus grammoptilus, Anampses lennardi, Chelmon marginalis, Epinephelus maculatus, Pomacanthus sexstriatus* and *Scarus ghobban.*

These 15 groups fall into four broad categories: widespread, regional, reef-associated and other. Note, however, that the groups making up each category may not be closely related (Fig. 3.8). For example, the five reef-associated groups do not cluster together, they comprise different species, and each is primarily associated with a single station.

## Tropical East Atlantic

The tropical and subtropical estuarine fish fauna of this area encompasses most of the west coast of Africa. More than 200 species have been recorded in the various open and blind estuaries and coastal lakes (Pillay, 1967b). The shelf is wide in many areas and the estuarine fauna extends into shallow marine waters. Most of the families and many of the genera in estuaries are the same as in the Indo–West Pacific, but in most cases the species are different.

The dominant families (and species that are widespread and important for fisheries in estuaries of the region) are the Ariidae *(Arius parkii)*, Bagridae *(Chrysichthys nigrodigitatus)*, Carangidae *(Caranx hippos)*, Cichlidae *(Sarotherodon melanotheron, Tilapia guineensis)*, Clupeidae *(Ethmalosa fimbriata, Ilisha africana, Sardinella madarensis)*, Elopidae *(Elops lacerta)*, Gerreidae *(Eucinostomus melanopterus, Gerres nigri)*, Haemulidae *(Pomadasys jubelini)*, Mugilidae *(Liza falcipinnis, L. dumerili* (extending from East Africa), *Mugil cephalus)*, Polynemidae *(Polydactylus quadrifilis)*, Sciaenidae *(Pseudotolithus senegalensis)* and Sphyraenidae *(Sphyraena afra)*.

The number of species in each system varies according to its size and type. It must again be remembered that not all systems have been sampled equally with a comprehensive suite of gear. Most of the open estuaries are species rich, for example the Senegal estuary has 133 species (Diouf *et al.*, 1991) and the Fatala estuary in Guinea 102 species (Baran, 1995). The coastal lakes vary considerably, with the Ébrié system in Ivory Coast having 153 species (Albaret, 1994) and the Lagos Lagoon system only 79 (Fagade and Olaniyan, 1974). Nevertheless there is a group of about 25 ubiquitous species that occur in all the estuaries of all types along the West African coast, the *peuplement de base* of Albaret and Écoutin (1990). This group can include up to eight species of Carangidae, six Sciaenidae, six Mugilidae, and several clupeids and ariids (Baran, 1995). In most West African estuaries, Clupeidae are the dominant family, especially in the area from Guinea to Senegal where they make up 61–85% of the number of individuals. However, the dominant species is not always the same: in Senegal it is *Ethmalosa fimbriata;* in Gambia and Guinea-Bissau, *E. fimbriata* and *Sardinella madarensis* are equally abundant; and in Guinea *Ilisha africana* is the most abundant species (Baran, 1995).

Examples of two systems that have been studied in detail are the coastal lake system of Lagos Lagoon in Nigeria and further west the open Fatala estuary of Guinea. Their fish faunas are summarized below.

### Lagos Lagoon, Nigeria

The fish community of this large system was documented in a series of papers by Fagade and Olaniyan (1972, 1973, 1974), Ikusemiju and Olaniyan (1977) and earlier by Williams (1962). A total of 79 species of 34 families have been recorded from the system. The great seasonal changes in the salinity regime of Lagos Lagoon, with periods of high and low salinity, has led to the classification of the fishes into three groups according to their seasonal distribution.

Group 1. Species in this group occur in the system throughout the year and can tolerate the great change in salinities between the dry and wet seasons (p.35). Most fall into the marine migrant category and include the commercially important clupeid *Ethmalosa fimbriata*. The remainder are freshwater species: *Chrysichthys nigrodigitatus, Sarotherodon melanotheron* and *Tilapia guineensis*.

Group 2. This group includes species that are found in the system only between December and May (mainly dry season) when the salinity fluctuates between 0.5‰ and 28‰. They are all marine migrants and all juveniles.

Group 3. There are 17 species in this group but they are usually few in number and are found only when salinities fall below 1‰. They are primarily freshwater species such as *Schilbe mystus*, *Clarias lazera*, *Lates niloticus* and Mormyridae.

### Fatala estuary, Guinea

The fishes of this open estuary have been the subject of an exhaustive study by Baran (1995). The Fatala River is 190 km in length, the estuarine portion of which is about 60 km long. Rainfall in the area is high (about 3.5 m per year) and salinity varies seasonally according to the wet (June–October) and dry seasons (November–May). There are extensive mangrove areas in the estuary and the banks are fringed almost entirely by three species of *Rhizophora* and *Avicennia africana*.

A total of 102 species have been recorded from the Fatala estuary and these are listed in Table 3.8 together with their affinities – marine migrants, estuarine or freshwater – and which species and genera also occur in the Indo–West Pacific or tropical West Atlantic. There is a strong freshwater influence in the Fatala estuary. In terms of species the Cichlidae comprise nearly 10% of the fish fauna. Altogether freshwater families, including the Characidae, Bagridae, Mormyridae, Schilbeidae, Centropomidae and Notopteridae, make up 24% of species. Estuarine species make up another 30% of the species, including 12 species of Gobiidae and Eleotridae, while the marine migrant category form the remaining 46%. In terms of numbers of individuals and biomass, the fish community is dominated by the sciaenid *Pseudotolithus elongatus* and the clupeid *Ilisha africana*. The relative abundances and biomasses of the top 25 species are shown in Fig. 3.9. The estuary is dominated numerically by large numbers of relatively small Clupeidae, but in terms of biomass the sciaenids, polynemids and mugilids are more important.

The number of species and genera common to the Fatala estuary (tropical East Atlantic) and the Indo–West Pacific and tropical West Atlantic are indicated in Table 3.8. Only five species are common to all three regions (one of which, *Mugil cephalus*, is circumtropical), nine are common to the Indo–West Pacific and the tropical East Atlantic, and twelve to the tropical East and West Atlantic.

Baran recognized seven ecological subdivisions of fishes in the Fatala estuary (Fig. 3.10). Subdivisions 2–5 fall into the marine migrant category and subdivisions 6 and 7 fall into the freshwater category.

1. A strictly estuarine group, in which adults and juveniles occur in the estuary and spawning takes place within the estuary. In this group are the Gobiidae, Syngnathidae and some species of Mugilidae, as well as *Ethmalosa fimbriata*.

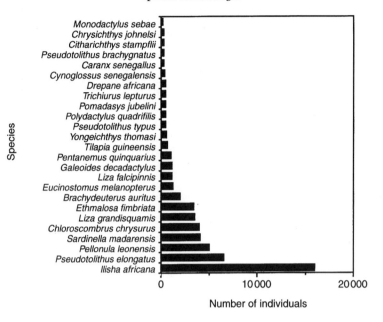

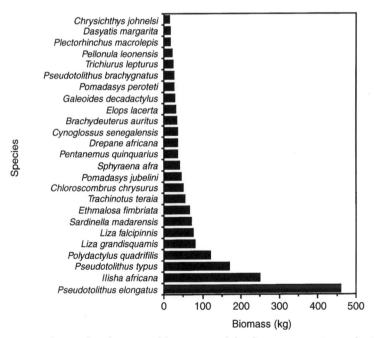

**Fig. 3.9** Relative abundances and biomasses of the dominant species in the Fatala estuary, Guinea, West Africa (redrawn from Baran, 1995).

**Table 3.8** Fishes found in the Fatala estuary, Guinea, West Africa, together with their category (M, marine migrant; E, estuarine; F, freshwater) and occurrence (G, genus present; S, species present) in the tropical Indo–West Pacific and tropical West Atlantic (modified from Baran, 1995)

| Family and species | Category | Indo–West Pacific | West Atlantic |
|---|---|---|---|
| Ariidae | | | |
| *Arius heudeloti* | M | G | G |
| *Arius latiscutatus* | M | G | G |
| *Arius parkii* | M | G | G |
| Bagridae | | | |
| *Chrysichthys johnelsi* | F | | |
| *Chrysichthys maurus* | E | | |
| *Chrysichthys nigrodigitatus* | E | | |
| Batrachoididae | | | |
| *Batrachoides liberiensis* | M | G | G |
| Belonidae | | | |
| *Strongylura senegalensis* | E | G | G |
| *Tylosurus crocodilus* | M | S | S |
| Bothidae | | | |
| *Citharichthys stampflii* | E | | G |
| Carangidae | | | |
| *Caranx hippos* | M | G | S |
| *Caranx senegallus* | M | G | G |
| *Chloroscombrus chrysurus* | M | | S |
| *Hemicaranx bicolor* | M | | |
| *Lichia amia* | M | S | |
| *Selene dorsalis* | M | | G |
| *Trachinotus teraia* | E | G | G |
| Carcharhinidae | | | |
| *Rhizoprionodon acutus* | M | S | G |
| Centropomidae | | | |
| *Lates niloticus* | F | G | |
| Characidae | | | |
| *Brycinus longipinnis* | F | | |
| *Brycinus macrolepidotus* | F | | |
| *Hydrocynus forskalii* | F | | |
| Cichlidae | | | |
| *Hemichromis bimaculatus* | F | | |
| *Hemichromis fasciatus* | E | | |
| *Sarotherodon caudomarginatus* | E | G | |
| *Sarotherodon melanotheron* | E | G | |
| *Tilapia brevimanus* | F | G | |
| *Tilapia buttikoferi* | F | G | |
| *Tilapia guineensis* | E | G | |
| *Tylochromis intermedius* | F | | |
| *Tylochromis leonensis* | F | | |
| Clupeidae | | | |
| *Ethmalosa fimbriata* | E | | |
| *Ilisha africana* | M | G | G |
| *Pellonula leonensis* | E | | |
| *Sardinella madarensis* | M | G | G |

**Table 3.8** Continued

| Family and species | Category | Indo–West Pacific | West Atlantic |
|---|---|---|---|
| Cynoglossidae | | | |
| *Cynoglossus monodi* | M | G | G |
| *Cynoglossus senegalensis* | E | G | G |
| Dasyatidae | | | |
| *Dasyatis margarita* | M | G | G |
| Echeneidae | | | |
| *Echeneis naucrates* | M | S | S |
| Eleotridae | | | |
| *Bostrychus africanus* | E | G | |
| *Eleotris daganensis* | E | G | G |
| *Eleotris senegalensis* | E | G | G |
| *Kribia kribensis* | E | | |
| Elopidae | | | |
| *Elops lacerta* | M | G | G |
| Ephippididae | | | |
| *Chaetodipterus lippei* | M | | G |
| *Drepane africana* | M | G | |
| Exocoetidae | | | |
| *Fodiator acutus* | M | G | |
| Gerreidae | | | |
| *Eucinostomus melanopterus* | M | | S |
| *Gerres nigri* | E | G | G |
| Gobiidae | | | |
| *Bathygobius soporata* | M | G | S |
| *Ctenogobius lepturus* | E | G | |
| *Gobioides ansorgii* | E | G | |
| *Gobionellus occidentalis* | E | | G |
| *Gobius rubropunctatus* | E | G | |
| *Periophthalmus barbarus* | E | G | |
| *Porogobius schlegelii* | E | | |
| *Yongeichthys thomasi* | E | G | |
| Gymnuridae | | | |
| *Gymnura micrura* | M | G | S |
| Haemulidae | | | |
| *Brachydeuterus auritus* | M | | |
| *Plectorhinchus macrolepis* | M | G | |
| *Pomadasys jubelini* | E | G | |
| *Pomadasys peroteti* | M | G | |
| Hemiramphidae | | | |
| *Hyporhamphus picarti* | E | G | |
| Hepsetidae | | | |
| *Hepsetus odoe* | F | | |
| Lobotidae | | | |
| *Lobotes surinamensis* | M | S | S |
| Lutjanidae | | | |
| *Lutjanus agennes* | M | G | G |
| *Lutjanus dentatus* | M | G | G |
| *Lutjanus endecacanthus* | M | G | G |
| *Lutjanus goreensis* | M | G | G |

Table 3.8   Continued

| Family and species | Category | Indo–West Pacific | West Atlantic |
|---|---|---|---|
| Monodactylidae | | | |
| *Monodactylus sebae* | E | G | |
| Mormyridae | | | |
| *Marcusenius thomasi* | F | | |
| *Mormyrops anguilloides* | F | | |
| *Petrocephalus tenuicauda* | F | | |
| Mugilidae | | | |
| *Liza dumerili* | M | S | |
| *Liza falcipinnis* | E | G | |
| *Liza grandisquamis* | E | G | |
| *Mugil bananensis* | M | G | |
| *Mugil cephalus* | M | S | S |
| *Mugil curema* | M | G | S |
| Notopteridae | | | |
| *Papyrocranus afer* | F | | |
| Ophichthidae | | | |
| *Myrophis plumbeus* | E | G | |
| Polynemidae | | | |
| *Galeoides decadactylus* | M | | |
| *Pentanemus quinquarius* | M | | |
| *Polydactylus quadrifilis* | M | G | |
| Rhinobatidae | | | |
| *Rhinobatos cemiculus* | M | G | |
| Schilbeidae | | | |
| *Schilbe micropogon* | F | | |
| Sciaenidae | | | |
| *Pseudotolithus brachygnatus* | M | | |
| *Pseudotolithus elongatus* | E | | |
| *Pseudotolithus epipercus* | M | | |
| *Pseudotolithus hostia* | M | | |
| *Pseudotolithus senegalensis* | M | | |
| *Pseudotolithus typus* | M | | |
| *Pteroscion peli* | E | | |
| Scombridae | | | |
| *Scomberomorus tritor* | M | G | |
| Soleidae | | | |
| *Synaptura cadenati* | E | G | |
| Sphyraenidae | | | |
| *Sphyraena afra* | M | G | |
| Syngnathidae | | | |
| *Enneacampus kaupi* | E | | |
| *Microphis brachyurus* | E | G | |
| Tetraodontidae | | | |
| *Ephippion guttifer* | M | | |
| *Lagocephalus laevigatus* | M | G | S |
| Trichiuridae | | | |
| *Trichiurus lepturus* | M | S | S |
| Zeidae | | | |
| *Zeus faber* | M | S | |

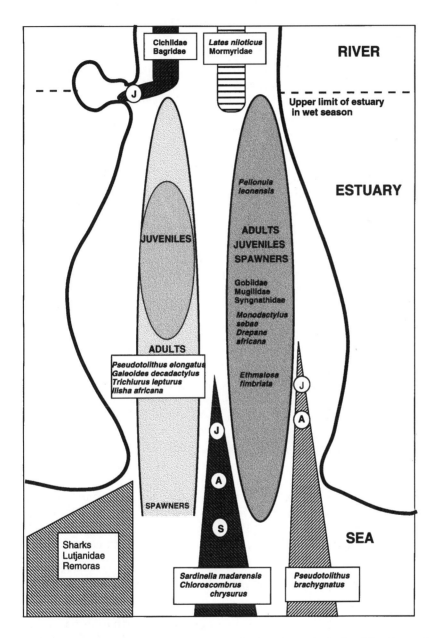

**Fig. 3.10** Diagrammatic representation of the seven ecological subdivisions of fishes in the Fatala estuary, Guinea, West Africa (J, juveniles; A, adults; S, spawners) (redrawn from Baran, 1995).

2. A group of euryhaline species, in which the juveniles use the estuary as a nursery ground, the adults are mainly coastal and all spawning takes place in coastal waters. Included here are the polynemid *Galeoides decadactylus* and the sciaenid *Pseudotolithus elongatus* as well as *Ilisha africana* and *Trichiurus lepturus*.
3. A group of less euryhaline species, in which juveniles and adults occur in the lower reaches of the estuary and spawning may take place around the mouth. Examples here are the clupeid *Sardinella madarensis* and the carangid *Chloroscombrus chrysurus*.
4. A small group in which the juveniles and adults both utilize estuaries, but spawning takes place in the sea. *Pseudotolithus brachygnatus* falls into this subdivision.
5. Occasional marine visitors to the estuary include sharks, lutjanids and remoras.
6. A freshwater group, in which the juveniles and adults utilize the estuary, but reproduce mainly in fresh water. This includes the numerous cichlids and the bagrid catfishes.
7. A group of occasional freshwater visitors to the upper reaches, usually juveniles of such species as *Lates niloticus* and Mormyridae.

### Comparisons with the East African coast

The overall number of species and the proportion of estuarine species in West African systems is similar to that of East Africa, but there are two important contrasts. Firstly, the number and penetration of freshwater species into West African systems is greater, in terms both of number of species and of numbers of individuals. This may be due, in part at least, to the higher rainfall in West Africa and hence the amount of fresh water entering the estuaries, resulting in very low salinities for much of the year. Secondly, in West Africa there is a greater similarity between the coastal fish fauna and that of the estuaries, than is the case in East Africa. This latter difference can be ascribed to the wide continental shelf and estuarine conditions that extend along much of the West African coast, when compared with the relatively narrow shelf and marked contrast between marine and estuarine waters in East Africa.

### Estuarine coastal waters of the tropical East Atlantic (West Africa)

As in the Asian and Australian regions of the Indo–West Pacific, there is an estuarine coastal zone and an assemblage of estuarine and coastal fishes that are commercially important. The shelf fauna of West Africa is divisible into a number of different assemblages that are correlated with depth and substratum type. A review and summary of the whole fauna is provided by Longhurst and Pauly (1987). The estuarine–coastal component extends to not more than 40 m depth and occurs over relatively soft, muddy bottoms. It is dominated by sciaenids such as *Pseudotolithus senegalensis, P. typus* and *Pteroscion peli*; ariid catfishes, particularly *Arius heudeloti, A. mercatoris* and *A. latiscutatus*; polynemids such as *Galeoides decadactylus* and *Pentanemus quinquarius*; the pomadasyid, *Pomadasys jubelini*; and several species of *Cynoglossus*. Some species are more estuarine than others, for example *Pseudotolithus typus* replaces *P. senegalensis* as conditions become more estuarine and *Pentanemus quinquarius* replaces *Galeoides decadactylus*. However, there is only a gradual change in conditions and an essentially estuarine fauna extends over much of the inner continental shelf.

### Tropical West Atlantic

This region extends from the Gulf of Mexico to northern South America and contains a full range of estuarine types and habitats. Estuarine waters extend into shallow coastal areas in parts of the Gulf of Mexico and particularly off Guyana in northern South America. The northern part of the Gulf of Mexico is barely subtropical, but is included here as its marine faunal affinities lie mainly with the rest of the Gulf. The fish faunas of the tropical West Atlantic are more diverse than those of the East Atlantic, but perhaps not as diverse as those of the Indo–West Pacific. Nevertheless, most larger systems contain more than 100 species. As in other regions, the numbers and composition of the fish community in an estuary depend upon its size, physical characteristics and range of habitats. A small system such as the Tortuguero estuary in Costa Rica has about 70 species (Gilbert and Kelso, 1971); in the Gulf of Mexico the large coastal lake of Terminós Lagoon has 122 species (Yáez-Arancibia *et al.*, 1985), whereas the hypersaline coastal lake of Laguna Madre in Texas has only about 70 species (Hedgpeth, 1967), and the huge Mississippi delta region on the margins of the subtropics has 208 species (many of which are North American freshwater species) (Deegan and Thompson, 1985). In South America, about 100 species have been recorded in the large Orinoco estuary system (Cervigón, 1985).

The dominant families are the Ariidae, Carangidae, Clupeidae, Engraulididae, Gerreidae, Mugilidae and Sciaenidae, but the dominant species vary considerably, and many subtropical Gulf of Mexico species do not occur in tropical South America, being separated by the reefs and sandy grounds of the Caribbean (Longhurst and Pauly, 1987). Large numbers of sciaenids (in terms both of species and of individuals) are present in the West Atlantic and form the basis for important commercial fisheries. Cichlidae are a significant part of the community in the upper reaches of many systems, although not in the marginally subtropical areas of the northern Gulf of Mexico, such as the Mississippi delta.

Well-studied estuarine fish faunas include those of the Tortuguero (Costa Rica), the Sinnamary (French Guiana), the Orinoco (Venezuela), Terminós Lagoon (Mexico) and Laguna Madre (USA), and these are used here to illustrate some of the characteristics of the estuarine fish communities of the subtropical and tropical West Atlantic. Some information is also provided for comparative purposes on the fish faunas of the Mississippi delta region.

### Tortuguero estuary, Costa Rica

The fishes of this Central American estuary (p.23) have received considerable attention and the following account is based upon the work of Gilbert and Kelso (1971) and Nordlie and Kelso (1975). The 30 families and 70 species recorded from the estuary are shown in Table 3.9 together with their category as marine migrant, estuarine or freshwater. As with other tropical estuaries, the marine migrant component makes up the greatest proportion of species (58%), followed by equal numbers of estuarine and freshwater species (21%). These proportions are similar to those for the Fatala estuary in West Africa (pp.82–88), another medium-size open estuary. The Tortuguero is also similar to the Fatala in that it has a pronounced wet and dry season, with resulting changes in salinity, and this strongly influences the composition of the fish fauna. In the wet season a number of freshwater species spread throughout the system, but in the dry season they are replaced by larger numbers of marine visitors. The three main categories of fishes can also be subdivided in a similar way with eight subdivisions. As before, subdivisions 2–5 fall into the marine migrant category. Subdivisions 6–8 fall into the freshwater category.

1. A strictly estuarine group comprising mainly small species of the families Bothidae, Eleotridae, Gobiidae, Soleidae and Syngnathidae. These are present in the system in both the wet and dry seasons. Some species have very specific habitat preferences, for example the eleotrid *Dormitator maculatus* almost always occurs in among the roots of the floating water hyacinth, *Eichornia crassipes*, and the goby *Gobiosoma spes* in thick growths of filamentous algae.
2. Marine migrants that utilize the estuary both as adults and juveniles, but spawn in the sea. Included here are *Mugil curema*, *Centropomus pectinatus*, *Caranx hippos* and *Caranx latus*.
3. Marine migrants that utilize the estuary primarily as a nursery ground and the juveniles of which are common in both wet and dry seasons.

The most abundant species are *Lutjanus jocu, Pomadasys crocro, Centropomus parallelus* and *C. undecimalis.*

4. Marine visitors that are found occasionally in the lower reaches as adults and juveniles. *Arius melanopus, Lutjanus griseus, Scomberomorus maculatus* and *Sphyraena gauchancho* are good examples, as well as the sciaenids *Larimus breviceps* and *Menticirrhus americanus.*

5. A catadromous species of mullet, *Agonostomus monticola,* spawns in the sea and the juveniles spend time in the estuary on their way to the fresh water of mountain streams; they return to the sea again to spawn.

6. Freshwater species that utilize the estuary extensively throughout the year, both as juveniles and adults, such as the cichlids *Cichlasoma citrinellum, C. friedrichsthali* and *C. maculicauda* and the poeciliid *Poecelia mexicana.*

7. Freshwater species that utilize the estuary extensively during the wet season when salinities are very low. Included here are *Cichlasoma centrarchus* and the characids. *Cichlasoma centrarchus* prefers heavy submergent vegetation in contrast to the other species of *Cichlasoma* listed above that occur in more open areas.

8. Occasional freshwater visitors such as *Heterotilapia multispinosa, Belenesox belizanus* and *Phallichthys amates* can be found in the upper reaches, usually in the wet season.

The number of genera and the few species common to the Tortuguero estuary and the East Atlantic (West Africa) and Indo–West Pacific faunas is shown in Table 3.9. Only six species are also found in West Africa while three (*Carcharhinus leucas, Echeneis naucrates* and *Elops saurus*) occur in the Indo–West Pacific. Many of the same genera are found in the tropical West and East Atlantic with the exception of the freshwater genera and the sciaenids.

### Sinnamary estuary, French Guiana

Eighty-three species from 35 families have been recorded from this medium-size estuary (pp.22–23). (Boujard and Rojas-Beltran, 1988) but only 11 of these are also found in the Tortuguero estuary. Estuaries of this region of South America are dominated by siluriiform catfish, sciaenids and engraulids. In the Sinnamary there are four families and 16 species of catfish (Table 3.10), 10 species of Sciaenidae, and 10 species of Engraulididae, none of which occur in the Tortuguero estuary of Central America. Only five species also occur in the tropical East Atlantic and four in the Indo–West Pacific.

About 60% of fishes fall into the marine migrant category, 27% are

*The fishes of tropical estuaries*

**Table 3.9** Fishes found in the Tortuguero estuary, Costa Rica, together with their category (M, marine migrant; E, estuarine; F, freshwater) and occurrence (G, genus present; S, species present) in the tropical Indo–West Pacific and tropical East Atlantic (modified from Gilbert and Kelso, 1971; Nordlie and Kelso, 1975)

| Family and species | Category | Indo–West Pacific | East Atlantic |
|---|---|---|---|
| Ariidae | | | |
| *Arius melanopus* | M | G | G |
| *Bagre filamentosus* | E | G | G |
| Batrachoididae | | | |
| *Batrachoides gilberti* | M | G | G |
| *Batrachoides surinamensis* | M | G | G |
| Belonidae | | | |
| *Strongylura marina* | M | G | G |
| *Strongylura timuca* | M | G | G |
| Bothidae | | | |
| *Citharichthys spilopterus* | E | | G |
| *Citharichthys uhleri* | F | | G |
| Carangidae | | | |
| *Caranx hippos* | M | G | S |
| *Caranx latus* | M | G | G |
| *Oligoplites palometa* | M | | S |
| Carcharhinidae | | | |
| *Carcharhinus leucas* | M | S | S |
| Centropomidae | | | |
| *Centropomus parallelus* | M | | |
| *Centropomus pectinatus* | M | | |
| *Centropomus undecimalis* | M | | |
| Characidae | | | |
| *Astyanax fasciatus* | F | | |
| *Brycon guatemalensis* | F | | |
| *Hyphessobrycon tortuguerae* | F | | |
| Cichlidae | | | |
| *Cichlasoma centrarchus* | F | | |
| *Cichlasoma citrinellum* | F | | |
| *Cichlasoma friedrichsthali* | F | | |
| *Cichlasoma maculicauda* | F | | |
| *Heterotilapia multispinosa* | F | | |
| Dasyatidae | | | |
| *Himantura schmardae* | M | G | G |
| Echeneidae | | | |
| *Echeneis naucrates* | M | S | S |
| Eleotridae | | | |
| *Dormitator maculatus* | E | | |
| *Eleotris amblyopsis* | F | G | G |
| *Eleotris pisonis* | F | G | G |
| *Gobiomorus dormitator* | E | | |
| *Leptophilypinus fluviatilis* | E | | |
| Elopidae | | | |
| *Elops saurus* | M | S | S |
| Engraulididae | | | |
| *Anchoa lamprotaenia* | M | | |
| *Anchoviella elongata* | M | | |
| Gerreidae | | | |
| *Diapterus olisthostomus* | M | | |

Table 3.9 Continued

| Family and species | Category | Indo–West Pacific | East Atlantic |
|---|---|---|---|
| *Diapterus plumieri* | M | | |
| *Diapterus rhombeus* | M | | |
| *Eucinostomus argenteus* | M | | G |
| *Eucinostomus pseudogula* | M | | G |
| Gobiidae | | | |
| *Awaous tajasica* | F | G | |
| *Bathygobius soporata* | M | G | S |
| *Evorthodus lyricus* | E | | |
| *Gobionellus boleosoma* | E | | G |
| *Gobionellus fasciatus* | E | | G |
| *Gobionellus pseudofasciatus* | E | | G |
| *Gobionellus spes* | E | | G |
| Haemulidae | | | |
| *Pomadasys crocro* | M | G | G |
| Hemiramphidae | | | |
| *Hyporhamphus roberti* | M | G | |
| Lutjanidae | | | |
| *Lutjanus griseus* | M | G | G |
| *Lutjanus jocu* | M | G | G |
| Megalopidae | | | |
| *Megalops atlanticus* | M | G | G |
| Microdesmidae | | | |
| *Microesmus carri* | M | G | G |
| Mugilidae | | | |
| *Agonostomus monticola* | M | G | |
| *Mugil curema* | M | G | S |
| Ophichthiidae | | | |
| *Myrophis punctatus* | E | G | G |
| Poeciliidae | | | |
| *Belenesox belizanus* | F | | |
| *Phallichthys amates* | F | | |
| *Poecilia mexicana* | F | | |
| Sciaenidae | | | |
| *Bairdiella ronchus* | M | | |
| *Larimus breviceps* | M | | |
| *Menticirrhus americanus* | M | | |
| *Micropogon furnieri* | M | | |
| Scombridae | | | |
| *Scomberomorus maculatus* | M | G | G |
| Soleidae | | | |
| *Achirus lineatus* | E | G | |
| *Trinectes paulistanus* | E | | |
| Sphyraenidae | | | |
| *Sphyraena gauchancho* | M | G | G |
| Syngnathidae | | | |
| *Coleotropis blackburni* | M | | |
| *Melaniris chagresi* | E | | |
| *Oostethus lineatus* | E | | |
| *Pseudophallus mindi* | M | | |
| Tetraodontidae | | | |
| *Sphoeroides testudineus* | M | G | G |

**Table 3.10** Siluriiform catfishes of the Sinnamary and Orinoco estuaries, their category (M, marine migrant; E, estuarine; F, freshwater), recorded salinity ranges and abundance in the Sinnamary estuary (****, in >10% of catches; ***, in 5–10% of catches; **, in 1–5% of catches; *, <1% of catches). Data from Boujard and Rojas-Beltran (1988), Cervigón (1985) and Cervigón *et al.* (1992)

| Family and species | Category | Salinity | Abundance |
|---|---|---|---|
| Ariidae | | | |
| *Arius couma* | E | 0–25 | ** |
| *Arius grandicassis* | M | 5–36 | *** |
| *Arius herzbergii* | M | 5–50 | *** |
| *Arius parkeri* | E | 5–25 | ** |
| *Arius passany* | E | 5–25 | ** |
| *Arius phrygiatus* | E | 0–25 | * |
| *Arius proops* | M | 5–50 | **** |
| *Arius quadriscutis* | E | 5–25 | **** |
| *Arius rugispinis* | E | 5–25 | * |
| *Bagre bagre* | M | 5–36 | *** |
| *Cathorops spixii* | M | 5–50 | **** |
| Auchenipteridae | | | |
| *Pseudauchenipterus nodosus* | E | 0–25 | **** |
| Loricariidae | | | |
| *Hypostomus gymnorhynchus* | F | 0–5? | * |
| *Hypostomus ventromaculatus* | F | 0–5? | * |
| *Hypostomus watwata* | F | 0–5? | * |
| Pimelodidae | | | |
| *Pimelodus blochii* | F | 0–5? | * |

estuarine and 13% freshwater. The relative abundance of the catfishes and their habitat preferences are shown in Table 3.10. The most abundant species are *Arius herzbergii, A. proops, A. quadriscutis, Cathorops spixii* and *Pseudauchenipterus nodosus* which are either marine or estuarine. The freshwater Loricariidae and Pimelodidae are less common, being confined mainly to the upper reaches of the estuary.

The abundant sciaenids are *Cynoscion acoupa, Macrodon ancylodon, Stellifer rastrifer* and *S. microps*, while the most common engraulids are *Lycengraulis batesii, Pterengraulis atherinoides* and *Anchovia clupeoides*. Other abundant species include the mullet *Mugil curema* and the tetraodontid *Colomesus psittacus*. The latter is an estuarine species and is replaced in coastal areas by the more marine *Sphoeroides testudineus*. Elasmobranchs are also well represented in the Sinnamary estuary, with five species of carcharhinid sharks recorded (the most abundant are *Carcharhinus leucas* and *C. limbatus)*, two rays and two species of hammerhead sharks (Sphyrnidae). Marine species with juveniles that use the estuary as a nursery area are *Caranx hippos, Chloroscombrus chrysurus, Hemicaranx amblyr-hynchus, Selene vomer, Trachinotus cayennensis, Chaetodipterus faber* and *Trichiurus lepturus.*

## Orinoco system, Venezuela

The fish fauna of this much larger estuary and delta is very similar to that of the Sinnamary although somewhat more diverse (Cervigón, 1985). There are fewer sharks but more rays (including *Dasyatis guttata, D. geijskesi, Himantura schmardae* and *Gymnura micrura*) as well as the sawfish *Pristis pectinata.* There are also more sciaenids (12), clupeids (4) and centropomids (4). The siluriiform catfish fauna is the same although more freshwater species are encountered in the upper reaches.

## Estuarine coastal waters

The coastal waters of Venezuela, French Guiana, Surinam, Guyana and north-east Brazil, stretching from Trinidad to the Amazon, are character-ized by a vast input of low-salinity water. Estuarine conditions exist along much of the coast and an essentially estuarine fish fauna is found to depths of at least 25 m and up to 60 m in some areas (Cervigón, 1985; Lowe-McConnell, 1987). The importance of the fishery potential of the region led to a number of detailed studies of the fishes, particularly of the Sciaenidae (Lowe-McConnell, 1962, 1966, 1987) and Engraulididae (Cervigón, 1969; Whitehead, 1973).

There are two distinct fish communities in the coastal zone, corre-sponding to those off the West African coast, but extending deeper because of the greater depth penetration of tropical water off South America (Longhurst and Pauly, 1987). The first is in the inshore 'brown-water' zone over soft muds, extends to depths of about 20 m, and is referred to as the 'brown-fish' zone (Lowe-McConnell, 1966, 1987) because the fish are mostly dark. This community is dominated by ariid catfishes such as *Arius grandicassis, Bagre bagre* and *Cathorops spixii;* stingrays such as *Dasyatis guttata* and *D. geijskesi,* flatfish such as *Achirus achirus* and *Symphurus* spp., the spadefish *Chaetodipterus faber,* the threadfin *Polynemus virginicus,* and about 12 species of sciaenid including *Macrodon ancylodon, Stellifer microps, S. rastrifer, Micropogonias furnieri, Cynoscion virescens* and *Nebris microps.* The second community is over firmer muds from about 20 to 60 m but is still influenced by reduced salinities and relatively high turbidities. This 'golden-fish' zone is dominated by sciaenids, some of which extend from the 'brown-fish' zone. As the name implies, most fish of this community are of a lighter colour. Although ariids, dasyatids and others are still present, they are less numerous than the sciaenids, which include *Micropogonias furnieri, Cynoscion virescens, Nebris microps, Larimus breviceps, Ctenosciaena gracilicirrhus* and *Polyclemus brasiliensis*. Carangids such as *Vomer setapinnis, Hemicaranx amblyrhynchus* and *Trachinotus cayennensis* are also abundant in the 'golden-fish' zone. At about 50 m the estuarine influence largely ceases and the 'silver-fish' zone begins – this is dominated by carangids, grunters, pomfrets and Spanish mackerel.

West of Trinidad along the Caribbean coast the situation is different; there is little freshwater influence on the coastal waters, and estuarine fishes are confined mainly to estuaries. The dominant coastal families are typically those such Carangidae, Ephippididae, Lutjanidae and Trichiuridae. However, in the far west of Venezuela and in Colombia, estuarine influences once more dominate the Caribbean coastal zone and the shallow-water fish community is very similar to that of the north-east of South America, with at least 35 species common to both areas. This similarity does not extend to the Gulf of Mexico, where estuarine coastal species are almost completely different from those of South America, although the Sciaenidae, Ariidae, Clupeidae and Engraulididae are still among the dominant families (Cervigón, 1985).

### Terminós Lagoon and Campeche Sound, Gulf of Mexico

The large coastal lake called Terminós Lagoon in the southern Gulf of Mexico (p.10, Fig. 2.1) has a fish fauna that is closely associated with that of the adjacent marine area known as Campeche Sound. The various fish communities are among the best studied of the region and comprehensive data can be found in Chavance *et al.* (1984), Yáñez-Arancibia *et al.* (1980, 1985a,b) and Yáñez-Arancibia and Sánchez-Gil (1986). Because of the close relationship between the fishes of this coastal lake and the estuarine coastal waters, they are treated together here.

A total of 122 species have been recorded from Terminós Lagoon, 82 species from the Puerto Real Inlet of the lagoon, 47 species from the Carmen Inlet of the lagoon, and 149 species from the coastal area of Campeche Sound. The relationships and overlaps between the fish communities of each of these areas is illustrated in Fig. 3.11 using the 56 species with the highest numerical abundances and greatest biomasses. In the area encompassed by the lagoon and the two inlets, 31 species make up 90% of the number of fishes, of which the most numerous are the catfishes *Arius felis* and *A. melanopus*, the gerreid *Eucinostomus gula*, the sciaenid *Bairdiella chrysoura* and the tetraodontid *Sphoeroides testudineus*. Both *Arius felis* and *Eucinostomus gula* are also abundant in Campeche Sound. Species that are most numerous only in the lagoon include the gerreid *Diapterus rhombeus*, the sciaenids *Micropogonias undulatus* and

*Cynoscion nebulosus*, and the batrachoidid *Opsanus beta*. There are also suites of species that are shared between the lagoon and the two different inlets (Fig. 3.11) as well as a number of species that are most numerous only in Puerto Real Inlet. In Puerto Real Inlet, 21 species make up 93% of numbers of fish and include Clupeidae, Haemulidae, Lutjanidae, Sciaenidae and Sparidae. In Carmen Inlet, 11 species constitute 81% of total numbers, but all are shared with the lagoon or with Campeche Sound. In Campeche Sound, 31 species make up 93% of the community by numbers and the dominant species are *Arius felis, Eucinostomus gula, Harengula jaguana, Chloroscombrus chrysurus, Synodus foetens, Diplectrum radiale, Syacium gunteri, Prionotus punctatus* and *Polydactylus octonemus*. The Campeche Sound fish fauna is a mixture of a tropical shallow-water sciaenid community together with various gerreids, haemulids and upeneids more characteristic of lutjanid and sparid communities (Longhurst and Pauly, 1987).

In the Terminós complex, Yáñez-Arancibia *et al.* (1980, 1985) have divided the fishes into seasonal visitors, occasional visitors and residents. The first two of these correspond to the marine migrant category and some of the third to the estuarine category. Approximately 45% of species are seasonal visitors, 45% occasional visitors and 10% residents. Hence in general terms, the proportion of marine migrants is higher than in other systems. The proportions of each of these groups in the different parts of the Terminós complex are shown in Fig. 3.12. The seasonal visitor group contains mainly species the juveniles of which utilize the Terminós system as a nursery and return to the sea as adults or subadults; it includes species such as *Cynoscion nebulosus, Lutjanus griseus* and *Archosargus probatocephalus*. It is thought that most immigration of juveniles is through the Puerto Real Inlet because the greatest numbers of juveniles are found in this area and the eastern parts of the main lagoon. There is then a gradual westward movement following the prevailing currents, with most emigration taking place through Carmen Inlet (Fig. 2.1). The greatest numbers of juveniles occur in the rainy season in July and August.

Occasional visitors do not show any definite patterns and include a wide range of euryhaline marine species, particularly sciaenids, lutjanids, clupeids, serranids and elasmobranchs. The resident group comprises those species that are found all the time in Terminós Lagoon and complete their life cycle within the system, but it does not correspond exactly with the estuarine category because at least three of the common resident species (*Arius felis, Eucinostomus gula* and *Sphoeroides testudineus*) also occur in Campeche Sound. If these are excluded from the estuarine category, then the dominance of marine species in the system is overwhelming. Relatively few freshwater species occur in Terminós but

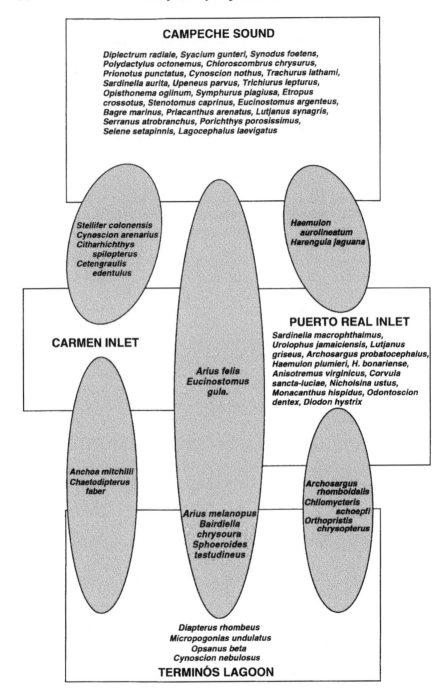

**CAMPECHE SOUND**

*Diplectrum radiale, Syacium gunteri, Synodus foetens, Polydactylus octonemus, Chloroscombrus chrysurus, Prionotus punctatus, Cynoscion nothus, Trachurus lathami, Sardinella aurita, Upeneus parvus, Trichiurus lepturus, Opisthonema oglinum, Symphurus plagiusa, Etropus crossotus, Stenotomus caprinus, Eucinostomus argenteus, Bagre marinus, Priacanthus arenatus, Lutjanus synagris, Serranus atrobranchus, Porichthys porosissimus, Selene setapinnis, Lagocephalus laevigatus*

*Stellifer colonensis Cynoscion arenarius Citharhichthys spilopterus Cetengraulis edentulus*

*Haemulon aurolineatum Harengula jaguana*

**CARMEN INLET**

**PUERTO REAL INLET**

*Sardinella macrophthalmus, Urolophus jamaiciensis, Lutjanus griseus, Archosargus probatocephalus, Haemulon plumieri, H. bonariense, Anisotremus virginicus, Corvula sancta-luciae, Nicholsina ustus, Monacanthus hispidus, Odontoscion dentex, Diodon hystrix*

*Arius felis Eucinostomus gula.*

*Anchoa mitchilli Chaetodipterus faber*

*Archosargus rhomboidalis Chilomycteris schoepfi Orthopristis chrysopterus*

*Arius melanopus Bairdiella chrysoura Sphoeroides testudineus*

*Diapterus rhombeus Micropogonias undulatus Opsanus beta Cynoscion nebulosus*

**TERMINÓS LAGOON**

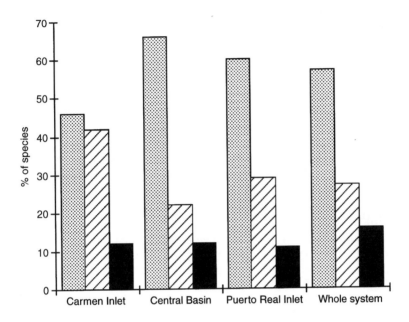

**Fig. 3.12** Relative proportions of occasional visitors (light-stipple), seasonal visitors (hatched) and resident fish species (black) in the Terminós Lagoon complex (data from Yáñez-Arancibia *et al.*, 1985).

*Cichlasoma urophthalmus, C. fenestratum* and *Cyprinodon variegatus* are listed by Yáñez-Arancibia *et al.* (1985).

There are strong seasonal changes in the species composition and diversity in the Terminós system, in large part due to the immigration of juveniles. Biomasses are highest in the rainy season (June–September) and in the northern storm season (October–March). This corresponds with maximum river discharge, reduced salinities and increased turbidities, as well as higher nutrient levels. Highest densities of fish are found in the *Thalassia* seagrass and *Rhizophora* mangrove areas.

The dominant fishes of Campeche Sound are shown in Fig. 3.11. They could be divided into those that occur year round and those that occur

**Fig. 3.11** Relationships and overlaps of the 56 dominant species in the Campeche Sound and Terminós Lagoon complex of southern Mexico (data from Yáñez-Arancibia *et al.*, 1985).

only at one or more seasons. The latter are largely related to the migrations to and from Terminós Lagoon. Of the total of 149 species recorded
in Campeche Sound, 49% are euryhaline and have a close ecological
relationship with Terminós Lagoon. The fishes of this estuarine coastal
area also fall into two groups associated with the two zones offshore from
Terminós Lagoon (pp.25–26). The shelf in this area can be divided into
western and eastern sections based on the influence of outflows from
Terminós Lagoon. The west, which receives the lagoonal discharge, has
high turbidities, lime-clay sediments and high organic content, whereas
the east is more marine in character with clearer water, calcareous
sediments and seagrasses growing on the bottom. Those species associated
with the more turbid western zone are generally more elongated, flattened
or silvery, whereas those of the quieter, clearer water of the eastern zone
have a greater variety of shapes and colours (Yáñez-Arancibia *et al.*,
1985b). In this respect, the zones are perhaps analogous with those off
the Guiana and West African shelves, except that here the change in
conditions takes place perpendicular to the shore rather than parallel to
it. These zones do, however, emphasize and reinforce the importance of
the influence of 'estuarine' conditions and habitats in determining the
species composition of the coastal fish fauna.

### *Laguna Madre, Texas, USA*

This usually hypersaline coastal lake system supports an abundant and
varied fish fauna (Hedgpeth, 1967) but is liable to extremes of both
temperature and salinity (pp.37–38) which have a marked influence on
the species composition. Approximately 70 species, many of them typical
of the subtropical region of the Gulf of Mexico, have been recorded, of
which the most numerous are the sciaenids, *Sciaenops ocellata*, *Pogonias
cromis* and *Cynoscion nebulosus*, the mullet *Mugil cephalus*, the cyprinodonts
*Fundulus grandis*, *F. similis* and *Cyprinodon variegatus*, several anchovies
*Anchoa* spp., the silverside *Menidia berylina*, the pinfish *Lagodon rhomboides*
and the tenpounder *Elops saurus*. Most can tolerate salinities of 50‰ to
60‰ but above this the number of species declines: 67 species have been
found in salinities up to 35‰, 52 up to 45‰, 19 up to 60‰ and 10 up
to 75‰ (Gunter, 1967). Both *Cynoscion nebulosus* and *Pogonias cromis*
have been recorded up to 75‰ and the pupfish *Cyprinodon variegatus* to
142‰ (Hedgpeth, 1967). The division of the fish fauna into the three
categories used previously in this chapter is not meaningful here. Many
could be classed as marine migrants but as Gunter (1967) states, most of
the common species in Laguna Madre are very euryhaline and are species
characteristic of estuarine areas rather than marine areas, having been
recorded elsewhere in salinities down to fresh water. Many species which

elsewhere spawn in the sea, spawn in Laguna Madre, not being constrained by low salinities, as is usually the case in estuaries. However, most do not spawn in salinities above 45‰, although the atherinid *Atherinops affinis* can spawn at 72‰. Among the commercially important sciaenids, *Cynoscion nebulosus* and *Pogonias cromis* spawn in Laguna Madre whereas *Sciaenops ocellata* returns to the sea to spawn.

### Mississippi delta region, USA

The estuaries and lakes of this vast delta lie outside the tropics and subtropics in terms of minimum water temperatures (winter temperatures can fall as low as 5–7 °C), but part of the fish fauna consists of subtropical species of the Gulf of Mexico so is mentioned briefly here. In a review of the fish communities of the Mississippi delta, Deegan and Thompson (1985) list 208 species of 69 families. The dominant families are Sciaenidae (14 species), Centrarchidae (a freshwater North American family) (14), Carangidae (12), Gobiidae (10), Cyprinodontidae (8), Clupeidae (7), Ictaluridae (freshwater catfishes of North America) (7) and Cyprinidae (7). However, in keeping with a more temperate regime, over 90% of fish numbers are contributed by only about seven species: *Anchoa mitchilli*, *Micropogonias undulatus*, *Cynoscion arenarius*, *Brevoortia petronus*, *Arius felis*, *Chloroscombrus chrysurus* and *Leiostomus xanthurus*. The various estuaries and lakes making up the delta vary in the composition of their fish faunas. Marine migrants range from a low of about 35% of species in the low-salinity waters of Atchafalaya Bay to 87% in Lake Pontchartrain (salinities up to 27‰). However, the bulk of these are juveniles using the estuaries as nursery grounds. The proportions of truly estuarine species are 25% and 5% for Atchafalaya and Pontchartrain respectively, and for freshwater species 38% and 8% (Deegan and Thompson, 1985). Estuarine coastal waters extend about 200 km from the Mississippi delta and the fish community is very similar to that already described for Campeche Sound. It is dominated by subtropical sciaenids with associated genera such as *Chloroscombrus*, *Vomer*, *Selar*, *Dasyatis*, *Chaetodipterus*, *Trichiurus* and *Bagre* (Longhurst and Pauly, 1987).

## Tropical East Pacific

The estuarine fish faunas of the tropical East Pacific contain relatively few forms from either the tropical Indo–West Pacific or Atlantic regions. Many of the genera are the same as those of the West Atlantic, but few species are shared. Many areas of the tropical South East Pacific (west coast of South America) are influenced by upwellings and cool winds and currents. Hence the tropical nature of the fauna of, for example, Peru, is restricted to low latitudes ( < 3°S) and much of the remainder has warm temperate affinities (Hildebrand, 1946). The tropical estuaries of Central America are less affected and are more comparable with those of other equatorial regions.

Three different types of system are chosen here to illustrate the main features of the fish communities of tropical East Pacific estuaries: the small open estuary of the Rio Claro in Costa Rica, the seasonally closed coastal lakes of Guerrero State of Mexico, and the estuarine coastal waters of Jiquilisco Bay, El Salvador.

### Rio Claro, Costa Rica

The fish community of this small (30 km) estuary and river in the Corcovado National Park was monitored and studied for a period of 8 years (Lyons and Schneider, 1990). Tidal influences normally only reach about 1 km from the mouth, although raised salinities penetrate for about 5 km. The maximum surface salinity recorded was 6‰ and during the wet season, salinities in the lower reaches vary between 1‰ and 5‰. The Rio Claro has a depauperate freshwater fish fauna and a very low number of primary freshwater species. Almost all fishes are marine or estuarine in origin and many are diadromous. A total of 22 species have been found and these are listed in Table 3.11 together with their category, relative abundance and whether they occur in the tropical East or West Atlantic and Indo–West Pacific regions.

The most abundant species throughout the system was the goby *Sicydium salvini*. Other common and widespread species were the gerreid *Eucinostomus currani*, the grunter *Pomadasys bayanus*, the centropomid *Centropomus nigrescens*, the mullets *Agonostomus monticola* and *Mugil curema* and the snapper *Lutjanus novemfasciatus*. The majority of species (64%) belong to the marine category, one of which, *Mugil curema*, is catadromous, and only three species (14%), *Astyanax fasciatus*, *Brachyraphis rhabdophore* and *Pseudophallus elcapitanensis*, are freshwater. The remaining five species are amphidromous – that is, they spawn in fresh water but their larvae drift into the estuary or sea where the juveniles grow for a time before returning to fresh water.

Although the fish fauna of this little system is depauperate, Lyons and Schneider (1990) point out that the species composition and abundances have been very persistent over time. Distance from the sea is one of the major determinants of the distribution of fishes in the Rio Claro because those species that spend part of their life in the sea decline in abundance upstream. The freshwater and diadromous species do not decline in numbers upstream and their distribution is controlled mainly by habitat preferences.

**Table 3.11** Fish species of the Rio Claro, Costa Rica, together with their category (M, marine migrant; E, estuarine; F, freshwater, a, amphidromous; c, catadromous), relative abundance (****, abundant, >1000 caught; ***, common, >100 caught; **, present, >10 caught; *, rare, <10 caught) and their occurrence (G, genus present; S, species present) in the tropical Indo–West Pacific, tropical East or tropical West Atlantic (modified from Lyons and Schneider, 1990)

| Species | Category | Abundance | West Atlantic | East Atlantic | Indo–West Pacific |
|---|---|---|---|---|---|
| *Caranx* sp. | M | ** | G | G | G |
| *Centropomus nigrescens* | M | *** | G | | |
| *Astyanax fasciatus* | F | ** | G | | |
| *Eleotris picta* | M, a | ** | G | G | G |
| *Gobiomorus maculatus* | F, a | ** | G | | |
| *Hemieleotris latifasciatus* | F, a | * | | | |
| *Eucinostomus currani* | M | *** | G | G | |
| *Gobiesox potomias* | M? | ** | G | | |
| *Awaous transandeamus* | F, a | ** | G | G | G |
| *Bathygobius andrei* | M | * | G | G | G |
| *Gobionellus sagittula* | M | ** | G | G | |
| *Sicydium salvini* | F, a | **** | G | | |
| *Pomadasys bayanus* | M | *** | G | G | G |
| *Lutjanus argentiventris* | M | *** | G | G | G |
| *Lutjanus colorado* | M | * | G | G | G |
| *Lutjanus novemfasciatus* | M | *** | G | G | G |
| *Agonostomus monticola* | F, c | *** | S | | G |
| *Mugil curema* | M, c | *** | S | S | G |
| *Brachyraphis rhabdophore* | F | * | | | |
| *Pristis pectinata* | M | * | S | S | S |
| *Pseudophallus elcapitanensis* | F | * | G | | |
| *Sphoeroides annulatus* | M | ** | G | G | G |

## The coastal lakes of Guerrero State, Mexico

There are about ten small, shallow (0.5–2.5 m deep), mangrove fringed coastal lakes in Guerrero, on the Pacific coast of Mexico that are seasonally isolated from the Pacific Ocean. From August to November they are open to the sea through a usually narrow channel (salinities 15–34‰); from November to May they become isolated from the sea when the channel closes and evaporation exceeds freshwater input (salinities > 35‰); and from May to August they are still isolated from the sea, but freshwater input exceeds evaporation, the lagoons gradually fill and finally open to the sea (salinities < 15‰). Their fish faunas were studied by Yáñez-Arancibia (1978), who recorded 105 species from 37 families.

In terms of species numbers the dominant families are Carangidae (11 species), Sciaenidae (8), Gobiidae (8), Gerreidae (7), Urolophidae (6), Engraulididae (5), Poeciliidae (5) and Haemulidae (4). Very few species are found during the entire year or throughout all the coastal lakes, but in general the most numerous species are the catfish *Galeichthys caerulescens*, the mullets *Mugil curema* and *Mugil cephalus*, the gerreids *Diapterus peruvianus* and *Gerres cinereus*, the clupeid *Lile stolifera*, the eleotrid *Dormitator latifrons*, the cichlid *Cichlasoma trimaculatum* and the goby *Gobionellus microdon*.

Freshwater species make up 14% of species, including four Gobiidae, three Eleotridae, two Cichlidae (one of which is *Oreochromis mossambicus* introduced from the Indo–West Pacific), five Poeciliidae and one characid. Only 6% of species are truly estuarine, of which three are ariid catfishes (*Galeichthys caerulescens*, *G. gilberti* and *Arius liropus*), two are gobies (*Gobionellus* spp.) and one a clupeid (*Lile stolifera*). The remaining 80% of species are marine. These can be divided into seasonal visitors (59%) and occasional visitors (21%) in the same way as in Terminós Lagoon in the tropical West Atlantic of Mexico. Juveniles of species that use the estuaries as nursery grounds make up 28% of the seasonal visitors and the dominant taxa are centropomids (2 species), carangids (7), gerreids (7), mugilids (2), bothids (3), soleids (4) and tetraodontids (2) as well as the circumtropical species *Albula vulpes* and *Chanos chanos*. The adults coming in as seasonal visitors (31%) are largely of different families from the juveniles, with clupeids (3 species), engraulids (5), lutjanids (3), haemulids (3) and sciaenids (7) the dominant families. The occasional visitors comprise six species of urolophid rays, several carangids, sciaenids and larger species such as the sawfish *Pristis pectinata*. Additional data on the community structure and biology of the fishes of the lagoons of eastern Mexico are provided by Warburton (1978, 1979).

The percentage of species in common with other areas of the East Pacific and tropical West Atlantic are shown in Fig. 3.13. Eleven species (10%) are endemic to the Pacific coast of Mexico but about 60% occur in estuaries along the tropical East Pacific coasts of Central and South America. In the West Atlantic, 16 species (15%) are found in the Gulf of Mexico and Caribbean (some of which may have colonized the east via the Panama Canal) while only 5 (5%) occur in the estuaries of South America. These last two groups include circumtropical species such as *Albula vulpes*, *Chanos chanos* and *Mugil cephalus*, which are the only three species the area has in common with the estuaries of the Indo–West Pacific. About three East Pacific species (e.g. *Caranx hippos* and *Mugil curema*) also occur in the tropical East Atlantic in addition to the three circumtropical species.

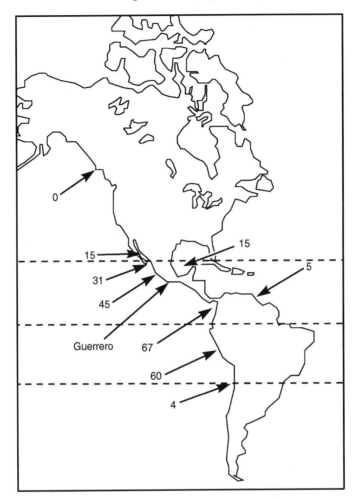

**Fig. 3.13** The location of the coastal lakes of Guerrero, western Mexico, and the percentage of species they have in common with other areas of the East Pacific and West Atlantic regions (data from Yáñez-Arancibia, 1978). Dashed lines denote the Tropic of Cancer, Equator, and Tropic of Capricorn.

### *Jiquilisco Bay, El Salvador*

This estuarine coastal area is a marine embayment of $121\,km^2$ fringed with mangroves and has an average depth of $7\,m$. It is a relatively homogeneous environment with a soft muddy substratum. Salinities vary from a high of 33‰ in the dry season (April) to a low of 23‰ in October

following peak rains in September. A total of 61 species from 27 families
were recorded by Phillips (1981) using a small bottom trawl. The
dominant families are Sciaenidae (8 species), Carangidae (7) and Haemulidae (5). In terms of numbers of individuals, however, the most
numerous species are the ariid *Galeichthys jordani*, the gerreids *Eucinostomus argenteus* and *Eugerres peruvianus*, soles *Achirus* spp., and two
species of anchovy, *Anchoa panamensis* and *Anchovia macrolepidota*.
The most abundant commercially important sciaenid is *Cynoscion phoxoce-
phalus*. Clupeidae and Engraulididae are more common in shallower
waters, ariids prefer deeper waters, and Sciaenidae occur throughout the
Bay. Among the flatfish, bothids are mainly found in deeper, less muddy
situations while cynoglossids occur in shallower muddy areas.

There is relatively little seasonality in species composition and diversity,
although the ranking of the various species differs slightly according to
the time of year. Bothidae are most common in January whereas
Cynoglossidae peak in May. All the species are essentially in the marine
category and both juveniles and adults of most are encountered in the
Bay.

Very few of the species occur in the West Atlantic, the notable exceptions being *Caranx hippos* and *Eucinostomus argenteus*. *Fistularia commersonni* is the only species in common with the Indo–West Pacific.

### 3.4   MAJOR SIMILARITIES AND DIFFERENCES IN THE FISH FAUNAS

The fish communities of tropical estuaries of all four zoogeographic regions
have many common characteristics. In almost all cases they are
dominated by fishes of marine origin – with more than half the number
of species, as well as the number of individuals, being contributed by
species that spend part of their lives in the sea. Given the 'estuarization' of
many tropical coastal areas this is not surprising, and perhaps the term
'marine' for many of these species is a misnomer, because many of them
do not occur outside estuarine coastal areas. It is mainly in the relative
proportions of freshwater and estuarine *(sensu stricto)* species that
estuaries differ, both between and within regions.

Freshwater species make a greater contribution in the tropical Atlantic
regions than in the Indo–West Pacific or East Pacific, particularly in
South America, where many of the very diverse fauna of silurid catfishes
penetrate estuaries. Similarly, in West Africa, various silurids and cichlids
make a significant contribution to estuarine communities. In East Africa
and Australia, however, freshwater species are usually insignificant
components of the estuarine fish fauna. The equatorial regions of South

East Asia have somewhat more freshwater species in estuaries than other areas of the Indo–West Pacific, but despite the diversity of the freshwater fish faunas of Borneo and Sumatra, relatively few live in estuaries. It is possible that higher-rainfall areas (e.g. eastern South America, West Africa, South East Asia), with their greater input of fresh water into estuaries, provide more suitable conditions for freshwater species entering estuaries, in contrast to the lower-rainfall areas of East Africa and Australia. Nevertheless, with regard to the proportions of freshwater species, it is probable that other factors, such as the greater diversity of possibly preadapted Siluriformes, are important.

The proportion of estuarine species varies somewhat between areas, but less so than in the case of freshwater taxa. Most are smaller species such as gobies, eleotrids and syngnathids and there is a higher degree of endemism.

The dominant taxa in each region are broadly similar, but there are some interesting contrasts. In all regions except the Indo–West Pacific, Sciaenidae are one of the dominant families. In the Indo–West Pacific, sciaenids are important in the equatorial regions of South East Asia but much less so elsewhere. This pattern follows that of the relative importance of freshwater species outlined above and may therefore be connected with greater rainfall and greater estuarization of coastal waters. In East Africa, for example, there are few estuarine coastal waters, and sciaenids, although present, are not a dominant component of the fauna. Except for the sciaenids, the other dominant families are broadly comparable across all regions. It is noteworthy that within the Indo–West Pacific, clupeids and engraulids are much more diverse and numerous in equatorial South East Asia than in other areas.

The species richness of fish in tropical estuaries is similar throughout the Indo–West Pacific but larger systems usually have more species than smaller ones; deep, open-water channels in the larger systems favour more of the larger species, particularly carangids and sharks, in addition to a higher number of occasional visitors. Comparisons between estuaries of the Indo–West Pacific and the tropical East and West Atlantic reveal that those of the Indo–West Pacific are usually more species rich, despite the large size of some Atlantic systems (e.g. Orinoco estuary, Terminós and Lagos Lagoons) (Table 3.12). Comparisons of overall fish biomasses in different tropical estuaries are difficult because there are few published figures, and because many of the results may not be comparable due to differences in methods. However, the scarce data (Table 3.13) suggest that fish biomass does not usually exceed $30\,\mathrm{g\,m^{-2}}$ and is most commonly between about 5 and $15\,\mathrm{g\,m^{-2}}$. These values are an order of magnitude higher than those of tropical non-estuarine seagrass areas in shallow waters (Blaber *et al.*, 1992).

**Table 3.12**  Numbers of species recorded from subtropical and tropical estuarine systems that have been comprehensively studied

| Region and system | Country | Type and size | No. of species | Reference |
|---|---|---|---|---|
| **Indo–West Pacific** | | | | |
| Alligator Creek | Australia | O, small | 150 | Robertson and Duke (1990) |
| Trinity | Australia | O, medium | 91 | Blaber (1980) |
| Embley | Australia | O, large | 197 | Blaber *et al.* (1990a) |
| Vellar Coleroon | India | O, large | 195 | Krishnamurthy and Jeyaseelam (1981) |
| Chilka | India | CL, large | 152 | Jones and Sunjansingani (1954) |
| Hooghly | India | O, large | 172 | Jhingran (1991) |
| Ponggol | Singapore | O, medium | 78 | Chua (1973) |
| Matang | W. Malaysia | O, medium | 117 | Sasekumar *et al.* (1994a) |
| Kretam Kechil | E. Malaysia | O, small | 44 | Inger (1955) |
| Purari | New Guinea | O, large | 140 | Haines (1979) |
| Pagbilao | Philippines | O, medium | 128 | Pinto (1988) |
| Solomon Islands | Solomon Islands† | O, small | 136† | Blaber and Milton (1990) |
| Morrumbene | Moçambique | O, medium | 113 | Day (1974) |
| Tudor Creek | Kenya | O, small | 83 | Little *et al.* (1988b) |
| Kosi | South Africa | CL, large | 163 | Blaber and Cyrus (1981) |
| St Lucia | South Africa | CL, large | 110 | Whitfield (1980d) |
| Mhlanga | South Africa | B, small | 47 | Harrison and Whitfield (1995) |
| **East Atlantic** | | | | |
| Lagos | Nigeria | CL, large | 79 | Fagade and Olaniyan (1972) |
| Fatala | Guinea | O, medium | 102 | Baran (1995) |
| Ebrié | Ivory Coast | CL, large | 153 | Albaret (1994) |
| Niger | Nigeria | O, large | 52 | Boeseman (1963) |
| **West Atlantic** | | | | |
| Itamaraca | Brazil | O, large | 81 | Paranagua and Eskinazi-Leça (1985) |
| Orinoco | Venezuela | O, large | 87 | Cervigón (1985) |
| Terminós | Mexico | CL, large | 122 | Yáñez-Arancibia *et al.* (1988) |
| Cienaga Grande | Colombia | CL, large | 114 | León and Racedo (1985) |
| Tortuguero | Costa Rica | O, small | 70 | Gilbert and Kelso (1971) |
| Sinnamary | French Guiana | O, medium | 83 | Boujard and Rojas-Beltran (1988) |
| **East Pacific** | | | | |
| Rio Claro | Costa Rica | O, small | 22 | Lyons and Schneider (1990) |
| Guerrero Lakes | Mexico‡ | CL, small | 105‡ | Yáñez-Arancibia (1978) |

* B, blind; CL, coastal lake; O, open.
† Incorporates 13 small estuaries – no one system more than 50 species.
‡ Sum of species for 10 small coastal lakes.

**Table 3.13** Biomasses (g m$^{-2}$) of fishes from various subtropical and tropical estuaries

| Region and country | Estuary or habitat | Fish biomass (g m$^{-2}$) | Reference |
|---|---|---|---|
| Indo–West Pacific | | | |
| Malaysia | Matang | 0.25–3.1 | Sasekumar *et al.* (1994a) |
| Australia | Alligator Creek | 2.5–29.0 | Robertson and Duke (1990) |
| | Embley | 5.0–16.0 | Blaber *et al.* (1989) |
| | Albatross Bay (inshore) | 5.0 | Blaber *et al.* (1995d) |
| | Albatross Bay (offshore) | 12.0–30.0 | Blaber *et al.* (1990b) |
| Solomon Islands | Small estuaries | 11.6 | Blaber and Milton (1990) |
| Tropical West Atlantic | | | |
| USA, Florida | Mangrove estuaries | 15.0 | Thayer *et al.* (1987) |
| Mexico | Terminós Lagoon | 0.4–3.4 | Yáñez-Arancibia *et al.* (1988) |
| Tropical East Pacific | | | |
| Mexico | Mangrove-lined canal | 7.9–12.5 | Warburton (1978) |

There are many physical and biological influences that control the distribution of fishes, and the roles played by the large array of interconnected factors that affect the distribution and abundance of fishes in tropical estuaries are discussed in the next chapter.

*Chapter four*

# Physical factors affecting distribution

## 4.1 INTRODUCTION

Although the factors that influence the distribution of estuarine fishes in cold and warm temperate estuaries are well understood (reviews: Green, 1968; McClusky, 1971; Dyer, 1973; Perkins, 1974; Burton and Liss, 1976; Barnes, 1980; Day *et al.*, 1989) much less information is available for tropical systems. In general, tropical estuaries are subject to greater extremes than their temperate counterparts (with the exception of temperature): very high and seasonal rainfall, regular droughts and extremes of salinity. They contain a greater wealth of species and hence the complexities of their biological interactions are greater. Both abiotic and biotic processes are important in relation to the distribution of their fishes. The balance between physical and biological influences varies from estuary to estuary and depends upon the location and type of estuary. In this chapter we examine the effects of the physical structure and hydrology on the distribution of fishes – such factors as food availability and predation are dealt with in Chapter 5.

Although tropical estuaries have a high species diversity relative to those of higher latitudes, the number of species is only a small proportion of those found in tropical seas. The total number of marine fish species in the Indo–West Pacific is approximately 10 000 (Nelson, 1984), but not more than about 600 species have been recorded in tropical Indo–West Pacific estuaries. Therefore, even after excluding deep-sea species, not more than 10% of species occur in estuaries. Also a large proportion of this 10% do not spend their entire life cycle in estuaries. Only those species that can tolerate, or are adapted to, the highly variable physical conditions, inhabit tropical estuaries. The relative importance of the range of biotic and abiotic factors that influence the distribution and life cycles of tropical marine fishes is central to an understanding of their ecology.

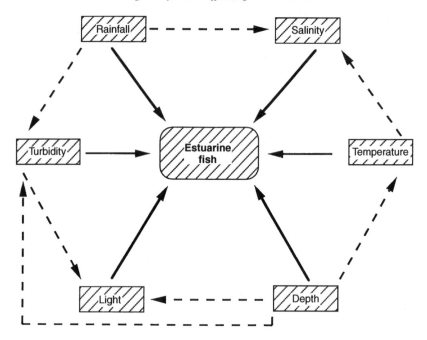

**Fig. 4.1**    Abiotic factors influencing the occurrence, distribution and movements of tropical and subtropical estuarine fishes. Solid arrows indicate direct influences on fishes; broken arrows denote indirect influences.

This in turn is the cornerstone of the knowledge necessary for any form of management, from fisheries to conservation.

Tropical estuarine fishes are subject to a complex matrix of interacting physical and biological factors that determine their occurrence, distribution and movement patterns. The abiotic influences are summarized in Fig. 4.1. Because of the close relationships and links between many of these parameters, it is often difficult to separate one particular factor from another. In most cases no one single factor can be identified as a key factor and certainly not generally for all species. However, the same groups of species often recur in similar environments and under similar physical regimes. Hence there is some degree of predictability about the structure of tropical estuarine fish communities, certainly within broad geographical regions. For this reason, and to better understand the role of the total environment with regard to tropical fish ecology, it is vital to examine as far as possible the relative importance of the different estuarine abiotic and biotic factors. In doing this we are concerned not only with trying to explain the reasons why particular species or suites of species occur in certain types of estuaries, or in broad geographical areas, but also with what influences their distribution within estuaries.

The majority of fishes in tropical estuaries have to be extremely tolerant of both short-term and long-term fluctuations in physical conditions. This is an important feature that distinguishes them from most marine fishes. For example, those few species that are able to tolerate hypersaline conditions in such systems as St Lucia in East Africa or Laguna Madre in Texas, are also those that are best able to tolerate very low or freshwater conditions (Chapter 3), not the many species that live in sea water of 35‰ but cannot tolerate its dilution. Tropical estuaries can be grouped in terms of the size of their physical fluctuations into a continuum from those that have been termed 'physically driven' systems (Dutrieux, 1991), characterized by large fluctuations in flow rate, salinity, turbidity and tidal range (e.g. large estuaries of Borneo), to those referred to as 'biologically driven' (Sanders, 1968; Dutrieux, 1991), where these factors are more constant and the physical environment apparently less harsh (e.g. many East African estuaries). The majority of estuaries fall between the two extremes.

## 4.2   HABITATS AND STRUCTURE

The physical structure of the environment is at least as important as the physical nature of the water itself in determining the structure and species composition of estuarine fish faunas. The main elements of this structure in tropical estuaries are the vegetation and substrata. The two are often related to each other.

### Vegetation

The structural vegetation of subtropical and tropical estuaries can be divided into four groups. Salt-marshes are not included as they are mainly temperate to warm temperate in distribution.

### *Mangroves*

Mangroves are a characteristic feature of most subtropical and tropical estuaries. There are more than 80 species of mangrove trees and the species composition and growth form of the trees and forests is dependent on a variety of physical and chemical parameters. For more details on mangrove floristics, zonation, ecology and physiology the reader is referred to one of the many texts on the subject (MacNae, 1968; Hutchings and Saenger, 1987; Robertson and Alongi, 1992).

**Fig. 4.2**  Dense mangrove forests line the Embley estuary in the Gulf of Carpentaria, Australia.


Mangrove forests can vary from a narrow fringe along the banks of an estuary to dense stands covering many square kilometres (Fig. 4.2). Monospecific stands of *Avicennia* spp. are characteristic of large intertidal regions of many subtropical estuaries (Fig. 4.3) and may also form a narrow fringe along tropical systems (Fig. 2.6). Some mangrove forests in the tropics have a complex zonation, may contain upwards of 17 species of trees (Hutchings and Saenger, 1987) and are very dense with many prop roots (Fig. 4.4). The specialized mangrove palm *Nypa fruticans* is mainly found along the channel margins in the upper and middle reaches of South East Asian and Australasian estuaries where the salinity is between 1‰ and 10‰ (Fig. 2.5) (Hutchings and Saenger, 1987; Robertson *et al.*, 1991).

Most mangroves are associated with soft, muddy substrata and accretionary shorelines. Their main influence on fishes is the physical structure they provide. Their pneumatophores, prop roots, trunks and fallen branches and leaves make a complex habitat not only for fishes, but for a host of invertebrates, both on the trees and burrowing in the mud. A rich epiflora of algae and diatoms is often found on the mangroves and

**Fig. 4.3** Open *Avicennia marina* forest in subtropical Moreton Bay, Queensland, Australia.

associated substrata. Because mangroves usually occur in shallow inter-tidal areas of deposition, with quiet waters, muddy substrata, variable turbidities and a rich fauna and flora, their effects on fish are inextricably linked with these other factors. The whole suite of mangroves plus associated biotic and abiotic conditions makes them one of the core habitats of tropical estuaries. The structural complexity of the mangrove environment must therefore be considered when comparing the fish faunas of different mangrove systems. Similarly, because of these inherent structural differences, reports of the ways in which fish utilize a mangrove, and any assessment of the value of mangrove systems to fishes, must take account of the structural nature of each individual system. Nevertheless, despite marked structural differences between mangrove systems in different estuaries, their fish communities share many common features.

The structural significance of mangroves for fishes is well demonstrated by the studies of Thayer *et al.* (1987) in Florida. They showed that the prop root habitat of *Rhizophora mangle* in the Everglades National Park in Florida was of major importance to a wide variety of fishes. The biomass

**Fig. 4.4**  Thick, tall mangrove forest with a dense network of prop roots in the Matang Forest Reserve, Malaysia.

and number of fishes in the prop roots (Fig. 4.5) was much greater than in adjacent seagrass areas. Juveniles of many species, including the commercially important snapper *Lutjanus griseus*, feed mainly in the prop root habitat, while their adults use the prop roots as shelter and forage into adjacent areas to feed.

In small Solomon Island estuaries, Blaber and Milton (1990) found differences in the species composition according to the species of mangrove, whether the channels are blocked or choked by fallen mangrove branches and the type of substratum (Table 3.7). The differences in mangrove trees may be related to tidal inundation and soil type (Percival and Womersley, 1975) or a variety of other factors such as seed predators. The hard-substratum estuary with an abundance of mangrove tree debris is inhabited mainly by fish species that apparently need the cover or structure provided by the debris. Pomacentridae and some species of Apogonidae and Gobiidae predominate. These species are largely absent from the soft-substratum estuaries where Gobiidae, including burrowing species, are dominant. Some species, such as *Apogon hyalosoma*

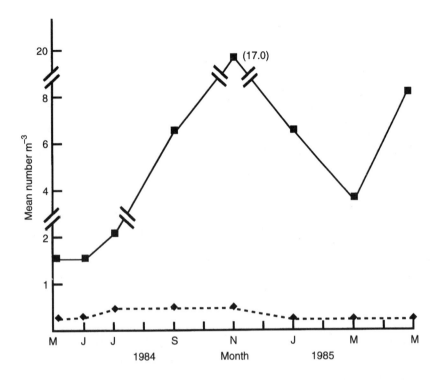

**Fig. 4.5** Numbers of fishes among the prop roots of *Rhizophora mangle* (solid curve) compared with numbers in adjacent seagrass beds (broken curve) in south Florida (data from Thayer *et al.*, 1987).

and the hemiramphid *Zenarchopterus dispar*, are not site-attached and are equally common in 'hard' and 'soft' estuaries.

It has been demonstrated that there are extensive tidal and nocturnal movements of fishes into mangroves. In an experiment to trap all the fishes moving out of a mangrove area on the falling tide in the Embley estuary in northern Australia, Blaber *et al.* (1989) isolated $9167\,m^2$ with a stake net, and trapped $647.3\,kg$ of fish of 39 species giving a biomass estimate of $70.6\,g\,m^{-2}$. Most of the fish were larger than $10\,cm$ SL. The catch was dominated by four teleosts and two species of Dasyatidae (stingrays), which made up 76% of overall weight. Numerically the most abundant species were *Arius proximus*, *Drepane punctata* and *Gerres abbreviatus* (Table 3.4).

The relative importance of mangroves and the dependence or otherwise of fish on them is discussed in Chapter 7.

*Seagrasses*

Aquatic macrophytes do not occur in the majority of tropical estuaries, being more characteristic of shallow inter-reef and coastal marine areas away from estuaries. However, where they are found in estuaries they exert a marked influence on the fish fauna. There are a number of species that grow in estuaries, some of marine affinity such as *Enhalus acoroides*, others from fresh water such as *Eichornia crassipes* (Fig. 4.6). Detailed information on the taxonomy and ecology of tropical seagrasses can be found in Larkum *et al.* (1989) and Brouns and Heijs (1985). As with mangroves, each species is usually associated with a particular set of conditions.

The seagrass areas in many areas of the subtropics and tropics, particularly in the Caribbean, East Africa, South East Asia and the South Pacific, consist mainly of small beds usually within coral reef lagoons. In contrast, many of the seagrass beds of tropical northern Australia extend over hundreds of kilometres of open coastline, often shoreward of soft, muddy substrata, and are not always associated with coral reefs (Poiner *et al.*,

**Fig. 4.6** Dense growth of *Eichornia crassipes* (water hyacinth) in the Tongati estuary, KwaZulu-Natal, South Africa.

1989). Here they sometimes extend into estuaries and the plant species are primarily marine. Where they occur in Central and South American estuaries, aquatic macrophytes are mainly freshwater (e.g. *Eichornia*); and in East Africa and South Asia they are freshwater and brackish-water species such as *Potamogeton pectinatus, Zostera* spp. and *Ruppia* spp.

In contrast to mangroves, which are usually intertidal, most seagrasses are subtidal. They vary from short sparse beds (e.g. *Halophila ovalis*) to dense stands of 1–2 m (e.g. *Enhalus acoroides*). Most are associated with low turbidities and sandy substrata in areas of low flow and small tidal range. In a study of the fishes of the coastal seagrass areas of Groote Eylandt in northern Australia, Blaber *et al.* (1992) distinguished four main seagrass habitats:

- tall, dense seagrass areas in depths between 1 and 3 m; these seagrass beds consist mainly of *Cymodocea serrulata, C. rotunda, Enhalus acoroides* and *Thalassia hemprichii*;
- tall, dense seagrass areas in depths < 1 m; these beds are dominated by *C. serrulata* and *E. acoroides*;
- short and more sparse seagrass areas in depths between 1 and 4 m; these consist mainly of *Halophila ovalis, Halodule uninervis* and *T. hemprichii*;
- short and sparse seagrass in depths of < 1 m, and partly intertidal; these are dominated by *H. uninervis* and *H. ovalis*.

Although the Groote Eylandt area is not estuarine, the species composition in the different types of seagrass and the ways the shallow water coastal fishes use the seagrass beds are relevant to similar, though smaller areas, in estuaries.

Shallow and intertidal sparse seagrass areas are dominated by several ubiquitous gobies (e.g. *Drombus triangularis* and *Yongeichthys nebulosus*) and small juvenile Gerreidae, Lutjanidae and Sillaginidae. In subtidal shallow areas, species compositions of larger juveniles and adults vary significantly between types of seagrass. Short seagrass sites are dominated by adult and juvenile Hemiramphidae and juvenile Lethrinidae, Gerreidae and Leiognathidae; tall, more dense seagrasses are populated mainly by juveniles of Leiognathidae, Lethrinidae and Gerreidae as well as by Teraponidae and the herbivorous Siganidae. Mixed seagrass areas are dominated by Sillaginidae, Gerreidae and Mugilidae. Although species composition changes according to seagrass or habitat type, a number of species, particularly *Gerres*, are common to all areas. Dionne and Folt (1991) have shown experimentally that macrophyte growth form has a marked effect on fish foraging and habitat associations, and Brewer and Warburton (1992) showed that in seagrass habitats the vulnerability of prey is linked to their distribution within the habitat. This suggests that differences in species composition in the different types of seagrass may be due to differences in food availability as well as shelter. In addition to the above species compositions, it was also shown that an array of larger species (e.g. sharks and carangids) move into shallower waters, particularly at night, in a similar manner to that demonstrated for mangrove areas. The majority of these are more abundant in tall seagrass than short seagrass, and more abundant in short seagrass than in open areas. These movements are probably related to feeding on the smaller resident and juvenile fishes that show the same patterns of abundance.

The species composition of fishes in tropical seagrass beds is partly influenced by the nature of adjacent habitats. Baelde (1990) found that the faunas of seagrasses near mangroves and coral reefs differed; mangrove seagrasses act mainly as nursery areas for small juveniles, and coral seagrasses as foraging areas for larger species that shelter in the coral by day. In this respect it is noteworthy that in an extensive seagrass bed, consisting mainly of *Enhalus acoroides*, adjacent to mangroves in the lower reaches of the Embley estuary in the Gulf of Carpentaria, Australia, where Blaber *et al.* (1989) recorded 53 species of fish, most were small species or juveniles ( < 10 cm SL); *Epinephelus suillus, Pelates quadrilineatus, Siganus canaliculatus, Apogon ruppelli, Lutjanus russelli* and *Monacanthus chinensis* contributed most to biomass (49%). The overall mean biomass was only $0.484 \, g \, m^{-2}$ (very much less than found in mangroves – see above).

In Lake St Lucia in KwaZulu-Natal, South Africa, four species of macrophytes are of importance: *Potamogeton pectinatus, Ruppia maritima, R. spiralis* and *Zostera capensis* (Ward, 1982). *Potamogeton pectinatus* is the least salt tolerant and declines at salinities above about 20‰, *Ruppia maritima* is sparse and only found in very shallow waters, while *R. spiralis* and *Z. capensis* are the most common species throughout the lake system. The areas covered by these macrophytes fluctuate depending on lake levels and salinities, but substantial areas of South and North Lake (Fig. 2.1) have a thick vegetation. The macrophytes support large numbers of epiphytic algae and a diverse invertebrate fauna and hence provide important feeding as well as shelter habitats for a wide variety of juvenile fishes. In addition, adults of estuarine groups such as Syngnathidae and Gobiidae occur mainly in the vegetated areas.

In the similar Lake Chilka in India, there is an abundance of aquatic macrophytes in the north and west parts of the lake. Here the dominant species are *Potamogeton pectinatus, Najas falcioulata* and *Halophila ovalis*. As in Lake St Lucia, these weed beds contain mainly juveniles or small species of syngnathids, teraponids and gobiids.

In the Tortuguero estuary in Costa Rica, the most important rooted macrophytes are species of *Najas, Potamogeton* and *Utricularia* which grow on the mudflats in the lower reaches. They likewise provide habitats for juveniles and adults of small species. However, the most abundant macrophytes in this estuary are the free-floating species that come in from tributary rivers. These include *Eichornia crassipes*, *Hydrocotyle* sp. and *Salvinia* sp. Mats of the water hyacinth, *E. crassipes*, are by far the most abundant but their occurrence varies according to salinity and river flow. This species has been introduced to Africa and has become established in many smaller estuaries in KwaZulu-Natal where it is considered a noxious weed. In a study of the blind and polluted Tongati estuary, Blaber *et al.* (1984) found that the levels of ammonia (from treated sewage effluent) and soluble reactive phosphorus were sufficient to promote rapid and sustained growth of *E. crassipes* (Fig. 4.6). This plant covered almost the entire surface of the estuary. The roots of the *Eichornia* supported a diverse fauna, mainly of insects and their larvae as well as crustaceans and polychaetes. The biomass of this fauna was higher in most months than that of the zoobenthos. Insect larvae were the most important food for most of the depauperate fish fauna, which included *Ambassis productus*, *Gilchristella aestuaria*, *Pomadasys olivaceum*, *Terapon jarbua* and *Tilapia rendalli*.

### Reed beds

Extensive beds of the reed *Phragmites* spp. (Fig. 4.7) occur in the estuaries of East Africa, often adjacent to mangroves. Many square kilometres of African coastal lakes also have such reed beds. Apart from providing shelter for juveniles and small adults of many species of fish, they usually harbour high biomasses of epifauna and epiflora. In the blind Mhlanga estuary of KwaZulu-Natal (Fig. 2.1), the most important components of the epifauna are the amphipod *Corophium triaenonyx* and the polychaete *Ficopomatus enigmatica* (Whitfield, 1980a), both important prey items of small fishes. The epiflora is grazed by the sparid *Rhabdosargus holubi* and several species of mullet. Detritus plays a dominant role in the energy flow in many estuaries and Whitfield (1980a) showed that the input of detritus into the Mhlanga estuary from the growth and decay of *Phragmites* beds has a significant impact on the detritus pool.

Other reed species that commonly occur in estuaries are *Juncus* spp. and *Scirpus* spp. Both are usually associated with areas of freshwater seepage.

### Swamp forests

Extensive, often seasonally flooded forests are frequently found adjacent to, or contiguous with, the upper reaches of tropical estuaries (Fig. 4.8). Large areas of such shallow, usually freshwater habitats are present in tropical South America, parts of Borneo and northern Australia. It is from

**Fig. 4.7**  *Phragmites* reeds lining a subtropical estuary.

these habitats that many of the freshwater species can move into the upper reaches of estuaries. However, in addition to this they are also the main juvenile habitat for a number of marine or estuarine species. Among the best known of these is *Lates calcarifer* (barramundi in Australia, seekap in South East Asia); the adults spawn in shallow coastal waters and the juveniles move up estuaries and into fresh waters where they live in the shallow, usually clear waters of swamp forests. After the wet season, they move back into the estuaries to continue their life cycle in the estuarine and marine environment.

### Substrata

Like the vegetation in estuaries, the type of substratum is interrelated to several other physical factors, particularly current speeds, tides and vegetation. It is also related to the overall catchment in terms of geology and human land use. The latter is especially important where there is much agriculture, forestry or soil erosion: much of the sediment from this may be deposited in estuaries.

**Fig. 4.8** Swamp forest in the upper reaches of the Baram River drainage in Sarawak, Borneo.

The type of substratum has a marked effect on the species composition and distribution of fishes in tropical estuaries. Even small amounts of rocky reef within an estuary can enable many additional species to live there, as described for Kosi estuary in East Africa (pp.54–57). In the artificial rocky environments created by humans as harbour walls, or more recently in the large numbers of residential canal estates in South Africa, Australia and USA, the fish community may become dominated by planktivores or micro-carnivores (Morton 1989, 1992; Chapter 9).

Estuaries are, however, generally dominated by soft substrata. The distribution of fishes in the Embley estuary described in Chapter 3 (pp.65–74) indicates that the species composition is significantly different according to whether the bottom is muddy or sandy. Although such differences may also be related to other phenomena linked with substratum type, such as turbidity, in some cases they can be ascribed directly to substratum type. A good example here is the relationship between the grain size of the substratum and the differential distribution of mullet species in African estuaries. The diet of most species of mullet is similar

**Table 4.1** Particle size preferences of Mugilidae in KwaZulu-Natal estuaries classified according to the Wentworth scale (after Blaber, 1976, and Blaber and Whitfield, 1977a)

| Species | Sand grain size (mm) | Wentworth classification |
|---|---|---|
| *Liza tricuspidens* | 0.5  –1.0 | Coarse sand |
| *Liza alata* | 0.5  –1.0 | Coarse sand |
| *Liza dumerili* | 0.25 –1.00 | Medium and coarse sand |
| *Liza macrolepis* | 0.25 –0.5 | Medium sand |
| *Mugil cephalus* | 0.25 –0.5 | Medium sand |
| *Myxus capensis* | 0.125–0.5 | Fine and medium sand |
| *Valamugil buchanani* | 0.125–0.25 | Fine sand |
| *Valamugil cunnesius* | 0.125–0.25 | Fine sand |
| *Valamugil robustus* | 0.125–0.25 | Fine sand |
| *Valamugil seheli* | 0.125–0.25 | Fine sand |

and is determined mainly by the occurrence of particular food items on substrata of the preferred particle size of each species. In an investigation of 10 sympatric species of Mugilidae, Blaber (1976, 1977) showed that each species has a preferred particle size range (Table 4.1). The exact sizes consumed by each species do not remain the same but vary according to locality. Within any one estuary, the species of mullet avoid interspecific competition by selecting particles of different sizes. Also the fish are always graded in the same order with regard to particle size. A similar phenomenon was described by Albaret and Legendre (1985) for five species of Mugilidae in Ébrié Lagoon in West Africa.

There is a close relationship between sediment type and the fish species composition of estuarine coastal waters. The soft, muddier substrata of the shallow coastal waters off the Guianas in South America, off West Africa and in South East Asia have a characteristically estuarine fish fauna of groups such as sciaenids, polynemids and ariids. In their detailed study of the almost unexploited fish communities of the Gulf of Carpentaria in northern Australia, Blaber *et al.* (1994c) showed that the fish fauna of the Gulf could be split into three major groups plus a fourth group of 'misfits': widespread, regional, reef-associated and other (Chapter 3, pp.76–80). The widespread and abundant species (especially of the families Leiognathidae, Polynemidae, Haemulidae, Carangidae, Ariidae, Gerreidae, Priacanthidae and Sciaenidae) form the most ubiquitous group. Most of these species are also characteristic of shallow ( < 80 m), soft-bottom trawl grounds throughout South East Asia (Mohsin *et al.*, 1988; Pauly, 1988) and the wider Indo–West Pacific region (Longhurst and Pauly, 1987).

## 4.3  SALINITY

Fishes in almost all estuaries are subject to changes in salinity. In open estuaries such changes are usually diel and depend on tidal influence, whereas in blind estuaries during periods of closure the changes take place more slowly depending on freshwater inflow and evaporation. The most stable salinities are found in coastal lakes, but these too are subject to long-term salinity changes in response to rainfall and freshwater input. Both blind estuaries and coastal lakes may experience hypersaline conditions of as much as three times sea water (Forbes and Cyrus, 1993). Sudden reductions in salinity may also occur following heavy rains and flooding of rivers.

Most estuarine fishes are able to cope with salinity fluctuations, but their ability to do so varies from species to species and hence may influence their distribution. Salinities down to about 25‰ apparently pose few osmoregulatory problems for most tropical marine fishes. As shown in Chapter 2, salinities over vast areas of estuarine coastal waters in the tropics may decline in the wet season to as little as 20–25‰. In some areas such as the Bay of Bengal and parts of the South China Sea, salinities seldom rise above 30‰. The fish communities of such areas are highly diverse (Pauly, 1985), often contiguous with those of open estuaries, and include many species of dasyatid rays and carcharhinid sharks. The osmoregulatory abilities of estuarine fishes have been extensively studied and physiological details are available in a number of excellent texts (Brown, 1957; Green, 1968; Rankin and Jensen, 1993; Jobling, 1995).

Almost all fishes living in subtropical and tropical estuaries are very euryhaline and routinely able to cope with salinities from almost fresh water ( < 1‰) to at least 35‰. It is apparently easier for tropical than for temperate marine species to adjust to living in fresh water (Pauly, 1985). The salinity ranges in which 100 Indo–West Pacific fishes have been recorded in Africa are shown in Whitfield *et al.* (1981) and Whitfield (1996). Sixty-five of these are in the marine migrant category, and in terms of their minimum salinity tolerance, 16 of these have been recorded in fresh water, another 31 in salinities of 1–3‰, another 14 in salinities of 4–10‰, with only two elasmobranchs having a minimum of 17‰. Among the most tolerant species are *Elops machnata* (0–115‰), *Pomadasys commersonni* (0–90‰), *Monodactylus falciformis* (0–90‰), *Mugil cephalus* (0–90‰), *Rhabdosargus sarba* (1–80‰) and *Liza macrolepis* (1–75‰). Whitfield (1996) lists 27 species in the estuarine category of which only two, *Gilchristella aestuaria* (0–90‰) and *Ambassis natalensis* (0–52‰) have been recorded in hypersaline conditions. However, all the tropical species in the estuarine category have been recorded in fresh water or at 1‰. A similar listing was provided by Renfro (1960) for the

Aransas estuary in Texas, where salinities ranged from nearly fresh water to about 60‰. Of the 26 species recorded, nine are freshwater and were found only in salinities less than about 15‰; eight estuarine species occurred from fresh water to about 35‰; and the remaining nine species were tolerant of the full range of salinities. This last group contains the most frequently occurring and abundant species, one of which is *Mugil cephalus* and another two are freshwater species (*Lepiosteus spatula* and *Dorosoma cepedianum*). The effects of hypersaline conditions on the fish fauna of the Casamance River in Senegal are described by Albaret (1987). Here, due mainly to low rainfall, the salinity is greater than sea water throughout the system, with values increasing with distance from the sea. In the dry season the upper reaches exceed 100‰ and only the cichlid *Sarotherodon melanotheron* persists and even proliferates. With the advent of the wet season, salinities fall sufficiently to allow other species to penetrate upstream. These include *Elops lacerta*, *Ethmalosa fimbriata*, *Liza falcipinnis* and *Arius latiscutatus*.

Using the terminology of Myers (1949), most primary freshwater fishes are intolerant of salt water but many secondary groups (e.g. Cichlidae, Synbranchidae, Lepiosteidae, Cyprinodontidae and Poeciliidae) are relatively tolerant and some, such as the African cichlids *Oreochromis mossambicus* and *S. melanotheron*, can tolerate hypersalinities ( > 100‰, Whitfield *et al.*, 1981; Albaret, 1987).

Although there are considerable field data on the salinity tolerances of estuarine fishes, there has been relatively little experimental work on subtropical or tropical species. Such work can be important in helping to explain distribution, because there is usually a synergistic tolerance relationship between temperature and salinity which determines the temperature × salinity envelope in which a species can live (Day *et al.*, 1989). This is well illustrated by a study of three closely related species of Ambassidae in KwaZulu-Natal estuaries (Martin, 1988). Their differential distribution within estuaries is largely determined by the interaction between their salinity and temperature tolerances (Fig. 4.9). *Ambassis gymnocephalus* is restricted to estuary mouths with marine conditions; *A. productus* to the less saline ( < 10‰) areas in both closed and open estuaries, and *A. natalensis* occurs throughout open and closed estuaries in a wide range of salinities, but is most abundant in the middle reaches of open estuaries where the salinity is between 8‰ and 33‰. The first two species have similar upper and lower lethal temperature limits in sea water. However, at lower salinities the survival temperature range of *A. productus* is increased, whilst that of *A. gymnocephalus* is decreased so markedly that at 1.5‰ they cannot tolerate a change of more than 3 °C above or below 24 °C. This inability to tolerate temperature changes at low salinity probably excludes *A. gymnocephalus* from the middle and

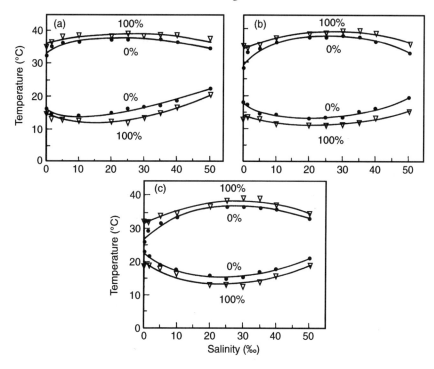

**Fig. 4.9** 0% and 100% mortality curves for adult Ambassidae exposed to different combinations of temperature and salinity for 118 h: (a) *Ambassis productus*; (b) *A. natalensis*; (c) *A. gymnocephalus*. Reprinted from *Journal of Fish Biology*, Vol. 33 (Supplement A), Interaction of salinity and temperature as a mechanism..., pages 9–15, 1988, by permission of the publisher Academic Press Limited, London.

upper reaches of estuaries. *Ambassis productus* on the other hand, is most common in low salinities (usually < 20‰) and can survive in fresh water. The third species, *Ambassis natalensis*, occurs throughout estuaries and coastal waters of south-east Africa and can tolerate lower temperatures in water above 20‰.

The synergistic effects of temperature and salinity occasionally result in extensive 'fish kills' in estuaries. In one such case in Lake St Lucia, KwaZulu-Natal, unseasonably high rainfall in the dry (cool) season reduced salinities throughout the lake system to below 3‰. When combined with water temperatures as low as 12 °C, this caused large-scale mortality among at least 11 normally very euryhaline species, including sciaenids, carangids, sparids, haemulids and mugilids (Blaber and Whitfield, 1977a).

In the coastal Lake Nhlange, part of the Kosi system of northern KwaZulu-Natal (Fig. 2.1), the changes in the fish fauna in response to gradual changes in salinity (Fig. 4.10) were a reduction in the diversity of marine species and an increase in numbers of individuals of a few freshwater species (Blaber and Cyrus, 1981). From 1978 onwards, the salinity decreased to 1‰ or less and a number of marine migrant species such as *Sphyraena barracuda* and *Scomberoides lysan* disappeared. There was, however, an increase in the numbers of freshwater cichlids (*Oreochromis mossambicus, Tilapia rendalli, Tilapia sparmanii* and *Pseudocrenilabrus philander*) and the freshwater catfish *Clarias gariepinus*. Nevertheless, even at salinities of 1‰ or less, the system still contained 23 estuarine or marine species (Table 4.2). The seasonal occurrence of juveniles and adults at low salinities in Lake Nhlange was thought to be due to the effect of interaction between temperature and salinity.

The artificial opening of the Ébrié Lagoon in the Ivory Coast, West Africa, showed how changes in the salinity regime influence the species composition of the fish fauna (Albaret and Écoutin, 1989). Prior to the opening in 1988, salinities were close to fresh water, but after opening, salinities increased to at least 20‰. The numbers of freshwater taxa such

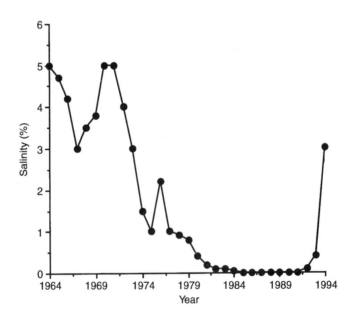

**Fig. 4.10**  Changes in salinity over a 30 year period in Lake Nhlange, KwaZulu-Natal, South Africa.

**Table 4.2**   Species of marine and estuarine teleosts occurring in salinities of less than 1‰ in Lake Nhlange, KwaZulu-Natal, South Africa, from 1977 to 1981 (J, juvenile; A, adult; −, absent)

| Species | Occurrence | |
|---|---|---|
| | Summer* | Winter† |
| *Chanos chanos* | − | A |
| *Elops machnata* | A | − |
| *Gilchristella aestuaria* | J A | J A |
| *Terapon jarbua* | J A | J A |
| *Caranx ignobilis* | − | J |
| *Caranx papuensis* | − | J |
| *Caranx sexfasciatus* | J | J |
| *Monodactylus argenteus* | J A | J A |
| *Gerres acinaces* | − | A |
| *Gerres filamentosus* | A | A |
| *Gerres rappi* | J A | J A |
| *Ambassis productus* | A | A |
| *Ambassis natalensis* | J A | J A |
| *Pomadasys commersonni* | A | A |
| *Acanthopagrus berda* | J | − |
| *Rhabdosargus holubi* | J | − |
| *Rhabdosargus sarba* | J A | J A |
| *Liza alata* | − | A |
| *Liza macrolepis* | − | A |
| *Mugil cephalus* | A | A |
| *Valamugil robustus* | − | A |
| *Sphyraena jello* | − | J |
| *Glossogobius giuris* | J A | J A |

* October–March.
† April–September.

as *Pellonula*, *Chrysichthys* and other silurids then decreased, and the system became dominated by marine species, some of commercial importance, such as *Ethmalosa fimbriata*, *Trachinotus teraia*, *Polynemus quadrifilis*, *Pomadasys jubelini* and *Liza grandisquamis*.

## 4.4   TURBIDITY

Relatively high water turbidity is usually one of the most noticeable characteristics of tropical estuaries. This is often enhanced by contrasts with adjacent fresh waters and the sea, both of which may be much clearer. Turbidity is defined as the optical property of a liquid which causes light to be scattered and absorbed, rather than transmitted in straight lines. It is thus a measure both of the suspensoid load and of properties of the suspensoids and liquid that cause light to be scattered and absorbed (Bruton, 1985). In a review on the effects of suspensoids on fish, Bruton (1985) states that studies on the effects of turbidity on fish have been hampered by inadequate measurement of turbidity. Only since the advent of the portable turbidimeter in about 1980 (measuring in nephelometric turbidity units – NTU) have accurate field measurements of turbidity become possible.

## Distribution in relation to turbidity

The importance of turbidity in relation to fish distribution was first recognized in the Great Lakes of the USA and Canada, where research showed that it may influence species composition and abundance (Doan, 1941; Langlois, 1941; Cordone and Kelly, 1961; Sorenson *et al.*, 1977). In the marine environment, Moore (1973) showed that the invertebrate fauna in kelp beds could be divided into clear-water, turbid-water and turbidity-indifferent species. It was not until Blaber and Blaber (1980) showed that turbidity was one of the major factors influencing the distribution of juvenile fishes in subtropical Moreton Bay, Queensland, Australia, that its relevance to estuarine fish was established.

In this study of the fishes of Moreton Bay, a large estuarine embayment on the Queensland coast, Blaber and Blaber (1980) divided the fishes into four groups according to the distribution of juveniles and adults.

Group 1 consists of those species in which juveniles and adults occurred in the same areas. It includes *Mugil cephalus, Polydactylus multiradiatus, Harengula abbreviata, Stolephorus carpentariae* and *Thryssa hamiltoni.*

Group 2 is the largest and contains those species where the juveniles occur only in estuaries and shallow turbid water, and the adults offshore in deeper water. It includes the carangids *Alepes djedaba, Caranx ignobilis* and *C. sexfasciatus* as well as various hemiramphids, mugilids and atherinids.

Group 3 consists of species in which the juveniles, like those of group 2, occur only in estuaries but the adults are widely distributed in a variety of habitats. It includes the sparid *Acanthopagrus australis* and the mugilids *Liza subviridis* and *Mugil georgii.*

Group 4 contains those species where the juveniles are restricted to shallow marine areas and the adults occur in deeper offshore waters. It includes several Gerreidae and *Rhabdosargus sarba.*

Variations in turbidity were correlated with the different distribution patterns of the fish, which could be divided into three categories: those tolerant of turbidity, those indifferent to turbidity (comparatively few), and those intolerant of turbidity. Juveniles of group 2 and group 3 fish occurred only in high-turbidity water during summer while their adults were found in low-turbidity waters. Recruitment of group 2 and group 3 fish took place in summer when estuarine turbidities were greatest and when turbidity gradients between estuaries and the deeper, more marine parts of Moreton Bay were greatest. During winter, juveniles that had not matured and moved to offshore areas, moved higher up the estuaries where turbidities remained high. Many adults that were restricted to the low-turbidity areas of the Bay in summer moved to the lower reaches of the estuaries in winter when turbidities were low and more uniform throughout Moreton Bay. At this time, many group 4 adults and juveniles that were restricted entirely to low-turbidity waters in summer were also able to spread throughout the Bay. Substratum type and depth are closely linked with turbidity. In summer, high turbidities are generated over the shallow, muddy substrata of estuaries by a combination of run-off and wind turbulence. It is difficult to isolate the effects of turbidity on fish, but the movement of adults into shallower estuaries in winter suggests that changes in turbidity may be more important than substratum or depth in Moreton Bay.

Cyrus (1988b,c) recorded a wide range of turbidities in 17 estuaries of KwaZulu-Natal, South Africa (0.5–1471 NTU): four of these were clear water (< 10 NTU), nine were semi-turbid (10–50 NTU), three were turbid (50–80 NTU) and one very turbid (> 80 NTU). If values of > 10 NTU are considered turbid (Swensen, 1978; Blaber and Blaber, 1980) then most subtropical and tropical estuaries range from turbid to very turbid, and most estuarine fishes are therefore turbid-water species. The shallow turbid-water areas of the world include those regions categorized as estuarine coastal waters (Chapter 2). These vast areas in, for example, South East Asia and West Africa, are inhabited by many of the same fishes as the estuaries. However, such turbid areas along the East African and east Australian coasts only occur in estuaries and hence many of the adults or juveniles using turbid areas are found only in estuaries. In a study of the distribution of fishes in Lake St Lucia, KwaZulu-Natal, in relation to turbidity (Cyrus and Blaber, 1987a), the fishes could be divided into intolerant or clear-water species (< 10 NTU), partially tolerant (< 50 NTU), widely tolerant (10–80 NTU) and indifferent (found at all turbidities). The results were supported by experimental work on the turbidity preferences of 10 species from St Lucia (Cyrus and Blaber, 1987b). Gradient tank experiments allowed the elimination of all environmental factors except turbidity. *Liza dumerili* was a clear-water species; *Liza macrolepis, Valamugil buchanani, Rhabdosargus sarba* and *Gerres filamentosus* preferred clear to partially turbid water ( < 50 NTU); *Monodactylus argenteus* preferred intermediate turbidities (10–80 NTU); and the remaining four, *Rhabdosargus holubi, Acanthopagrus berda, Pomadasys commersonni* and *Terapon jarbua*, were indifferent to turbidity. These patterns corresponded with the occurrence of the species in the field. Further research in the Embley estuary of northern Australia also showed that turbidity was a key factor in relation to fish distribution in tropical estuaries, with the same or similar species having the same turbidity ranges (Cyrus and Blaber, 1992).

Juveniles and adults of the same species, and fish of the same genus or family, often fall into different turbidity categories. Among the carangids, juvenile *Scomberoides lysan* (Fig. 3.1) are not tolerant of high turbidities while juvenile *Caranx ignobilis, C. sexfasciatus, Scomberoides commersonianus* and *Trachinotus blochii* have wide turbidity tolerances (Blaber and Cyrus, 1983). Adults of the last four species occur mainly in clear waters. Juvenile *Sphyraena barracuda* are clear-water species (Fig. 3.1) while juvenile *S. jello* are turbidity tolerant (Blaber, 1982).

Turbidities in coastal estuarine waters, although high in areas such as the Bay of Bengal or South China Sea, are usually more uniform than in open estuaries. Probably for this reason, turbidity plays a less significant role than depth or substratum in relation to variations in fish species distribution (Blaber *et al.*, 1994c, 1995a; Martin *et al.*, 1995). Nevertheless, the majority of fishes inhabiting such waters are those tolerant of high turbidities.

## The significance of turbidity

Turbidity is relevant to estuarine fishes in three main ways:

- it may afford greater protection for juvenile fish from predators;
- it is usually associated with areas where there is an abundance of food;
- it may provide a mechanism for migration to and from estuaries.

In terms of protection, turbidity may provide juvenile fish with a form of cover through a reduction in light intensity, visually obscuring prey species from their predators. Adults of many piscivorous fishes such as carangids and sphyraenids are visual feeders and occur mainly in clear water. In turbid waters they would have to get much closer to their prey before attacking and the prey can elude capture (Hecht and van der Lingen, 1992). Turbidity would help protect juvenile and small fish not only from fish predators, but also from piscivorous birds that rely on sight to locate their prey. Evidence from Lake St Lucia, KwaZulu-Natal suggests that clearer-water and surface-swimming species predominate in the diets of fish-eating birds (Whitfield and Blaber, 1978a, 1979b,c).

Turbid waters are usually associated with an abundance of invertebrate food and this may be an important factor attracting large numbers of juvenile fishes. In Lake St Lucia the biomass of invertebrate benthos is four times greater in muddy substrata with turbid water than in sandy substrata with clear water (Blaber *et al.*, 1983). In a review of feeding strategies in relation to turbidity, Hecht and van der Lingen (1992) conclude that turbidity has no effect on the size selection of prey. The feeding rate of visual predators such as *Elops machnata* is reduced at high turbidities, but that of benthic feeders, such as *Pomadasys commersonni*, is not. Turbid estuaries generally have higher densities of zooplankton than clear estuaries and hence more planktivores, particularly filter feeders. This is very marked in the estuaries of Borneo, where the highly turbid estuaries support large numbers of engraulids and clupeids whereas the clearer ones do not (Blaber *et al.*, 1997).

It has been postulated that turbidity gradients that exist between the sea and estuaries act as one of the orientation 'clues' (*sensu* Harden Jones, 1984) for postspawning juvenile fishes migrating into estuaries (Cyrus and Blaber, 1987b). Most spawning of marine migrant species takes place in the coastal zone during the wet season when river outflow is highest and turbidity gradients are steepest. Gradients such as that shown in Fig. 4.11 exist along most tropical estuaries and extend varying distances into the sea (section 2.6). Similar salinity gradients also exist from sea water to fresh water, but recent evidence suggests that salinity may not be a major clue for tropical estuarine species because many will also migrate along reversed salinity gradients into hypersaline areas (Blaber *et al.*, 1985; Albaret, 1987). There exists within most estuaries a vertical gradient from relatively low turbidity on the surface to high turbidity on the bottom. Under experimental conditions, larvae of species such as *Liza macrolepis* and *Mugil cephalus* avoid strong illumination and congregate in places of low light intensity or high turbidity (Liao, 1975; James *et al.*, 1983). To find maximum turbidities or lowest light intensities, mullet larvae in the mouth of an estuary must move towards the bottom and up the estuary. That they do this is well established in relation to their change-over from pelagic to benthic feeding (Blaber, 1987). In addition, however, this movement enables them to take advantage of the stronger tidal movement of water that occurs close to the bottom, particularly during periods of strong river flow. This relationship between turbidity and other mechanisms involved in the migration of juvenile mullet into estuaries is shown in Fig. 4.12.

The roles and relative importance of high turbidity, reduced salinity, high food availability, reduced predation and the presence of calm sheltered waters, as attractants to juvenile fish, and the question of estuarine dependence are discussed further in Chapter 7.

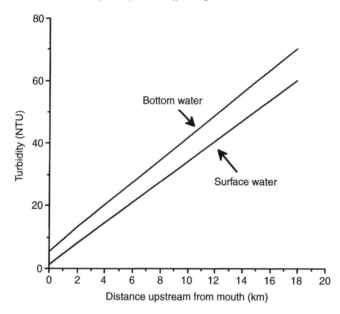

**Fig. 4.11**  Turbidity gradients from the mouth to 20 km up the estuary at Lake St Lucia, KwaZulu-Natal, South Africa.

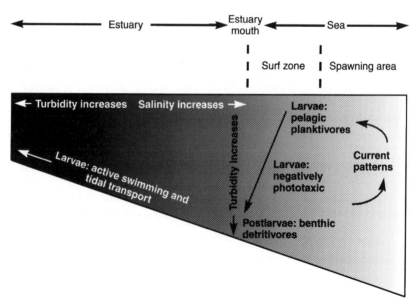

**Fig. 4.12**  Mechanisms involved in the migration of postlarval mullet into estuaries (redrawn from Blaber, 1987).

## 4.5   WATER MOVEMENTS

Included in this section are the currents, tides and wave action that cause circulation in estuaries. Water movement in open estuaries is usually controlled by tidal flow and riverine input, and also, particularly in coastal lakes and blind estuaries, by wind-induced water transport. Estuarine geometry and bathymetry have important modifying effects on all water movements and circulation patterns. The complex mechanisms involved in circulation of water in estuaries are comprehensively reviewed by Day *et al.* (1989) and discussion here is restricted to their influences on estuarine fishes.

The effects of water movements on tropical estuarine fishes can be divided into (1) those primarily related to currents and tides and (2) those that result from wave action and turbulence.

### Currents and tidal influences

Most water movement in tropical open estuaries is related to the tidal cycle with modifications imposed by the amount of freshwater run-off. During the wet season in equatorial regions, freshwater input may be sufficient to almost obscure the ebb and flow of the tide. Because tidal flow is linked with salinity and turbidity fluctuations and gradients, it is often not possible to separate their combined effects on fishes. Quinn and Kojis (1987) showed that in the small, open Labu estuary in Papua New Guinea, fish distribution was not related to the state of the tide; most species were not carried to and fro by tidal currents, but maintained their position with respect to the substratum. They postulate that in small estuaries with a tidal range of less than 1 m, tidal range and velocity have little influence on fish distribution. In contrast to this, however, the effects of high current speeds generated by tidal flow in estuaries with a large tidal range may be significant. In the Norman estuary in the Gulf of Carpentaria, Australia, which has a tidal range of nearly 4 m, current speed was strongly correlated with species composition when the influences of other physical variables were taken into account (Blaber *et al.*, 1994a). The relative abundance of some species in the shallow ( < 5 m) estuarine coastal waters of Albatross Bay, also in the Gulf of Carpentaria, varies according to the tide. *Arius proximus, Carcharhinus cautus, Hyporhamphus quoyi* and *Valamugil buchanani* are more abundant during spring tides (range > 2 m), whereas *Absalom radiatus, Nematalosa come, Polydactylus multiradiatus* and *Rhizoprionodon acutus* were most abundant over neap tides (range < 1 m). It is noteworthy, however, that the abundances of another 31 species were not correlated with tidal range (Blaber *et al.*, 1995a).

While it is evident that current speed can have a marked effect on the distribution of some species, the mechanisms are usually obscure. The cichlid *Oreochromis mossambicus* does not usually occur in the lower reaches of open estuaries, but may be abundant in the middle and upper reaches and in blind estuaries and coastal lakes throughout much of the Indo–West Pacific. It was shown by both field observations and experimental work (Whitfield and Blaber, 1979a) that, in spite of its euryhalinity and the presence of suitable foods, high current speeds may be one of the factors excluding this cichlid from the lower reaches of open estuaries. It requires calm, relatively still water for nesting (Fig. 3.2).

In most coastal lakes, wind-driven circulation patterns may be as important as tides, or more so. Most coastal lakes are large and shallow and the wind-generated seiche can alter water depths. For example, in Lake St Lucia in KwaZulu-Natal with a mean depth of < 1 m (pp.31–34), northerly winds drive water southwards, thus reducing the depth in North Lake and increasing it in South Lake (Taylor, 1982) and southerly winds vice versa. Persistent northerly winds can cause much of North Lake to become exposed or only a few cm deep. This affects fishes firstly directly, because many species move out of North Lake, and secondly indirectly, because an increase in the area of very shallow water renders them more vulnerable to predation by piscivorous birds.

## Wave action and turbulence

Estuaries are usually quiet-water areas with the calm conditions thought to be an important prerequisite for many small juvenile and adult fishes (Bruton, 1985; Day *et al.*, 1989). In type 2 estuaries (section 2.6), there is a great difference between wave action in the sea and its near absence in adjacent estuaries (Figs 2.7, 4.13). In such areas, the species in the surf zone of the sea are largely different from those in estuaries, or are adults the juveniles of which are found in estuaries (Lasiak, 1984, 1986; Ross *et al.*, 1987). The quiet waters of estuaries are more conducive to the growth of the macrophytes and marginal vegetation that provide habitats for estuarine fishes.

The effects of turbulence are felt more in large, type 1 (section 2.6) estuaries and their adjacent coastal waters. In Albatross Bay, Gulf of Carpentaria, and in the South China Sea, the inshore waters are usually more turbid than the deeper offshore waters, but they are generally less turbid than those of adjacent open estuaries. Salinities, although reduced in the wet season, are also more stable than in the estuaries. Conversely, the wind has a much greater effect than in the calmer estuaries, and during the north-west monsoon (wet season), the inshore waters may be very turbulent with much wave action and swell. The main effects of this

**Fig. 4.13** Aerial view of the mouth of the Kosi estuary, KwaZulu-Natal, South Africa, showing differences in wave action and turbulence inside and outside the estuary.

are to increase the numbers of sharks, polynemids and sciaenids – all turbid-water groups – in the nearshore zone.

In coastal lakes, wind-induced turbulence is a major cause of increased turbidity (Bruton, 1985) and hence indirectly affects fish distribution (Cyrus and Blaber, 1987a). It may also directly affect some species: the estuarine burrowing goby *Croilia mossambica*, a species endemic to the coastal lakes and a few blind estuaries of south-east Africa, is restricted to calm, shallow areas with a sandy bottom. Sand rather than mud is needed for their burrow construction. The minimum depth inhabited by *C. mossambica* on any particular shore is probably related to wave action. In the coastal lakes of Poelela, Nhlange and Sibaya the prevailing winds are from the north or south, and the north and south shores are consequently subject to more wave action than the east and west shores. On the shallow terraces on the east and west shores, *C. mossambica* are present from a depth of about 1 m, whereas on north- and south-facing shores they are not present at depths of less than 3.5 m. The absence of *C. mossambica* and their burrows from rippled areas of sand indicates that they do not inhabit areas of appreciable water movement. This excludes them from tidal estuaries where water movements occur, even if shallow sandy areas suitable for burrow building are present (Blaber and Whitfield, 1977b).

## 4.6 SYNTHESIS AND CONCLUSIONS

The relative importance of each of the factors described in this chapter in determining the distribution and abundance of estuarine fishes obviously varies from estuary to estuary. The permutations of their relative influences are almost as variable as estuaries themselves. The study by Martin *et al.* (1992) of the ichthyoplankton of the St Lucia coastal lake system of KwaZulu-Natal, South Africa, before and after severe cyclones provides some insight into the effects of physical changes on recruitment into estuaries. During 1982 and 1983, salinities in St Lucia ranged between 28‰ and 36‰. After Cyclone Demoina in January 1984 they dropped sharply to 14‰ on the ebb tide. Two weeks later a second cyclone, Imboa, caused salinities to drop to 6‰ on the ebb tide. The species composition of the ichthyoplankton (postlarvae and small juveniles) in terms of estuarine resident and marine migrant categories before and after the cyclones is shown in Fig. 4.14. Only species with a density greater than one fish per $100\,\text{m}^3$ were considered. The marine migrant species *Elops machnata, Leiognathus equulus, Johnius dussumieri, Ambassis gymnocephalus* and *Pomadasys olivaceum* all showed increases after the floods. This may be attributed to the successful migration into the estuary of postlarvae, possibly facilitated by increased turbidity or salinity gradients in the nearshore zone of the sea, caused by the massive outflows of fresh water. The large decreases in the numbers of the benthic estuarine residents *Gobius acutipennis* and *Croilia mossambica,* and the marine migrant *Solea bleekeri,* are attributable to the removal of existing sediments and deposition of new sediments laden with allochthonous litter and detritus. Such disturbances probably caused the temporary destruction of much of the habitat and food resource of these species. The reduction in numbers of the benthic marine migrant *Rhabdosargus sarba* may also be attributed to diminished food availability. There was a fifteen-fold increase in the numbers of the estuarine clupeid *Gilchristella aestuaria* immediately following the flood. The increased abundance of this species, which adapts well to low salinities (Blaber *et al.*, 1981), was probably linked to the large blooms of phytoplankton and zooplankton in the wake of the floods, as well as the more favourable lower salinities.

The Gulf of Carpentaria in tropical northern Australia provides some interesting contrasts in terms of its estuaries, and estuarine and coastal fish faunas, which enabled Blaber *et al.* (1994a) to try and assess how

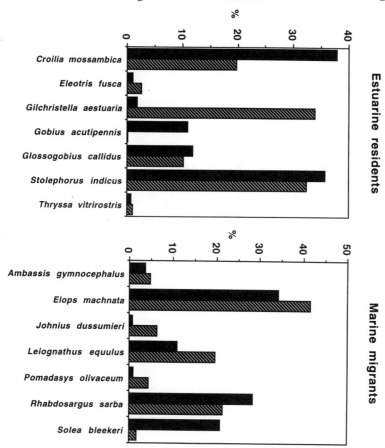

**Fig. 4.14** The percentage species composition of the estuarine resident and marine migrant components of the ichthyoplankton of Lake St Lucia before (black) and after (hatched) cyclones Demoina and Imboa, 1984 (data from Martin *et al.*, 1992).

species composition is affected by habitat types and physical phenomena. The sites chosen were the Embley and Norman estuaries and the coastal waters of Groote Eylandt (Fig. 4.15). These three sites have very different vegetation, substrata and hydrological conditions (Table 4.3). The Norman estuary has a narrow fringe of mangroves, soft muddy substrata, a large tidal range, high turbidity, extremes of salinity, and high current speeds (similar to many of the estuaries of Borneo and New Guinea); the Embley estuary has extensive mangrove forests, a variety of substrata and less extreme hydrological conditions; and the coastal waters of Groote Eylandt have thick seagrass beds, sandy substrata and relatively stable hydrological conditions.

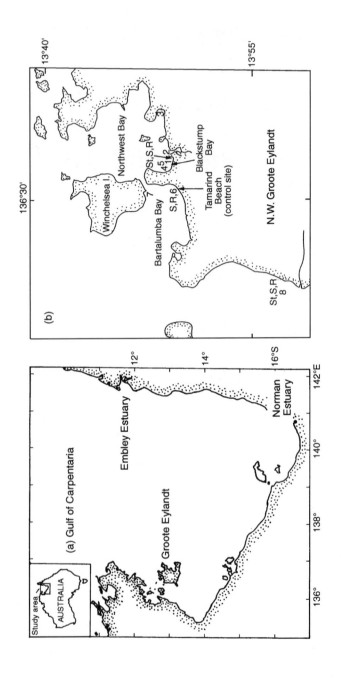

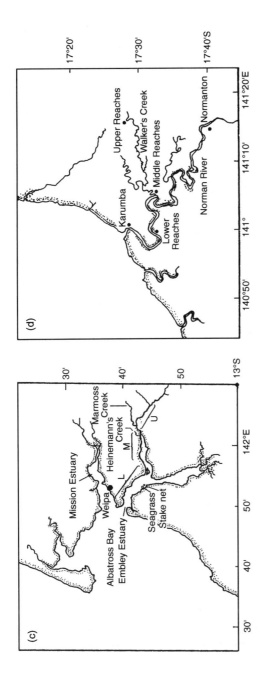

**Fig. 4.15** (a) The location of the Embley estuary, Norman estuary and Groote Eylandt study sites in the Gulf of Carpentaria; (b) Groote Eylandt; (c) Albatross Bay and the Embley estuary; and (d) the Norman estuary.

**Table 4.3** Main physical features of the Embley (E), Norman (N) and Groote Eylandt (G) study areas in the Gulf of Carpentaria, northern Australia (D, dominant type; m, minor; –, absent)

| Physical parameter | Type | Study areas | | |
|---|---|---|---|---|
| | | E | N | G |
| Habitat type | Open water up to 7 m deep | D | D | D |
| | Intertidal mud/sand <2 m | D | m | D |
| | Mangroves/ mangrove creeks | D | D | – |
| | Seagreass | m | – | D |
| Salinity (‰) | Strong sesonal (0–35) | – | D | – |
| | Seasonal (10–35) | D | – | – |
| | Marine (35) | – | – | D |
| Temperature (°C) | Maxima | 32 | 33 | 33 |
| | Minima | 25 | 19 | 24 |
| Turbidity (NTU) | Maxima | 55 | 2656 | 17 |
| | Minima | 0.2 | 9.5 | 1.3 |
| Current speed (m s$^{-1}$) | Maxima | 0.1 | 0.7 | <0.01 |
| Tides (m) | Minimum height | 0.1 | –0.1 | 0.2 |
| | Maximum height | 3.0 | 4.0 | 1.5 |
| | Maximum daily range | 2.6 | 3.8 | 1.5 |

As described in Chapter 3, the general species composition of tropical estuarine faunas is similar throughout the Indo–West Pacific (Blaber *et al.*, 1989; Robertson and Blaber, 1992). The majority of species characteristic of tropical estuaries also live in adjacent shallow, turbid areas of the sea. Along the East Australian and East African coasts (except in estuaries), where clear waters and coral reefs abound, such shallow areas are scarce, but they are common in South East Asia and West Africa. Blaber (1981a) and Longhurst and Pauly (1987) postulated that the fishes of tropical estuaries are better defined not as estuarine, but as characteristic of a particular set of conditions. The fish faunas of the three sites studied in the Gulf of Carpentaria include many of the species of non-coral-reef, inshore and estuarine areas of the tropical Indo–West Pacific. The sites encompass a wide range of habitats and physical conditions.

The sites are best compared by habitat and physical gradient (Fig. 4.16). Although some species occur at all sites, for all habitats (open water, beaches, seagrass, mangroves), species composition among the three sites differs markedly, and at least six distinct types of fish community can be recognized.

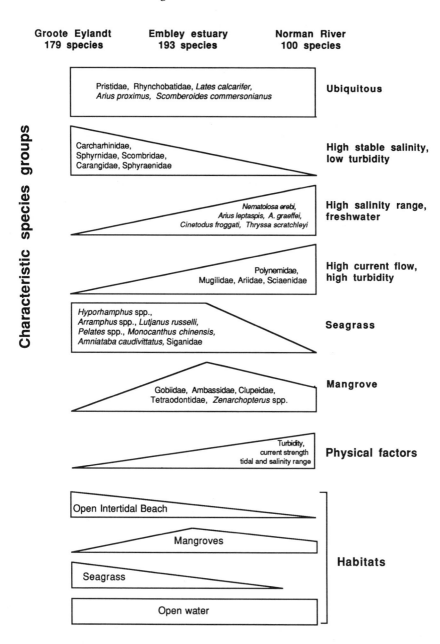

**Fig. 4.16** The abundance of different estuarine fishes in relation to habitats and physical gradients at Groote Eylandt and the Embley and Norman estuaries, showing six distinct community groups. Reproduced with permission from Blaber *et al.* (1994a).

1. A ubiquitous group occurs across all habitats and conditions – its most prominent species are *Scomberoides commersonianus*, *Lates calcarifer* and *Arius proximus*.
2. An open-water group of taxa characteristic of slow currents, stable high salinities and low turbidities includes Carcharhinidae, Carangidae, Scombridae, Sphyraenidae, Sphyrnidae and *Psammoperca waigiensis*. They are most abundant at Groote Eylandt, but were also common in the lower reaches of the Embley (Blaber *et al.*, 1989, 1992).
3. A contrasting open-water community lives in areas of high current flow and high turbidity in the Norman estuary and upper and middle reaches of the Embley; this group includes Polynemidae, Sciaenidae, some Mugilidae (e.g. *Liza subviridis*, *L. tade*) and most Ariidae.
4. A community characterized by taxa such as *Hyporhamphus quoyi*, *Arrhamphus sclerolepis*, *Monacanthus chinensis*, *Lutjanus russelli* (juveniles), Siganidae and *Amniataba caudavittatus* (juveniles) predominates in shallow ( < 2 m) seagrass areas of low turbidity, low current flow and relatively stable salinity.
5. Although a very wide range of species occurs in and adjacent to mangroves (Robertson and Blaber, 1992), there is a distinct community of smaller resident species characteristic of small mangrove creeks; members of the families Ambassidae, Gobiidae, Clupeidae and Tetraodontidae as well as *Zenarchopterus* spp. are included. This group is most diverse in the extensive mangroves of the Embley but is still evident in the narrow mangrove fringe along the Norman estuary.
6. A smaller group of species characteristic of low-salinity or freshwater conditions in the Norman and the upper reaches of the Embley (Blaber *et al.*, 1989) includes *Nematalosa erebi*, *Leptobrama mulleri*, *Thryssa scratchleyi* and the ariids *Arius graeffei*, *A. leptaspis* and *Cinetodus froggati*.

These analyses suggest that species composition and community structure may be strongly influenced by current speed and turbidity, and to a lesser extent, salinity. Turbidity has been shown to be an important determinant of species composition in estuaries (Blaber and Blaber, 1980; Cyrus and Blaber, 1987a), but the effects of current speed are not well understood. Current flow in tropical estuaries, such as the Norman, results largely from tidal range and freshwater inflow. Both are also correlated with highly variable salinities and high turbidities. In addition, habitat diversity and structure are important in relation to species composition; the Embley estuary, with a broad range of habitats, has more species than either Groote Eylandt or the relatively homogeneous Norman estuary. The influence of the structural heterogeneity of each habitat is important, particularly in mangroves (Thayer *et al.*, 1987); this is reflected by the greater number of species found in the complex forests of the Embley than in the more depauperate Norman estuary. It is thus probable that the fish communities of estuarine and inshore waters of the tropical Indo–West Pacific are determined largely by the interplay of the interrelated physical factors of current speed, turbidity and salinity, as well as by the structure of the habitat. This interplay produces communities characteristic of particular combinations of conditions such as those described here, some or all of which may occur in estuaries, but need not be confined to them.

*Chapter five*

# Feeding ecology and trophic structure

## 5.1 INTRODUCTION

The food and feeding ecology of estuarine fishes has long been a favourite subject for study. This is because information on diets, food availability and feeding behaviour of fishes is fundamental to an understanding of their community structure, their distribution patterns and their life history strategies. The collection of such fundamental data has also been vital to those concerned with effective fisheries management and conservation. For example, it has been important for the management of the commercially important tropical penaeid prawn fisheries in the Gulf of Carpentaria, Gulf of Mexico and Gulf of Arabia to know the levels of natural predation on prawns. Hence much research has been undertaken on the diets and feeding ecology of the fishes of these gulfs and adjacent estuaries, with particular reference to the relative importance of prawns in fish diets and the numbers eaten (Sheridan and Trimm, 1983; Sheridan *et al.*, 1984; Pauly and Palomares, 1987; Salini *et al.*, 1990, 1992, 1994; Brewer *et al.*, 1991; Smith *et al.*, 1991, 1992).

The overall energy pathways in estuaries are well summarized in such texts as Green (1968), Day (1981) and Day *et al.* (1989), and a wide range of topics related to feeding ecology are covered by Gerking (1994). The present chapter is confined to subtropical and tropical fishes and their feeding interrelationships with estuaries. Firstly, selected examples are used to illustrate the closely related topics of feeding specializations and conversely, dietary flexibility, as well as ontogenetic changes, with particular reference to their effects on fish distribution in estuaries. Secondly, the various feeding or trophic categories among the major groups of tropical estuarine fishes are surveyed; this is necessary because detailed information on diets is relevant to community structure and distribution patterns. Finally, we examine how the complexity of food webs and hence community structure may be influenced by the type of estuary, its state and its perturbations. Questions related to predation on fishes in estuaries, both by fishes themselves and by other organisms, are dealt with in Chapter 7.

## 5.2  FEEDING SPECIALIZATIONS

The extraordinary array of feeding adaptations and mechanisms in fishes allows them to exploit almost all available sources of food (for reviews of feeding mechanisms see e.g. Alexander, 1967; Norman and Greenwood, 1975). A rich variety of foods exist in tropical estuaries because of their diversity of habitats and their interfaces with freshwater, marine and terrestrial habitats. It is not surprising, therefore, that there is considerable scope for dietary specialization in the fish fauna. In this section a few examples of the adaptive significance of some of these dietary specializations are highlighted.

### Mugilidae

Grey mullet are one of the most numerous and characteristic families in subtropical and tropical estuaries. Almost all species are iliophagous (feeding on the organisms in and on the substratum by ingestion of the substratum) and their diet consists largely of benthic algae, bacteria, meiofauna and smaller epifauna. Ontogenetic changes in diet (pp.153–154) take place at a small size, and from a length of about 50 mm the diets of juveniles and adults are the same (Blaber, 1977; Albaret and Legendre, 1985). The prey of mullet are taxa that occur most abundantly in shallow and quiet waters, and in or on substrata with a relatively high organic content. In subtropical and tropical regions, these conditions are found mainly only in estuaries where there is a high input of organic detritus. The substratum in such systems is frequently covered with a carpet of benthic diatoms and other algae. Mullet are well adapted to take advantage of this situation. They feed by taking up the surface layer of the substratum or by grazing on submerged rock and plant surfaces (Odum, 1970). Most species ingest large quantities of inorganic particles together with food items. Blaber (1976, 1977) showed that the most important diet items of 10 species (Table 4.1) in south-east Africa are Foraminifera, flagellate Protozoa, pennate diatoms, unicellular and filamentous green algae, bluegreen algae, Ostracoda, and a variety of small animals such as *Assiminea* (a gastropod) and harpacticoid copepods. The diets of the various species are similar, but vary from estuary to estuary, and are probably largely determined by the occurrence of particular food items on substrata of the preferred particle size of each species. These particle size preferences are thought to reduce interspecific competition (p.124, Table 4.1). A very similar situation exists in West African estuaries for *Liza falcipinnis, L. grandisquamis, L. dumerili* and *Mugil cephalus* (Fagade and Olaniyan, 1973; Payne, 1976; Albaret and Legendre, 1985; Baran, 1995); and in South and South East Asian estuaries for *Liza macrolepis, L. parsia, L. subviridis, L. tade, L. vaigiensis, Mugil cephalus, Valamugil buchanani, V. cunnesius* and *V. speigleri* (Sarojini, 1954; Luther, 1962; Chan and Chua, 1979; Silva and De Silva, 1981; Wijeyaratne and Costa, 1987a,b,c, 1988, 1990).

The importance of bacteria in the diet of *Mugil cephalus* was demonstrated quantitatively by Moriarty (1976) in Australia, where bacteria supplied about 15–30% and diatoms 20–30% of the organic carbon in the stomach.

### Eleotridae

This family (gudgeons and sleepers) is very widely distributed in tropical estuaries of all regions of the world, where many of the species form part of the estuarine category of fishes. Although ubiquitous, little is known of the biology of most species and their relationships with the estuarine environment are unclear. A notable exception to this are the eleotrids of the Tortuguero estuary in Costa Rica (Chapter 2, p.23). Research by Nordlie (1981) on four species of eleotrid (*Eleotris amblyopsis, E. pisonis, Dormitator maculatus* and *Gobiomorus dormitor*) in the Tortuguero estuary indicates that three of the four species are closely associated with the mats of floating water hyacinth, *Eichornia crassipes*, while the fourth, *Gobiomorus dormitor*, is benthic and occurs mainly in sandy areas. *Eleotris amblyopsis* is found only under the mats of water hyacinth, and *Eleotris pisonis* and *Dormitator maculatus* occur mainly under the mats. This ecological association extends to their diets because stomach analyses have revealed that these species are partially herbivorous and that water hyacinth roots are an important part of the diet. These roots together with insect larvae form the bulk of the food. It was shown in south-east Africa (Blaber *et al.*, 1984; Chapter 4, p.121) that water hyacinth supports a diverse fauna of insect larvae among its roots so it is likely that the insect component of the diets of the eleotrids in the Tortuguero was obtained from among the roots.

### Lepidophagy

Lepidophagous fishes are those that feed by pulling scales off other living fishes. Scale eating is known in a number of characoid and cichlid genera from South America and Africa (Fryer *et al.*, 1955; Roberts, 1970), but is rare away from fresh waters. There are, however, a number of estuarine fishes that obtain most of their food in this curious way. *Terapon jarbua* (thornfish) is abundant in the estuaries of the Indo–West Pacific and feeds mainly on fish scales (Whitfield and Blaber, 1978b). Laboratory experiments showed that it is able to digest the scales which have a relatively high calorific value. They are taken only from living fish and usually from fish larger than *T. jarbua*. Slow-swimming species are favoured and target species in Lake St Lucia, South Africa, include the sparids *Acanthopagrus berda* and *Rhabdosargus sarba*, several species of mullet, the sciaenid *Argyrosomus hololepidotus*, the engraulid *Thryssa vitrirostris* and the cichlid *Oreochromis mossambicus*. *Terapon jarbua* always attack the lateral surface and scales are removed mainly from the posterior part of the body around the caudal peduncle.

In the tropical East Pacific, scale feeding is common in juveniles of at least seven species of ariid catfishes in the Gulf of Nicoya, Costa Rica. It is particularly evident in *Ariopsis seemani* (Szelistowski, 1989). As in *Terapon jarbua*, most target fish are larger than the juvenile ariids and include centropomids, sciaenids, bothids and clupeids. *Arius felis* from the Gulf of Mexico is also known to remove scales from *Mugil cephalus* (Hoese, 1966). Szelistowski (1989) states, "the high number of scales eaten per individual and the abundance of ariids in tropical East Pacific estuaries may make scale feeding an important form of parasitism in fishes from these areas." Evidently estuarine conditions of high water turbidity, cover among mangrove roots or seagrasses, and shallow water may also facilitate scale feeding, allowing the predator to get close to the target species without detection and disappear again once a scale has been obtained.

## 5.3   ONTOGENETIC CHANGES IN DIETS

The diets of most fish species change with growth. As larvae and postlarvae, the majority are planktivorous; the timing and extent of changes in food and feeding ecology varies from species to species and is often associated with changes in life style or habitat. For many species in tropical estuaries, the diet of juveniles is markedly different from that of the adults (Blaber and Blaber, 1980; J.H. Day *et al.*, 1981; J.W. Day *et al.*, 1989; Whitfield, 1996). In this case the diets of juveniles, and the ontogenetic changes themselves, may be specialized and directly related to living in estuaries. The actual timing of switches in diet usually relates to juveniles becoming subadults or adults and leaving estuaries (see also Chapter 3, pp. 47–52), or may be related to changes in morphology of jaws or teeth or size. For most species, larger fish eat larger prey and hence there will be a gradual change in the size of prey as the fish grows (Dall *et al.*, 1990).

In this section we use selected examples to examine how ontogenetic changes in diet enable species to utilize or maximize their use of estuarine habitats, and how this influences their distribution patterns.

### Sciaenops ocellata

The red drum is a commercially important sciaenid of the Gulf of Mexico and Florida. Spawning takes place in the sea near estuary mouths and the larvae or juveniles move into estuaries and sheltered waters. The juveniles live in a wide variety of estuarine nursery areas, particularly quiet coves and lagoons with seagrass over sand or mud bottoms, but they are also found in unvegetated areas. Much of this habitat variety is apparently related to their migrations as they grow and move from the deeper waters into quiet backwaters, and then at a length of about 200 mm back to deeper waters again. The following account of the ontogenetic changes in their diet is taken largely from the studies of Peters and McMichael (1987) in Tampa Bay and its estuaries.

The larvae feed mainly on copepods and copepod nauplii but postlarvae and juveniles (10–75 mm) switch to a diet of mainly mysids, amphipods and polychaetes, with mysids predominant in the smaller size classes. Larger juveniles (> 75 mm) feed on fish, crabs and shrimps – a diet that is continued into adulthood (Fig. 5.1). The ontogenetic changes can be related both to the size of the fish and to the various habitats. Polychaetes occur only in fish from non-vegetated areas while copepods are more numerous in juveniles from seagrass areas. The availability of fish and crabs taken by the larger sizes of red drum is much greater in deeper, more marine waters, whereas the shrimps (mainly carids) that feature in the diets of smaller fish are more abundant up the estuaries.

### Lates calcarifer

This large and commercially important centropomid is found in tropical and subtropical estuaries and coastal waters in South and South East Asia and Australia. In a comprehensive study of its diet in northern Australia, Davis (1985) showed that from a length of 4 mm to 1200 mm it is an opportunistic predator with an ontogenetic progression in its diet from microcrustaceans to macrocrustaceans to fish. Small *Lates calcarifer* (< 80 mm) feed mainly on copepods and amphipods. Such microcrustaceans are absent in fish longer than 80 mm where they are replaced by macrocrustaceans such as penaeids and palaeomonids. Fish and macrocrustaceans are equally important up to about 300 mm but the

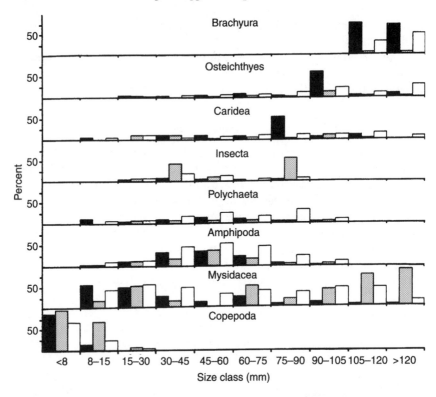

**Fig. 5.1** Composition of the diet of 15 mm size classes of *Sciaenops ocellata* from Tampa Bay, Florida (dark grey, percentage volume; light grey, percentage by numbers; white, percentage frequency of occurrence). Reproduced with permission from Peters and McMichael (1987). © Estuarine Research Federation.

proportion of fish increases thereafter until it represents more than 80% of the diet in larger fish. A wide variety of fish species are consumed with Mugilidae, Engraulididae and Ariidae among the most frequent. There is a strong correlation between the length of *Lates calcarifer* and that of its prey, with most prey fish being about 30% of the length of the predator.

Davis (1985) reports that the transition from eating prawns to eating fish entails spending more time feeding on pelagic species in mid-water or on the surface, rather than on benthic forms. This could result in greater exposure to predators, but predation may be a factor of decreasing importance as the fish attain a large size (with the possible exception of crocodiles – Chapter 7). The prey chosen by the smaller size classes of *L.*

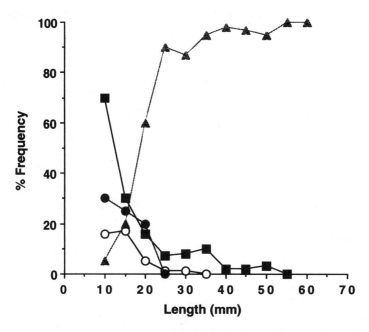

**Fig. 5.2** Ontogenetic changes with fish length in the diet of postlarval and juvenile mullet, *Liza macrolepis* (adapted from Blaber and Whitfield, 1977a). ○, Plankton; ■, vertically migrating plankton; ●, meiobenthos; ▲, microbenthos.

*calcarifer* may reflect the types of food available in areas where predation risk is reduced, such as in vegetated backswamps and shallow creeks in the upper reaches. Cannibalism is common in *L. calcarifer* (up to 11% of the diet in some areas), but is probably reduced by the different size groups living in different habitats. For example, the small juveniles occupy shallow habitats that are not accessible to larger conspecifics.

## Mugilidae

The diets of juvenile mullet undergo a series of ontogenetic changes between lengths of about 10 and 50 mm until the iliophagous habit is attained (pp.148–149). These changes were studied for 11 species in south-east African estuaries (Blaber and Whitfield, 1977a), all of which spawn in the sea and the young enter estuaries at a length of about 10 mm. In estuaries they change their feeding habits in the following sequence: zooplankton to zooplankton in the benthos (10–15 mm), zooplankton in the benthos to meiobenthos (10–20 mm), and meiobenthos to sand particles and associated microbenthos (15–25 mm). All species show a similar pattern of change and that for *Liza macrolepis*, a species widespread in the Indo–West Pacific, is illustrated (Fig. 5.2). Interspecific competition is probably reduced by the rapid switch to the adult iliophagous habit and because species enter estuaries at different times according to spawning periods. Once they have entered an estuary (Fig. 4.12), two important changes in the physical environment influence the feeding ecology of juvenile Mugilidae. Firstly, most tropical estuaries, especially those of type 2 (section 2.6), are shallow compared with the sea, and secondly, estuarine waters are less turbulent than those of the sea. Juvenile mullet in estuaries prefer water less than 1 m deep and because much of the estuarine zooplankton is in, on or near the bottom during the day (Perkins, 1974), the fish are able to capture zooplankton from the benthos. It is a short step from feeding on vertically migratory zooplankton at the benthic stage to ingesting meiobenthic fauna such as nematodes, polychaetes and oligochaetes. Once they are feeding on the benthos, the number of sand grains increases and the meiobenthos is abandoned in favour of microbenthic foods associated with sand grains.

It is probable that the change-over from planktonic feeding to the ingestion of microbenthos and sand could take place only in the shallow, quiet waters of estuaries. Under turbulent conditions in shallow waters it would be difficult for small juveniles to consume sand and associated organisms, and deeper waters support a lower density of food organisms on the substrata such as diatoms. Estuaries thus play a vital role in providing suitable conditions for juvenile mullet to change their feeding habit to that of the adults. It is significant that the change in diet takes place relatively soon after they enter estuaries. Mullet of all ages feed in estuaries and it may be only in such areas that conditions are suitable for substratum feeding.

### Toxotidae

Archerfish (genus *Toxotes*) inhabit mangrove estuaries and freshwater streams from New Caledonia and Solomon Islands in the Pacific, through northern Australia and New Guinea, to South East Asia and southern India (Allen, 1978). Their ability to shoot down prey from overhanging vegetation, by squirting them with a jet of water from the mouth (Lüling, 1963), has led to a great deal of research on the mechanisms involved. The biomechanics of spitting have been described and analysed (Smith, 1936, 1945; Myers, 1952; Elshoud and Koomen, 1985), the fine structure of the eyes and retinal pigments described (Braekevelt, 1985a,b), and the problems of accurate aiming and refraction at the air–water interface investigated (Dill, 1977). Lüling (1963) stated, however, that specialized water shooting is not the archerfish's primary means of food gathering, and that it depends largely on food it finds on or below the surface, thus prompting the question of the evolutionary or survival value of the spitting mechanism. Despite all of the above studies, no quantitative data could be found on the diet of any of the six species of *Toxotes*. The author had the opportunity to gather such data during research on the estuarine fishes of the Gulf of Carpentaria in northern Australia.

The diet of archerfish, *Toxotes chatareus*, was studied in two tropical mangrove estuaries in northern Australia. There are ontogenetic changes in the diet of *T. chatareus*: small fish (< 100 mm) eat no plant material, medium-sized fish (100–139 mm) ingest some plant material, while about a quarter of the intake of larger fish consists of parts of mangrove trees. There is a gradual reduction in the importance of green tree ants, *Oecophylla smaragdina*. In medium and large fish these are to some extent replaced by larger insects such as Orthoptera, but overall, with increasing fish size, arboreal insects are replaced by larger arboreal prey, particularly grapsid crabs. The various arboreal prey thus make up about half the diet of all size groups. This indicates that, contrary to previously published theory, spitting is indeed the main prey capture method. A smaller proportion of the diet of all size groups consists of animals captured in the water, such as amphipods, carids, penaeids and fish. The increasing incidence of plant material in the diet is curious; it is not digested and its intake could be accidental. If so, its origin can be explained in two possible ways: firstly, parts of mangroves incidentally falling onto the surface of the water are snapped up by the fish; or secondly, when the archerfish shoots at an arboreal crab or insect it may cause the dislodgment of pieces of bark or seeds, especially if it misses hitting the prey. The absence of mangrove material in small fish may be because their jet of water is insufficient to break off the plant material. Either of these scenarios suggests that the seizing of prey by the archerfish is a reflex action to material falling onto the water, and that the fish may be unable to discriminate between living and inanimate material. It has been shown that other fish species, which prey largely on insects captured from the water surface, are able to localize and identify prey from non-prey, by discriminating between the different surface wave frequencies set up by struggling insects as opposed to other objects (Bleckmann *et al.*, 1981; Hoin-Radkovsky *et al.*, 1984; Käse and Bleckmann, 1987). It would appear that *Toxotes* are not capable of such discrimination, or perhaps employ a response that is too rapid to use such a system. It is possible that the plant material is consumed deliberately, perhaps to supplement the diet in some unknown way or to aid in the digestion of arthropods. It is possible that the shooting method may also be employed for aquatic prey as suggested by Elshoud and Koomen (1985), but more research would be needed to verify this. Regardless of the answers to these questions, it is apparent that this mode of feeding can only be employed successfully in quiet water areas with overhanging vegetation with epifauna – conditions such as exist in the mangrove forests of the middle and upper reaches and tributary creeks of tropical estuaries.

### *Rhabdosargus holubi*

This sparid is endemic to south-east Africa where the juveniles occur mainly in estuaries and the adults in the sea. The diet of juveniles is more than 50% (often up to 90%) aquatic macrophytes with the remainder consisting mainly of small Crustacea (Blaber, 1974). The fish cannot, however, digest the vegetation eaten due to the absence of a cellulase or a method of breaking up the plant tissue. Therefore the plant material is passed out in an undigested state. Epiphytic diatoms and sessile Bryozoa are, however, removed from the plants in the stomach and are subject to digestion. Diatoms may form up to 50% of the weight of plants eaten. It would thus appear that during the juvenile phase of its life history, which is spent in estuaries, *Rhabdosargus holubi* is an omnivore with a marked preference for plant material covered with diatoms. The migration to the sea at maturity is accompanied by a change in the form of the teeth and a reduction in the amount of plant material in the diet. The tricuspid teeth of the juveniles are well adapted to cutting off pieces of macrophyte, but less useful for feeding on animal prey. As the species grows, the teeth become more molariform and *R. holubi* switches to a diet mainly of bivalve molluscs. Such molluscs are scarce in estuaries of the region, but aquatic macrophytes are abundant; conversely macrophytes and their diatom flora are scarce in the sea and bivalves abundant. It appears, therefore, that the food requirements of both the juveniles and adults largely determine when the species enters and leaves estuaries.

### 5.4   FLEXIBILITY IN DIETS

Although most fishes feed on a limited range of prey types, dietary flexibility is a feature of many families. Such flexibility confers important advantages in terms of both survival and mobility in estuaries where there may be large fluctuations in the type and numbers of prey available. There are usually, for example, very marked differences between the sorts of invertebrate prey living in coastal estuarine waters and those found in the low-salinity upper reaches of estuaries. In the following examples it is evident that, although particular groups may have specializations enabling them to feed on or capture particular prey, they are not limited

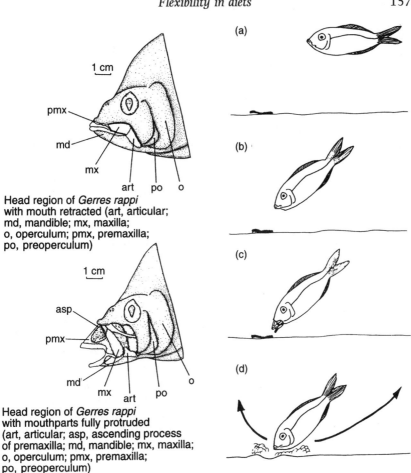

Head region of *Gerres rappi* with mouth retracted (art, articular; md, mandible; mx, maxilla; o, operculum; pmx, premaxilla; po, preoperculum)

Head region of *Gerres rappi* with mouthparts fully protruded (art, articular; asp, ascending process of premaxilla; md, mandible; mx, maxilla; o, operculum; pmx, premaxilla; po, preoperculum)

**Fig. 5.3** The protrusible jaw mechanisms of *Gerres rappi* and sequence of prey capture by *Gerres*: (a) orientation, (b) downward pivot, (c) protrusion of mouthparts and (d) return to horizontal. Reproduced with permission from Cyrus and Blaber (1982).

to these prey, and are able to switch to other foods if it is necessary or advantageous.

### Gerreidae

The Gerreidae (pursemouths) are a characteristic component of tropical and subtropical estuarine fish communities in all regions. They have evolved an efficient protrusible jaw structure and suction pressure mechanism that enables them to feed on a variety of prey on or just below the surface of the substratum (Cyrus and Blaber, 1982) (Fig. 5.3). In a study of the feeding ecology of five species of *Gerres* in the Kosi system of KwaZulu-Natal, Cyrus and Blaber (1983) showed that they feed largely on the siphon tips of *Hiatula lunulata,* an abundant bivalve. For this, prey suction is probably not used because only the distal 5 mm of siphon is taken and it is probable that the rapid contraction of the siphon, together with the biting motion of the fish, causes the siphon to break off before the mouth touches the substratum. In this estuary *Gerres* feed almost exclusively on bivalve siphon tips. However, benthic sampling at the sites where siphons are dominant in the diet of *Gerres* showed that they were not the most numerous food item present. Interestingly, because siphon tips regrow they are also a renewable food resource. They are positively selected for, perhaps because their energy value is higher than that of other foods. Despite this apparent feeding specialization in the Kosi system, however, on a worldwide basis, members of the family (including those species found in the Kosi system) feed on a wide variety of invertebrates, with polychaetes the single most important food of most species – from India (Prabhakara Rao, 1968) to Australia (Blaber, 1980) and from West Africa (Longhurst, 1957) to the Caribbean (Austin and Austin, 1971; Carr and Adams, 1973; Charles, 1975). The next most common food is whole bivalves (Fagade and Olaniyan, 1973; Albaret and Desfossez, 1988) followed by various Crustacea and Gastropoda. For most of these slower-moving taxa the suction pressure developed by the protrusible mouth and feeding action is more useful in feeding than the 'nipping' action used on bivalve siphons. *Diapterus* species in the tropical West Atlantic include a large quantity of aquatic macrophytes in their diets (Austin and Austin, 1971), but how these are fed upon has not been studied.

The Gerreidae thus possess a specialized morphology and feeding mechanism for eating slower-moving or small invertebrates, but they are also able to exploit, when available, a valuable food resource such as bivalve siphon tips, using a different feeding method. The Gerreidae serve to emphasize that for a full understanding of the relative significance of any feeding specialization, it is necessary to have information from throughout the range of a species, and from a variety of habitats.

### Ariidae

Fork-tailed catfishes of the family Ariidae occur throughout the shallow waters of the tropics, both in fresh water and the sea, with numerous euryhaline species in estuaries (Nelson, 1984; Kailola, 1990). Ariidae in the northern Australia/southern New Guinea region consist of a complex of closely related genera and species, all of relatively uniform coloration and external morphology. In a major review of the family, Kailola (1990) recognized about 35 species from New Guinea and Australia, of which 16 occur in tropical northern Australian waters. Thirteen of these catfishes are a prominent component of the estuarine and inshore fish communities of the Gulf of Carpentaria (Blaber *et al.* 1989, 1992) where as many as 10 species may be sympatric (Blaber *et al.*, 1994b). The number, shape and arrangement of the premaxillary and palatine teeth plates (Fig. 5.4) are important taxonomic characters (Kailola, 1990) and vary markedly from one species to another. The mandibular tooth band apposes the premaxillary band and is usually similar to it.

The diets of 13 species of ariid catfishes from the tropical waters of the Gulf of Carpentaria have been described and compared (Blaber *et al.*, 1994b). Fish were collected from two estuaries and inshore and offshore marine areas. Up to 10 species have been recorded from a single estuary. Although all are carnivorous and consume a variety of prey, diet analyses and statistical ordination reveal three feeding guilds: piscivores, polychaete-eaters and molluscivores. The diets of most species are similar between sites. There are strong relationships between dietary guild and the size and arrangement of the palatine teeth (Figs 5.4, 5.5). The piscivorous group of catfish (guild 1) have large mouths with relatively large multiple palatine tooth plates, either in a band or in a triangular pattern and armed with sharp recurved teeth. The primarily polychaete-feeding group (guild 2) have a variable mouth size, but it is usually smaller than that of guild 1 fish; their palatine teeth plates are fewer and smaller, and they have small, sharp recurved teeth. Guild 3 eat mainly molluscs, and have a small mouth and large posteriorly situated palatine plates with globular, truncated teeth. Paradoxically, morphological specializations contrast with opportunistic feeding behaviour and the ability of most species to cope with a range of prey. Overlaps in diet between species are probably reduced by differential distribution patterns within estuaries and different habitat preferences. The mouth-width and tooth-plate arrangements of ariids are suitable for dealing with broad classes of prey rather than specific items, conferring dietary flexibility. This probably optimizes the trade-off for most species between occupation of broad feeding niches and the ability to shift diet easily.

### *Bairdiella chrysoura*

In the large coastal lake system of Terminós Lagoon in Mexico there are pronounced wet and dry seasons (Chapter 2, p.37) which influence the availability of food for fishes. One of the most common species is the sciaenid *Bairdiella chrysoura*. It is present year-round in the system and is

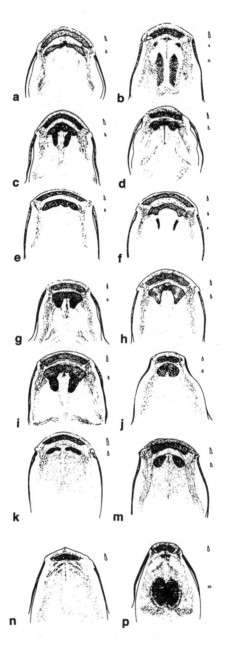

**Fig. 5.4** Palatine and vomerine tooth patterns of Ariidae from the Gulf of Carpentaria. Guild 1 (piscivores): a, *Hemiarius* sp. 1 (254 mm); b, 'Genus 1' *nella* (161 mm); c, *Arius bilineatus* (235 mm); d, *Arius graeffei* (295 mm); e, *Arius leptaspis*

essentially an opportunistic carnivore (Chavance *et al.*, 1984; Vega-Cendejas *et al.*, 1994). This is illustrated by the changes in its main foods between the wet and dry seasons. During the dry season its diet comprises by weight about 79% engraulids, 4% penaeid prawns and 3% tanaids, but in the wet, the proportion of engraulids falls to 30% and penaeid prawns disappear, to be replaced by 10% ariid catfishes, 33% carid prawns and 19% crabs. Such flexibility in diet is a general feature of many of the smaller species of sciaenids, and similar seasonal changes in diet between decapod Crustacea, particularly penaeid prawns, and fish have been recorded for seven species in the Matang system in Malaysia (Yap *et al.*, 1994) and for two species in the Gulf of Carpentaria, Australia (Brewer *et al.*, 1991).

## 5.5   TROPHIC LEVELS AMONG ESTUARINE FISHES

Although a great deal has been written about what fish eat, usually classified from a taxonomic standpoint, knowledge about how they get their food and whether they choose particular foods is perhaps equally important. In broad terms the diet is controlled by the relative sizes of predator and prey, the capability of the predator in terms of capture, handling and digestion, and the opportunity to encounter the prey. We have seen that many species appear to be opportunistic, in other words if they encounter a prey item, can catch it, manipulate it, swallow it and digest it, then it forms part of the diet. However, we also know that fish make choices, and that some foods are preferred, perhaps because of higher calorific value or because of other dietary requirements, as suggested for *Sillago analis* by Brewer and Warburton (1992). Much further research is needed on this topic for tropical estuarine fishes, but a sound knowledge of their diets is an essential prerequisite for a deeper understanding of their feeding ecology and behaviour. Most fishes are omnivorous to some extent, and their trophic level or category depends upon, and may vary according to, the relative proportions of different foods eaten. Hence the utility of the term 'omnivore' is questionable and no single omnivorous category has been included here.

---

(384 mm); f, *Arius mastersi* (389 mm); g,h, *Arius proximus* (337, 331 mm); i, *Arius thalassinus* (378 mm). Guild 2 (polychaete eaters): j, *Cinetodus froggatti* (325 mm); k, *Arius armiger* (212 mm); m, *Arius* sp. 3 (259 mm); n, '*Arius*' sp. 4 (242 mm). Guild 3 (molluscivores): p, 'Genus 1' *argyropleuron* (314 mm) (redrawn from Blaber *et al.*, 1994b).

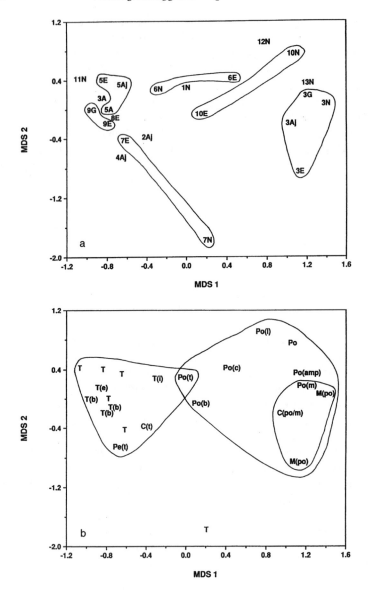

**Fig. 5.5** (a) Plots of MDS 1 versus MDS 2 resulting from the multidimensional scaling analysis of ariid catfish species groups based on the composition of their diet. Each plotted point represents a group based on the species of catfish (1–12), the sampling site (capital letter) and whether they were juvenile fish (j). Those occurring at more than one site are circled together. 1, *Arius armiger*; 2, *Arius bilineatus*; 3, 'Genus 1' *argyropleuron*; 4, 'Genus 1' *nella*; 5, *Arius thalassinus*; 6, *Arius graeffei*; 7, *Arius leptaspis*; 8, *Arius mastersi*; 9, *Arius proximus*; 10, *Arius*

### Herbivorous fishes

Herbivores are rare among tropical estuarine fishes, but many species incorporate plants, either algae or macrophytes, as part of their diet. Families that occur in estuaries and in which herbivory is recorded are Hemiramphidae (needlefish), Sparidae (breams), Siganidae (rabbitfish), Gobiidae (gobies), Cichlidae (cichlids), Monacanthidae (leatherjackets), Scatophagidae (scats) and Stromateidae (pomfrets).

Tropical estuarine herbivorous fishes can be divided into three groups that are not mutually exclusive: (a) partial herbivores, (b) those that are herbivorous for part of their life history, and (c) those that obtain most of their nutrition from plants.

Partially herbivorous taxa include some of the Hemiramphidae: *Hyporhamphus capensis* in South Africa feeds predominantly on aquatic macrophytes such as *Ruppia* and *Potamogeton* when they are abundant but switches to an animal diet (mainly amphipods and isopods) when plants are scarce (Coetzee, 1981). *Hyporhamphus melanochir* in Australia is a diurnal herbivore and nocturnal carnivore: it feeds on *Zostera* leaves during the day, but at night switches to animals such as amphipods that move into the water column. This feeding strategy appears to be an adaptation to diel changes in food availability and perhaps to the fish's metabolic requirements (Robertson and Klumpp, 1983). Another partial herbivore is *Monacanthus chinensis*, which lives mainly in seagrass beds in estuaries and includes a high proportion of seagrass in a mixed diet of small crustaceans and polychaetes (Warburton and Blaber, 1992).

Many of the Hemiramphidae, for example *Arrhamphus sclerolepis*, *Hemiramphus robustus*, *Hyporhamphus dussumieri* and *H. regularis*, feed almost entirely on aquatic macrophytes as adults. Their juveniles consume mainly planktonic animals such as amphipods, gastropod and bivalve larvae, copepods and isopods. Among the Sparidae, *Rhabdosargus holubi* (p.156) of South Africa is primarily herbivorous during the first year of life but adults are totally carnivorous (Blaber, 1974), whereas conversely, *Archosargus rhomboidalis* of the Gulf of Mexico is mainly herbivorous as an adult and carnivorous on small invertebrates as a juvenile

---

sp. 3; 11, *Hemiarius* sp. 1; 12, *Cinetodus froggatti*; 13, '*Arius*' sp. 4. A, Albatross Bay; E, Embley River estuary; G, Groote Eylandt; N, Norman River estuary. (b) Plots of catfish groups as in Fig. 5(a), except that they are represented here by their main prey category, and where applicable a secondary important prey in the diet. T, teleost prey; Po, Polychaeta; M, Mollusca; C, Crustacea; Pe, Penaeidae; b, Brachyura; e, Echinodermata; i, Insecta; amp, Amphipoda. Members of each of the three feeding guilds are circled together. Reproduced from Blaber *et al.* (1994b) with kind permission from Kluwer Academic Publishers.

(Chavance *et al.*, 1986; Vega-Cendejas *et al.*, 1994). The stromateid *Pampus argenteus*, common in estuarine coastal waters of South East Asia, is mainly herbivorous in the postlarval phase, but becomes a carnivore as it grows larger (Pati, 1983).

Two families that are primarily herbivorous are the Siganidae and Scatophagidae. There are 27 species of Siganidae and as their single row of flattened, close-set teeth suggests, they feed mainly on seaweeds (Woodland, 1990). However, only one species, *Siganus vermiculatus*, is exclusively estuarine; the remainder, although venturing occasionally into estuaries, live mainly on coral reefs, reef flats with seagrasses, or along rocky shores. *Siganus vermiculatus* is an Indo–West Pacific species that lives amongst mangroves and in and around the lower reaches of large estuaries. It is particularly common in Sri Lanka and south India in coastal lakes, but extends through South East Asia to the South Pacific. Woodland (1990) observed schools of young *Siganus vermiculatus* in the creeks and shallow turbid waters of the Rewa estuary in Fiji, feeding on fronds of seagrass and on algae that encrusted stones.

The Scatophagidae is a tropical Indo–West Pacific family occurring in estuaries and coastal waters from the east coast of India through South East Asia to Australia and the South Pacific (Nelson, 1984). The most widespread species is *Scatophagus argus*, a very euryhaline species that lives in coastal waters, estuaries and even in fresh waters. Its successful colonization of a variety of quiet-water habitats has been aided by its broad diet, for although primarily herbivorous (Barry and Fast, 1992), it includes zooplankton and macrobenthos in its diet (Datta *et al.*, 1984). Another scatophagid, *Selenotoca multifasciata*, which occurs in northern Australian estuaries, has a narrower food spectrum, but eats a wide variety of plants, including filamentous algae and macrophytes (Lee *et al.*, 1993). Both *Siganus vermiculatus* and *Scatophagus argus* are being used in aquaculture in South and South East Asia and have potential as cage culture species (Gargantiel, 1982; Barry *et al.*, 1993).

Many of the species that consume unicellular algae and diatoms, such as the iliophagous species considered in the next section, may be regarded as partially herbivorous. Few, however, are sufficiently selective to feed solely on plants. One that does is the goby *Pseudapocryptes dentatus*, a mudskipper of the Arabian Gulf. Studies at the Fao estuary show that it feeds solely on benthic phytoplankton, particularly the pennate diatoms *Pleurosigma* and *Navicula*, scraped from the surface of the mudflats (Sarker *et al.*, 1980). Such mudskippers, so characteristic of nearly all tropical mangrove areas, vary considerably in their feeding ecology and behaviour (MacNae, 1968). *Boleophthalmus boddaerti* of South East Asia is another herbivore that skims algae from the surface of the mud (it also has the long intestine of a herbivore, quite unlike the shorter intestine of other mudskippers); *Scartelaos viridis* is omnivorous, scraping up mud and associated algae and microfauna; species of the genus *Periophthalmus* are usually carnivores, eating mainly small Crustacea and errant polychaetes, and very little plant material; and the larger *Periophthalmodon* species are mainly piscivores (MacNae, 1968; Milward, 1974).

### Iliophagous species

This group feeds on benthic algae such as diatoms, microfauna including Foraminifera and Flagellata and to a lesser extent smaller meiofauna, as well as indeterminate particulate organic matter usually of plant origin, sometimes referred to as detritus (in essence all the small organisms in or on the surface layer of the substratum and associated organic matter). To feed, they ingest relatively large volumes of sand or mud, digest the food material and pass out the inorganic particles. They are sometimes referred to as 'detritus feeders' but this term is imprecise and unsatisfactory because much of the food is not detritus. The importance of estuaries as detritus traps has been emphasized by many authors (Bruton, 1985; Day *et al.*, 1989; Whitfield, 1996) and the biomass and productivity of benthic microalgae and microfauna is often very high in tropical estuaries, particularly in mangrove areas and on mudflats adjacent to them (Hutchings and Saenger, 1987). It is therefore not surprising that iliophagous fishes are among the most numerous species in subtropical and tropical estuaries. The dominant and most speciose group is undoubtedly the Mugilidae, the feeding ecology of which has already been described (pp.123–124, 148–149, 153–154). Mullets are not, however, the only iliophagous group. Two other important iliophagous species in Indo–West Pacific estuaries are the milkfish, *Chanos chanos*, and the cichlid *Oreochromis mossambicus*. The diet of *C. chanos* has received much attention in relation to its role as a premier aquaculture species in South East Asia, but relatively little is known of its feeding ecology under natural conditions throughout its extensive geographic range. In a study of the iliophagous species in St Lucia, South Africa, Whitfield and Blaber (1978c) showed that a suite of mullet species, *C. chanos* and *O. mossambicus* shared the following food resources: centric and pennate diatoms, filamentous algae, particulate organic matter, Foraminifera, Gastropoda (*Assiminea bifasciata*) and various small Crustacea. Hiatt (1944) found that mullet and milkfish compete for food under artificial conditions in Hawaiian fish ponds, feeding on the same species of diatoms and bluegreen algae. However, detailed comparisons of the diets of the iliophagous species in St Lucia revealed that although there is some overlap, there are also important differences. Whereas *O. mossambicus* feed predominantly on epiphytic pennate diatoms, mullet and *C. chanos* consume

mainly centric diatoms attached to sand grains; *C. chanos* eat more crustaceans than the other groups, mainly small vertically migratory zooplankters that are on the substratum surface during the day; and *O.
mossambicus* takes less animal material than any of the mullet or *C.
chanos*. There is a higher incidence of sand grains in the stomachs of
mullet than in either *O. mossambicus* or *C. chanos*, suggesting that their
feeding methods differ. This could result in differential selection of prey,
but foods that are very abundant in St Lucia, such as the tiny gastropod
*Assiminea* (up to $488\,000\,m^{-2}$, Boltt, 1975), are swallowed by all, regardless of feeding methods.

A similar degree of overlap and partitioning of food resources also
occurs in West African estuaries and coastal lakes, where apart from the
five or so mullet species, at least two cichlids, *Oreochromis melanotheron*
and *Tilapia guineensis*, are iliophagous and feed mainly on algal fragments,
diatoms and particulate organic matter (Fagade and Olaniyan, 1973;
Baran, 1995).

The role of unidentifiable particulate organic matter (detrital aggregate)
in diets in tropical estuaries is obscure, but it may sometimes form a
considerable proportion of the diets of iliophagous species. In a study of
*Oreochromis mossambicus* in a coastal lake in South Africa, Bowen (1979)
showed that detrital aggregate which formed an important component of
the diet is generally low in protein, but that the protein content is higher
in shallow areas and declines with increasing depth. He suggests that this
nutritional constraint may account for the nearshore distribution of 'detritivorous' cichlids and the paucity of exclusively 'detritivorous' species of
fishes.

### Meiofauna feeders

Meiofauna is usually defined as benthic metazoa that pass through a
$500\,\mu m$ sieve but are retained on one of $100\,\mu m$. They have largely been
overlooked as a significant food of estuarine fishes in the tropics and
subtropics (Gee, 1989). In a recent review, however, Coull *et al.* (1995)
state that many bottom-feeding juvenile fish pass through an obligatory
meiobenthos (particularly copepod) feeding stage. They go on to suggest
that copepod dominance of this feeding may not be real because non-
chitinous meiofauna, such as nematodes, may be digested rapidly and not
be visible in guts. In studies in subtropical Moreton Bay, Queensland,
Warburton and Blaber (1992) and Coull *et al.* (1995) showed that
meiofauna was important in the diets of seven species: two whitings,
*Sillago maculata* and *S. ciliata*; three gobies, *Favonigobius exquisitus*, *Amoya*
sp. and *Valenciennia longipinnis*; the teraponid *Pelates quadrilineatus*; and
the ponyfish *Leiognathus moretoniensis*. Juvenile *Sillago maculata* ( < 43 mm)

and juvenile and adult *Favonigobius exquisitus* fed almost exclusively on harpacticoid copepods. In an experimental study using tanks with a substratum of mangrove mud, Coull *et al.* (1995) showed that although juvenile *Sillago maculata* reduced nematode abundance in the substratum by 54% and copepod abundance by 56%, no nematodes could be detected 2–3 h later in the guts of the fish – presumably because they are fully digested. For fishes such as juvenile *Sillago maculata* and the various gobies that live in intertidal mangrove areas, the meiofauna clearly provide most of their nutrition. Further research is obviously required in subtropical and tropical estuaries to determine the relative significance of this trophic level.

### Macrobenthos feeders

Included in this category are those species that feed primarily on macro-benthic invertebrates. It is not a tight grouping because many species, particularly in the Sciaenidae and Carangidae, feed on both invertebrates and fishes. Also there is frequently an ontogenetic progression from an invertebrate diet to one of fish (section 5.3). There are, however, a number of families that feed mainly on invertebrate benthos. Good examples are the Haemulidae (grunters) and the Sparidae (breams) as well as the Gerreidae (pursemouths, pp.157–158).

*Pomadasys* species (Haemulidae) are a common and often prominent part of the fish fauna in estuaries of the Indo–West Pacific and tropical East Atlantic. *Pomadasys commersonni* of South and East African estuaries is a large species attaining 92 cm. It feeds on polychaetes, bivalves, crabs and prawns when small, but from a length of about 20 cm preys mainly on burrowing bivalves such as *Solen* and the burrowing prawns *Upogebia* and *Callianassa*. The fish uncovers these burrow-dwelling invertebrates by blowing holes in the sand with its protrusible snout (Day *et al.*, 1981). *Pomadasys kaakan* occurs throughout estuaries and coastal waters of the Indo–West Pacific and feeds mainly on smaller crabs and stomatopods, polychaetes and molluscs (Day *et al.*, 1981; Salini *et al.*, 1990). Another smaller Indo–West Pacific species, *Pomadasys maculatus*, is characteristic of estuarine coastal waters throughout the region and has a similar diet with the addition of penaeid prawns (Brewer *et al.*, 1991).

In the tropical East Atlantic region, *Pomadasys jubelini* is common in most West African estuaries. In Lagos Lagoon and the estuaries of Guinea its diet consists of hermit crabs (without shells), penaeid prawns, stomato-pods and bivalves (Fagade and Olaniyan, 1973; Baran, 1995), with the addition of polychaetes in the Sierra Leone estuary (Longhurst, 1957). All *Pomadasys* species also take small quantities of fish but they are not a large part of the diet.

Most sparid breams have strong molariform teeth suitable for crushing hard-shelled prey such as crabs and molluscs. One of the most widespread estuarine species in the Indo–West Pacific is *Acanthopagrus berda*. In African estuaries, juveniles eat mainly amphipods and tanaids while adults feed on bivalves, gastropods, crabs, penaeid prawns and small amounts of macrophytes and fish (Day *et al.*, 1981; Kyle, 1988; Harrison, 1991). In a north Queensland estuary, Beumer (1978) found that the diet of *A. berda* consists chiefly of crustaceans but that the composition of this group changes from month to month, with Natantia, Reptantia, Cirripedia and Copepoda dominant at different times. Nematodes, polychaetes, insects and molluscs formed the remainder of the invertebrates. Small amounts of fish and filamentous algae were usually present in the diet. A similar species, *Acanthopagrus australis*, is endemic to the estuaries of eastern Australia. Its diet also consists of a wide variety of invertebrates as well as small but significant quantities of filamentous algae and macrophytes (Pollock, 1982).

*Rhabdosargus sarba* is another sparid in which the juveniles and subadults are very common in the estuaries of south-east Africa, India and Australia. In Africa its diet consists mainly of a mixture of aquatic macrophytes, bivalves, gastropods, crabs and amphipods (Blaber, 1984). The diet is similar in Australia (Thomson, 1959) and in Lake Chilka in India (Patnaik, 1973). In Africa *Rhabdosargus sarba* is sympatric with two endemic congeners, *R. holubi* and *R. auriventris*. Their diets differ in the proportions of plant material eaten. *Rhabdosargus holubi* ingests large quantities of plant material from which it digests the epiphytic diatoms (p.156) together with associated amphipods and polychaetes; *R. auriventris* is almost entirely carnivorous, taking bivalves and Crustacea. Therefore, despite occurring in mixed-species shoals, the three *Rhabdosargus* species exhibit differences in feeding ecology. Both *R. sarba* and *R. holubi* consume plants but treat them differently, the former gaining nutriment from the plants and the latter only from the epiphytes. *Rhabdosargus holubi* consume fewer of the bivalves that form the main invertebrate food of *R. auriventris*. Bivalves form an important part of the diets of adults of all three species in the sea. It is possible that in the divergence of these three closely related species there has been evolution in the juveniles, either towards herbivory or towards bivalve feeding. Juveniles of *R. sarba* are the most omnivorous, and therefore perhaps ancestral to the more localized *R. holubi* and *R. auriventris*, which have become more specialized in their feeding ecology. Juveniles of all three species have nevertheless retained the opportunistic and flexible diet necessary for living in estuaries, as evidenced by their consumption of a wide range of invertebrate benthic fauna.

## Plankton feeders

The numbers and species diversity of planktivorous fishes in subtropical and tropical estuaries varies according to the biomass and composition of the plankton as well as the physical characteristics and location of the estuary. There are more species of estuarine clupeoids in the Indo–West Pacific and tropical West Atlantic than in the tropical East Atlantic or tropical East Pacific. Planktivorous fishes can be divided into those that filter feed – straining the plankton out of the water through gill rakers – and those that select individual zooplankters. Most of the Clupeidae and Engraulididae fall into the former group, while those that prey on individual plankters are usually smaller species or juveniles from a wide range of families. Although the filter feeders are the dominant planktivores in the sea and some larger estuaries, those that select individual prey may be equally as abundant in smaller estuaries. The general biological features of clupeoids of tropical seas are comprehensively reviewed by Longhurst and Pauly (1987), who divide them into predators, zooplankton feeders, omnivores that can take both zooplankton and phytoplankton, and phytoplankton feeders.

The size of plankton retained by filter feeders is mainly controlled by the width of the gap between the gill rakers, and the length of, and gap between, the gill raker denticles – the whole forming a filtering basket. These not only vary according to species but also with the size of the fish. The strong relationships between these parameters and the prey retained have been described for six clupeoids in the St Lucia and Kosi coastal lake systems of south-east Africa (Table 5.1) (Blaber, 1979; Blaber and Cyrus, 1981). Generally the narrower the gill raker and denticle gaps, the smaller the sizes of prey taken. The only species in St Lucia that feeds mainly on phytoplankton is *Hilsa kelee*, which has relatively narrow gill raker denticle gaps. Most planktivores in St Lucia feed on the abundant estuarine calanoid *Pseudodiaptomus*, and at small sizes various veliger and zoae larvae. However, *Thryssa vitrirostris* begins life as a particulate feeder on individual *Pseudodiaptomus*, then switches to filter feeding on the same calanoid, before reverting at large sizes to individual selection of small fish prey.

Some of the most economically important fishes of tropical estuaries are filter-feeding planktivores: for example, *Ethmalosa fimbriata* (bonga) of West Africa, *Brevoortia patronus* (menhaden) of the Gulf of Mexico, and *Tenualosa ilisha* (hilsa) of the Bay of Bengal and Ganges delta. The fisheries for these species are described in Chapter 9 so only features of their feeding ecology are given here.

*Ethmalosa fimbriata* ( = *dorsalis*) occurs in estuaries and estuarine coastal waters along most of the West African coast, with concentrations in and

**Table 5.1** Measurements of mean gill raker gap, mean gill raker denticle gap and mean gill raker denticle length in six species of filter-feeding teleosts from the Lake St Lucia and Kosi systems, South Africa (–, no denticles). After Blaber (1979), Blaber and Cyrus (1981)

| Species | Standard length (cm) | Gillraker $\bar{X}$ gap (μm) | Denticle $\bar{X}$ gap (μm) | Denticle $\bar{X}$ length (μm) | Main prey |
|---|---|---|---|---|---|
| Gilchristella aestuaria | 4 | 103 | 68 | 26 | Pseudodiaptomus, mysids, postveligers, zoae and copepodids |
| | 2 | 51 | – | – | |
| Hilsa kelee | 23 | 238 | 44 | 66 | Phytoplankton, Pseudodiaptomus, Acartia, Amphipoda |
| | 10 | 168 | 87 | 40 | Phytoplankton – centric diatoms |
| | 7 | 185 | 50 | 26 | Phytoplankton – centric diatoms |
| | 3 | 91 | 60 | 7 | Postveligers, bivalve spat, copepodids |
| Thryssa vitrirostris | 15 | 285 | 104 | 263 | Small fish |
| | 5 | 220 | 63 | 38 | Pseudodiaptomus, mysids |
| | 2 | 210 | 70 | 20 | Pseudodiaptomus and copepodids |
| Stolephorus commersonii | 5 | 76 | 46 | 27 | Pseudodiaptomus, bivalve spat, zoae, cyclopoids, postveligerse |
| Sardinella melanura | 10 | 320 | 109 | 23 | Ostracods, juvenile penaeids |
| | 5 | 115 | 112 | 22 | Calanoids, ostracods, zoae |
| | 4 | 105 | 90 | 15 | Calanoids, ostracods, zoae |
| | 3 | 105 | 53 | 12 | Calanoids, ostracods, zoae |
| Spratelloides delicatulus | 4 | 111 | 77 | 21 | Calanoids, ostracods, veligers, bivalve spat |

around most estuaries. It is primarily a non-selective phytoplankton filter feeder with an array of very fine gill rakers (denticle gaps as little as 40 μm) that enable it to retain large centric diatoms that form the bulk of its food. Changes in the composition of its food closely follow the succession of species in the phytoplankton (Bainbridge, 1963). It is rarely found in clear water and has onshore–offshore movements related to the tidal cycle and its food. During neap tides, when diatom blooms occur in estuaries, *Ethmalosa* move into the estuaries to feed. On spring tides, when the turbidity and amount of suspended silt increase in the estuaries, they move to the nearshore zone to feed (Longhurst and Pauly, 1987). The position of *Ethmalosa* in the planktonic food web is discussed in section 5.6.

Longhurst and Pauly (1987) describe *Brevoortia patronus* of the Gulf of Mexico as having a very similar feeding ecology to that of the more tropical *Ethmalosa*. Like *Ethmalosa* it has a fine gill raker apparatus allowing it to feed on larger phytoplankton, particularly diatoms. There is also a seasonal movement of *Brevoortia* into estuaries of the Gulf of Mexico in relation to phytoplankton blooms.

The hilsa, *Tenualosa ilisha*, of the Bay of Bengal is an anadromous species that ascends most of the larger rivers of the region. It is a non-selective filter feeder with closely set fine gill rakers (Hora, 1938; Hora and Nair, 1940; Jones and Sujansingani, 1951) and takes a very wide variety of plankton, both plant and animal. It occurs in the sea, in estuaries – including coastal lakes such as Chilka – and far into fresh waters. Given its extensive range, both geographical and in terms of habitats, it is not surprising that almost all planktonic organisms have been recorded at one time or another in its diet (Pillay and Rosa, 1963; Raja, 1985). This flexibility is probably the key to its widespread distribution (Fig. 8.1). Pillay (1958) states that stomachs of spawning fishes in the rivers often contain mud mixed with food organisms, indicating that it can also feed near the bottom and may be relatively tolerant of gill raker clogging. A closely related species, *Tenualosa toli*, which has a much more restricted distribution in the estuaries and coastal waters of Borneo (Fig. 8.1), feeds only on calanoid copepods with small quantities of other zooplankton (Blaber *et al.*, 1996).

The planktivorous group defined by Longhurst and Pauly (1987) as predators (non-filter-feeding) is well represented in tropical estuaries and includes a number of clupeoid genera. *Ilisha africana* is an economically important fish of the estuaries and coastal waters of West Africa that feeds mainly on larger zooplankton, penaeid prawns and small fish in the water column (Fagade and Olaniyan, 1973; Whitehead, 1985; Baran, 1995). Other species of *Ilisha* are important in the estuarine artisanal fisheries of Asia but their diets have not been studied. The engraulid genera *Coilia* and *Setipinna* are numerous in tropical Asian estuaries and important in artisanal fisheries, but likewise little information is available on their feeding ecology. All are surface to midwater predators. According to Jones and Menon (1952), the diet of *Coilia dussumieri* in India consists of copepods, prawn larvae, stomatopod larvae, mysids, isopods and *Sagitta* – in other words the larger elements of the zooplankton. *Setipinna tenuifilis* is widespread in South and South East Asia and its diet in the Hooghly estuary in India also comprises larger zooplankton such as small prawns, copepods and fish larvae (Babu Rao, 1962).

Apart from clupeoids, the smaller juveniles of species of many families are selective zooplankton feeders for relatively short periods. Small estuarine species that are planktivorous include the ambassids (perchlets or glassies) of the Indo–West Pacific. Most species of this family feed on animals taken in the water column or from the surface such as cladocerans, ostracods, copepods, insects and small fish (Allen and Burgess, 1990). The diet varies according to habitat: *Ambassis interruptus* of the Sepik estuary in Papua New Guinea feed mainly on insect larvae and insects from the surface of the water as well as ostracods and cladocerans (Coates, 1990); *Ambassis productus*, *A. natalensis* and *A. gymnocephalus* of East African estuaries are all primarily zooplanktivores taking mainly Crustacea, but whereas *A. gymnocephalus*, which occurs only in higher salinities near the mouth (> 30‰), has a narrow food spectrum, the others have a broader diet, particularly *A. productus*, which lives mainly in the upper reaches and has a high proportion of insects in its diet. All species feed to some extent on larval and postlarval fishes (Martin and Blaber, 1983). It is probable that *Ambassis* species are the dominant zooplanktivores in the smaller, less turbid estuaries of the Indo–West Pacific where filter feeders are scarce, and where, because of their abundance and their predation on fish larvae and fry recruiting into such estuaries, they may have a significant impact on estuarine fish communities.

### Piscivorous species

Piscivorous fishes are the top predators apart from other vertebrates (Chapter 7) in subtropical and tropical estuaries. Although most larger fishes are piscivores, many smaller species, as we have seen earlier, incorporate fish in their diets. For many there is an ontogenetic progression towards piscivory (pp.150–156). The main piscivorous families in estuaries are Belonidae, Carangidae, Carcharhinidae, Centropomidae, Elopidae, Megalopidae, Sciaenidae and Sphyraenidae (Fig. 5.6).

The Belonidae (garfishes, longtoms) are surface predators and species

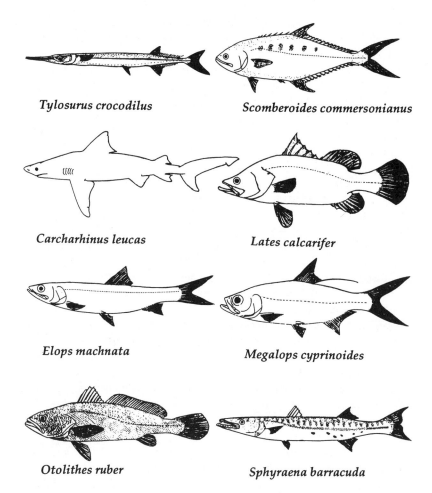

*Tylosurus crocodilus*

*Scomberoides commersonianus*

*Carcharhinus leucas*

*Lates calcarifer*

*Elops machnata*

*Megalops cyprinoides*

*Otolithes ruber*

*Sphyraena barracuda*

**Fig. 5.6** Representative examples of important piscivorous species in Indo-West Pacific tropical and subtropical estuaries. Not to scale.

such as *Strongylura strongylura, S. leiura* and *Tylosurus crocodilus* – abundant throughout the estuaries and coastal waters of the Indo–West Pacific – feed on pelagic species such as *Thryssa, Stolephorus* and hemiramphids (Whitfield and Blaber, 1978d; Salini *et al.*, 1990).

Many species of carangids occur in tropical estuaries, particularly as juveniles and subadults. *Caranx hippos* is common in West African estuaries where it feeds mainly on juvenile clupeids (particularly *Ethmalosa*), gerreids and cichlids (Fagade and Olaniyan, 1973). In tropical Australian estuaries the queenfish, *Scomberoides commersonianus*, is the most numerous carangid and feeds on a very wide range of species. In the Embley estuary 33 species of teleosts have been recorded in its diet (Salini *et al.*, 1990), with Mugilidae, Clupeidae (mainly *Sardinella* and *Anodontostoma*), Engraulididae and Hemiramphidae the most numerous. In a study of 17 species of carangids in KwaZulu-Natal estuaries, Blaber and Cyrus (1983) found that only juveniles and subadults were present. Among the juveniles, *Caranx ignobilis* and *C. sexfasciatus* are mainly piscivorous while *C. melampygus*, *C. papuensis* and *Scomberoides lysan* filter feed on mysids and juvenile penaeids (pp.48–49). However, subadults of all species are primarily piscivorous. Most Carangidae are clear-water species and hunt visually. In clearer-water estuaries they may form the dominant piscivorous group, together perhaps with Sphyraenidae, but in more turbid systems, particularly in South and South East Asia, are not among the major piscivores.

Carcharhinid sharks occur in large numbers in the lower reaches of many estuaries and in coastal waters of the Indo–West Pacific, particularly in South East Asia and Australia. They are apparently less numerous in other regions although the reasons for this are obscure. Most are juveniles or subadults. One species, *Carcharhinus leucas* (bull or Zambezi shark) penetrates to fresh water and occurs commonly in most tropical estuaries around the world. The diets of this species and 10 others from the estuaries and coastal waters of the Gulf of Carpentaria in northern Australia have been described by Salini *et al.* (1990, 1992). A very wide range of prey are eaten in an opportunistic and density-dependent manner, and there is little evidence of any ecological partitioning of the food resource between most species.

Several large centropomids are important piscivores in the estuaries of the tropical West Atlantic in South and Central America. *Centropomus constantinus*, *C. undecimalis* and *C. parallelus* occur from the Gulf of Mexico to the estuaries of Brazil and Venezuela, and to the coastal lakes of Colombia where there is an additional three species (Cervigón, 1985; León and Racedo, 1985; Yáñez-Arancibia *et al.*, 1985). All are primarily piscivorous on a wide range of small and mainly benthic fishes. The only estuarine centropomid of the Indo–West Pacific (excluding the ambassids which are sometimes included in the family) is the commercially important *Lates calcarifer*. At larger sizes, *Lates* is almost entirely piscivorous and a significant predator of many smaller species in South and South East Asian and Australian estuaries (pp.151–153).

A single species of *Elops* occurs in estuaries in each of the tropical regions: *Elops machnata* in the Indo–West Pacific, *E. lacerta* in the tropical East Atlantic, *E. saurus* in the tropical West Atlantic, and *E. affinis* in the tropical East Pacific. All have a similar feeding ecology and as adults are pelagic piscivores, preying particularly on clupeids such as *Ethmalosa* in

West Africa (Fagade and Olaniyan, 1973) and *Thryssa* in East Africa (Whitfield and Blaber, 1978d). Similarly, *Megalops cyprinoides* of the Indo–West Pacific and *M. atlanticus* of the tropical East and West Atlantic feed on a variety of small pelagic species. In a detailed study of *Megalops cyprinoides* in the Sepik estuary in Papua New Guinea, Coates (1987) found that small fishes from beneath macrophyte mats, such as eleotrids, cichlids and ambassids, are favourite prey.

Among the Sciaenidae it is the larger species and individuals that are piscivorous. Genera from Indo–West Pacific estuaries include *Argyrosomus*, *Otolithes*, *Otolithoides* and *Protonibea*. They are among the most important piscivores in turbid estuaries, in contrast to the Carangidae, and are particularly common in larger South East Asian systems. They feed mainly on benthic fishes. In the tropical West Atlantic the very diverse sciaenid fauna of the estuaries and coastal waters of northern South America includes *Macrodon ancylodon*, *Cynoscion virescens* and *C. jamaicensis*, the adults of which feed mainly on fish (Lowe-McConnell, 1962). Curiously, although some of the abundant estuarine sciaenids of the West African coast, such as species of *Pseudotolithus*, include some fish in their diets, they are not primarily piscivorous (Fagade and Olaniyan, 1973; Baran, 1995).

The Sphyraenidae (barracudas) are almost entirely piscivorous from a very small size (pp.48–49). Like the carangids they are not common in more turbid estuaries, but are abundant predators of a great variety of species in clearer estuaries and coastal waters (Fagade and Olaniyan, 1973; Blaber, 1982; Salini *et al.*, 1990; Brewer *et al.*, 1995).

Many marine piscivorous species visit the lower reaches of estuaries, probably to prey on the often abundant small fishes. Good examples include the scombrid mackerels, *Scomberomorus maculatus* of West Africa and *S. semifasciatus* of northern Australia, both of which are frequent estuarine visitors.

## 5.6 FOOD WEBS

One of the distinguishing features of tropical communities, and estuarine fishes are no exception, is their high species diversity and the complexity of their trophic structure and interrelationships. Despite the amount of information available on the diets of many tropical estuarine fishes, there are relatively few well-worked examples of food webs as they relate to fishes. As discussed in Chapter 3, the number of fish species in an estuary is related to a suite of factors, biotic and abiotic. The trophic structure of tropical estuarine fish communities is a result of the influences of these factors, the type of estuary, as well as the state or condition of the estuary.

Here we examine how the trophic structure of the fish community is influenced by the type of estuary, its state and its perturbations. Most generalized estuarine food webs are driven by phytoplankton or detrital production, or a combination of the two (De Sylva, 1985).

In terms of numbers of species, many estuarine systems in the Indo–West Pacific appear to have approximately similar proportions of species at each broad trophic level (Table 5.2). Most of the variation is in the planktivorous and iliophagous species, probably reflecting whether or not a particular system is phytoplankton or detritus driven. Numbers of piscivorous species are higher in estuarine coastal waters or large open estuaries such as the Embley where sharks and other larger predators can occur.

Two very different coastal lakes in south-east Africa provide an interesting comparison. St Lucia is shallow and turbid with a high species diversity ($\sim 110$ species – pp.57–58), whereas Lake Nhlange is oligotrophic with clear water and a relatively low species diversity ($\sim 30$ species). Outlines of their food webs are shown in Figs 5.7 and 5.8. In terms of biomass their trophic structures are different; St Lucia has more piscivores and planktivores while Nhlange has a higher proportion of iliophagous species and macrobenthic carnivores. The differences can partly, at least, be attributed to the high plankton productivity in St Lucia (Blaber, 1979) compared with Nhlange (Allanson and van Wyk, 1969). This results in high numbers of planktivorous fishes in St Lucia and the piscivores that feed upon them. Iliophagous and benthic invertebrate feeders have the same relative proportions of biomass in each of these systems although the number of species in each of the groups is higher in St Lucia. From these data it is apparent that the St Lucia fish food web is both phytoplankton and detritus driven, while that of Nhlange is detritus driven with very little phytoplankton production.

Where large mangrove forests line tropical estuaries, as in much of Asia and Australia, the detritus chain is probably the most important. The Sundarbans area in India and Bangladesh is one of the largest mangrove ecosystems in the world and Sarkar (1993) has shown that this ecosystem consists of several basic subsystems linked together and is a good example of a coupled system that achieves a balance between physical and biotic components and a high rate of biological productivity. The mangroves provide an important nutrient input and primary energy source for the tropical estuaries of the Sundarbans area through the export of decomposable organic matter (mostly in the form of leaf detritus) into adjacent coastal water. In a deterministic model of a food chain in this area, Sarkar (1993) postulated that a micro-organism pool lives on the detritus of mangrove litter and has its invertebrate predators. The growth rates of the microbes are assumed to be controlled by the input of detritus and hence also indirectly that of the predators. The low

**Table 5.2** Percentage contribution of each trophic level to the total number of species ($n$) in 11 Indo–West Pacific estuarine systems

| Estuary | Type* | Country | Trophic level | | | | | $n$ | Reference |
|---|---|---|---|---|---|---|---|---|---|
| | | | Herbi-vorous | Ilio-phagous | Plankti-vorous | Benth.† | Pisci-vorous | | |
| Trinity | O | Australia | 5 | 9 | 27 | 36 | 15 | 55 | Blaber (1980) |
| Ponggol | O | Singapore | 5 | 3 | 33 | 40 | 19 | 78 | Chua, Thia-Eng (1973) |
| Morrumbene | O | Moçambique | 5 | 4 | 22 | 50 | 19 | 113 | Day (1974) |
| Embley | O | Australia | 5 | 4 | 15 | 50 | 26 | 197 | Blaber et al. (1989) |
| Nhlange | CL | South Africa | 3 | 20 | 24 | 33 | 20 | 30 | Blaber (1978) |
| St Lucia | CL | South Africa | 2 | 11 | 16 | 54 | 17 | 110 | Whitfield (1980a) |
| Chilka | CL | India | 1 | 11 | 18 | 53 | 19 | 152 | Pillay (1967a), Jhingran (1991) |
| Tongati | B | South Africa | 6 | 25 | 16 | 34 | 19 | 32 | Blaber et al. (1984) |
| Mhlanga | B | South Africa | 2 | 28 | 11 | 48 | 11 | 46 | Harrison and Whitfield (1995) |
| Trinity Bay | EC | Australia | 5 | 12 | 13 | 38 | 28 | 60 | Blaber (1980) |
| Albatross Bay | EC | Australia | 3 | 4 | 17 | 48 | 28 | 115 | Blaber et al. (1995a) |

*O, open estuary; CL, coastal lake; B, blind estuary; EC, estuarine coastal waters.
† Macrobenthic invertebrate feeders.

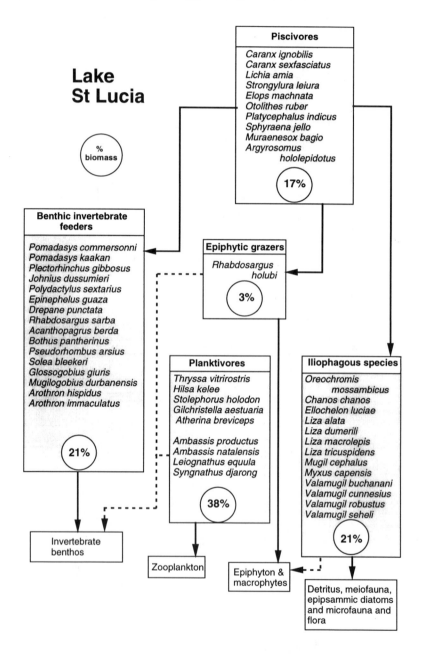

**Fig. 5.7** Outline of the fish community food web of Lake St Lucia, KwaZulu-Natal, South Africa. Solid arrows denote major pathways; broken arrows signify minor pathways.

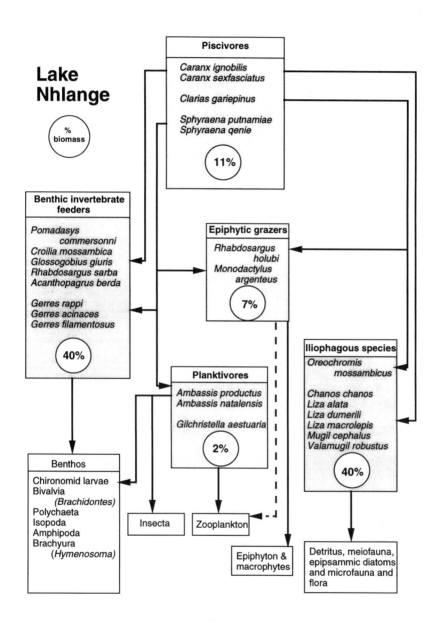

**Fig. 5.8** Outline of the fish community food web of Lake Nhlange, KwaZulu-Natal, South Africa.

production of phytoplankton is assumed due to high turbidities although production of benthic diatoms may be high. Such diatoms, however, usually enter the food web through the iliophagous chain.

Based on work in Cochin Backwater, an open estuary on the west coast of India, Qasim (1970) postulated that in shallow tropical estuaries an increase in such benthic production available to predators compensates for the lack of zooplankton. In shallow coastal waters adjacent to mangroves in the Straits of Malacca, Ong and Sasekumar (1984) found that detritus in terms of broken-down vegetation is a component of the diet of at least a third of the fish species. In marked contrast to this, the large, 'physically driven', deep, open estuaries of Borneo, such as the Lupar (p.16), have few iliophagous species, but a high diversity of plankti-vores and benthic invertebrate feeders (Fig. 5.9). The high flow rates prevent much accumulation of detritus – much of which remains suspended in the water column and may enter the food web via the very diverse and abundant plankton.

In other coastal waters, complex food webs mainly involving piscivory and predation on benthic invertebrates have been described for the Guyana shelf (Lowe-McConnell, 1962), the West African shelf (Longhurst and Pauly, 1987) and the Gulf of Mexico (Yáñez-Arancibia *et al.*, 1985).

Such trophic outlines as are described in this chapter can only give a general indication of the structure of a food web. Detailed studies reveal an astonishing complexity that varies from estuary to estuary. The pelagic food web at St Lucia involves relatively few species, but when examined in detail, an elaborate network of relationships is evident with different size classes of the anchovy *Thryssa* having different feeding strategies (Fig. 5.10). The planktivores of Lagos Lagoon in West Africa are different from those in St Lucia but form part of a similarly complex food web (Fig. 5.11).

Comparisons between the trophic structures of fish communities of a degraded blind estuary and one in a natural state in KwaZulu-Natal, south-east Africa, show the sorts of changes induced by anthropogenic perturbations (Fig. 5.12). The Tongati estuary is affected by sewage effluent, industrial waste and agricultural run-off. There is no epifauna or flora on its *Phragmites* reed beds and almost no zooplankton. Much of the surface is covered with the alien water hyacinth *Eichornia* (p.121). In contrast, the Mhlanga estuary forms part of a nature reserve, has an abundant *Phragmites* epifauna and flora and ten times the biomass of zooplankton. Both systems are dominated by iliophagous fishes that make up about 90% of the biomass. The benthic invertebrates in the polluted estuary are less diverse but polychaetes and chironomid larvae are still present and support a low biomass of benthic feeders. Surprisingly, the planktivores in Tongati constitute a greater proportion of the biomass

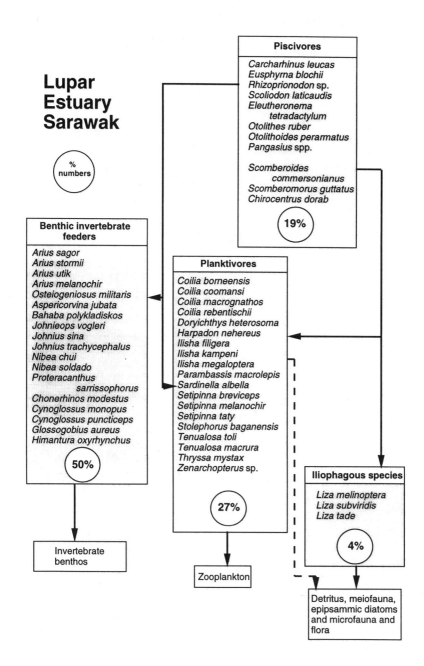

**Fig. 5.9** Outline of the fish community food web of the Lupar, a large open estuary in Sarawak, Borneo.

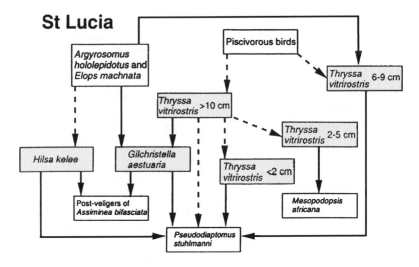

**Fig. 5.10** The pelagic food web in Lake St Lucia, KwaZulu-Natal, South Africa. Planktivorous species are shown in grey boxes, invertebrate prey in small print (adapted from Blaber, 1979).

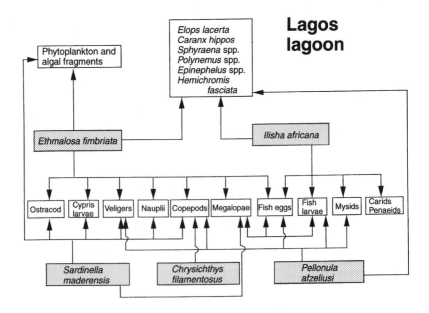

**Fig. 5.11** The pelagic food web in Lagos Lagoon, Nigeria. Planktivorous species are shown in grey boxes (adapted from Fagade and Olaniyan, 1973).

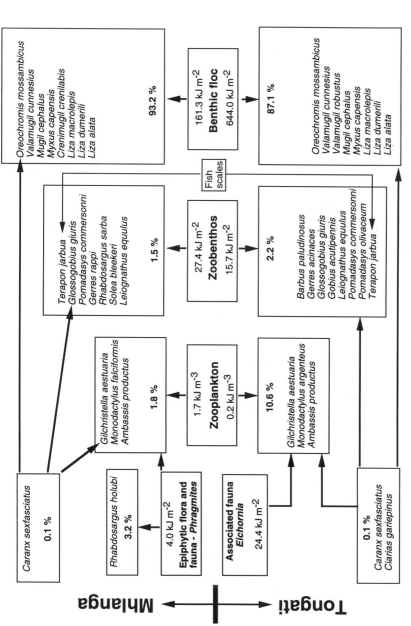

**Fig. 5.12** Contrasting fish community food web of degraded (Tongati) and 'natural' (Mhlanga) blind esturies in KwaZulu-Natal, South Africa, in terms of per cent biomass of fishes and standing stock (energy) of foods (data from Blaber *et al*, 1984, and Whitfield, 1980a).

than they do in Mhlanga despite the paucity of zooplankton. In Tongati they have switched to feeding on the abundant epifauna associated with *Eichornia* roots, emphasizing once again the value of dietary flexibility in tropical estuaries. The proportion of piscivores in these blind estuaries is low (0.1%) and may relate more to the size of the systems and lack of access to the sea, than to their degree of degradation. In the degraded Tongati estuary, the food chains involving zooplankton and zoobenthos are impoverished, probably owing to the greater sensitivity of these prey organisms to environmental perturbations compared with benthic floc and microorganisms. The data suggest that such polluted systems may, however, remain viable through the iliophagous and detritus chains under a variety of stresses.

The effects of defoliation of mangrove estuaries carried out in Vietnam during the war (1965–1975) was the subject of an intensive study (De Sylva, 1985). Unfortunately the results were inconclusive because of the lack of comparative pre-war data on the estuaries. However, the studies did show that in a non-defoliated estuary, the food web does not vary much between wet and dry seasons among the benthic carnivores and piscivores, but that during the wet season the primary source of energy is detritus, and in the dry season it is phytoplankton.

# Chapter six

# *Reproduction and spawning*

## 6.1 INTRODUCTION

The fish communities of tropical and subtropical estuaries consist of species that spend parts of their lives in estuaries, either as juveniles or adults, or both; species that are entirely confined to estuaries; species that are in transit to or from sea water or fresh water; and species that are merely visiting. Not surprisingly, their reproductive cycles vary greatly, and have evolved as part of their diverse life history strategies. In this chapter we are concerned with the ways in which estuarine fishes reproduce, where they spawn, spawning stimuli, over what periods of time they spawn, and under what conditions. These questions are relevant to recruitment into estuaries, to estuarine dependence, and to the abiotic and biotic characteristics of particular estuaries. For general information on fish reproduction the reader is referred to the many detailed texts on reproduction in fishes, such as Hoar and Randall (1969), Hoar *et al.* (1983a,b), Potts and Wootton (1984), Wootton (1990) and Jobling (1995). As with most aspects of fish ecology, most information about reproduction of fishes comes from studies of temperate species. The reproductive biology of the vast majority of tropical species has never been studied, and inferences drawn from temperate species may, at best, be misleading. Most of the data collected on tropical estuarine species consist of basic biological parameters such as size at maturity and fecundity, together with information on spawning seasons and areas. Relatively little is known about reproductive behaviour, spawning stimuli or the relationships between reproductive strategy and environment. The absence of seasonal changes in temperature or daylength (photoperiod) – stimuli that may be very important in controlling the reproductive cycles of temperate species – the lack of predictability in such factors as river flow, salinity and isolation from the sea, all often combine to necessitate unique and sometimes complex reproductive strategies.

## 6.2  MODES OF REPRODUCTION

The variety of life history traits in tropical estuarine fish communities involves a wide array of reproductive strategies. For instance: most species have separate sexes but some are hermaphrodite; some lay eggs while others are livebearers; and most release their eggs into the water, but some exhibit parental care. The mode of reproduction shown by each species is an integral part of its lifestyle, and cannot be considered in isolation from its biological needs, such as food, or its ecology in relation to, for example, habitat type or predator avoidance. A summary of the reproductive characteristics of a selection of the fishes discussed in this chapter is shown in Table 6.1.

### Single sex or hermaphrodite

Most tropical estuarine fish are gonochoristic, that is, they have separate sexes and each individual remains the same sex throughout its life. However, hermaphroditism is widespread among teleosts (Yamamoto, 1969; Chan and Yeung, 1983) and exists in a variety of forms controlled by both extrinsic and intrinsic factors (Reinboth, 1980; Ross, 1990). As Jobling (1995) points out, the fact that sex change occurs naturally in a number of species points to a sexual lability in fishes that is not seen in other vertebrate groups. Hermaphroditism in fishes is of two basic types: either synchronous, where functional ovaries and testes are present in the same individual, or successive, when sex change takes place. The latter is the most frequent type and may be protandrous, when males change to females and hence the older and larger fish are female, or protogynous, where the larger, older individuals are males and the fish start life as females. Protandry has been found in species of at least 8 families and protogyny in 14 families. Hermaphroditism in fishes has interested many researchers and the known number of hermaphroditic species is increasing all the time (Sadovy and Shapiro, 1987). The evolution of hermaphroditism has been reviewed by Ghiselin (1969), Warner (1978) and Ross (1990), and the hypotheses concerning its adaptive significance discussed by Warner (1988). It is most frequent in coral reef fish and deep sea fish. In the former, a clear-water environment, the primary determinants of sex-change mechanisms are social organization and mating systems that depend on the distribution of limited habitat and food resources for mainly sedentary species (Ross, 1990). In the darkness of the deep sea, where fish densities are low and intraspecific contacts few, most hermaphroditic fishes are synchronous hermaphrodites. This has adaptive significance because self-fertilization is possible or the meeting of two individuals can lead to the fertilization of two batches of eggs (Jobling, 1995).

In subtropical and tropical estuaries, fish densities may be high but the water is seldom clear and many species are migratory or very mobile. The complex behavioural and social systems that operate among sedentary coral reef species are largely absent and hermaphroditism is uncommon. Synchronous hermaphroditism is unknown in tropical estuarine species, but successive hermaphroditism does occur in some common and economically important species of the families Centropomidae, Clupeidae, Platycephalidae, Serranidae (most *Epinephelus* species are protogynous hermaphrodites) and Sparidae (mainly protandrous). In the two last-named families, sex inversion is the rule for most species (Yamamoto, 1969; Chan and Yeung, 1983), but in the others it is less predictable and there is little consistency, even within genera. To illustrate the possible adaptive significance of successive hermaphroditism, and to show that where it occurs, it may be a key part of the life history, details of three estuarine hermaphroditic species are given below.

## *Lates calcarifer* (barramundi or sea bass)

This large species of centropomid supports extensive commercial, aquacultural and recreational fisheries in the Indo–West Pacific region. It occurs in coastal waters, estuaries and fresh waters from western India, around Sri Lanka to the Bay of Bengal, and through the whole of South East Asia to eastern Papua New Guinea and northern Australia (Grey, 1987). It has a complex life history (Fig. 6.1) and is a protandrous hermaphrodite (Moore, 1979). Garrett (1987) reviewed the reproductive biology, from which the following account is largely taken. The gonads are dimorphic and a complete reorganization of gonad structure and function takes place after sex inversion, probably under the influence of hormones. *L. calcarifer* spawn as males for several years before sex inversion. The sex reversal is initiated as the testes ripen for the last time, and the change to ovary takes place rapidly within a month of spawning. The change to female usually takes place at about 7 years of age and a length of about 800 mm, but is apparently more related to age than to length. The length at which sex change occurs varies somewhat across its extensive geographic range, due probably to habitat, food and genetic differences. Moore (1980) postulated that protandrous sex reversal in *L. calcarifer* allows the larger and more successful fish (the females) to make the greatest contribution to the gene pool of a particular population. Movement to spawning sites and gonad maturation takes place at the end of the dry season in estuarine waters (28–36‰), usually near the mouth. The eggs and larval stages have a narrower range of salinity and temperature tolerance than the adults (Russell and Garrett, 1983). There is a tidal-based monthly cycle of spawning, with postlarvae entering coastal swamps on spring tides as the wet season progresses. Spawning at this time ensures that the juveniles can enter and remain in the shelter of flooded backswamps and floodplains until they dry up during the early part of the dry season (Fig. 6.1). There is thus a strong link between the reproductive cycle of *L. calcarifer* and the monsoon pattern in Australia and Asia. The fecundity of *L. calcarifer* is among the highest of any teleost fish, with estimates of from 0.6 to $2.3 \times 10^6$ eggs kg$^{-1}$ of body weight (Garrett, 1987). It is reported that some individuals spawn all at once whereas others may be multiple spawners. The high fecundity may be a response to the late onset of female reproductive function in the population and the relatively low numbers of females in the overall population (Moore, 1980; Davis, 1984). This implies that most recruitment is derived from small numbers of large female fish (Garrett, 1987) and makes the species very vulnerable to overfishing (Chapter 8). Sex reversal in *L. calcarifer* may on the one hand allow the most successful individuals (those that have survived to large size) to contribute most to the gene pool, but it results in a need for very high fecundities in the low number of large surviving fish.

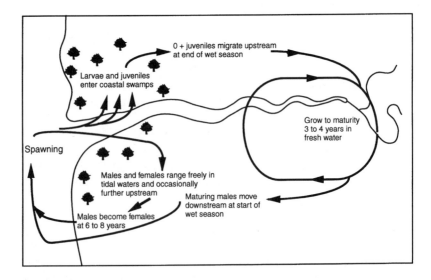

**Fig. 6.1**  The life cycle of *Lates calcarifer* (adapted from Grey, 1987).

### *Acanthopagrus australis* (yellowfin bream)

The yellowfin bream is a common sparid of the estuaries of subtropical and tropical eastern Australia. It is of importance in both commercial and recreational fisheries and its biology has been intensively studied in Moreton Bay (Pollock, 1982, 1984, 1985; Pollock *et al.*, 1983), a large estuarine embayment in southern Queensland. It spawns in winter over 2 to 3 months (Fig. 6.2). It is a protandrous hermaphrodite and three types of gonad have been recognized: ovotestis, ovary and testis. The ovotestis is invariably immature, and its presence in *A. australis* indicates that this fish undergoes sex reversal as a normal part of its life cycle. This ovotestis is typical of sparids in which the testis and ovary occur in different zones. Histological and structural study of the ovotestis shows that all fish have previtellogenic cells in the ovarian zone but only juvenile and male fish have developing spermatogenic cells in the testis. Testicular structure is of the unrestricted spermatogonial type (lobular) with germinal cysts containing synchronously developing germ cells along the tubules. As

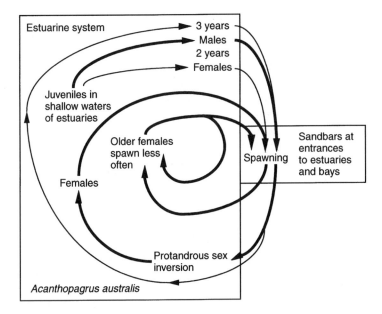

**Fig. 6.2** The life cycle of *Acanthopagrus australis* in Moreton Bay, Queensland, Australia. The majority of juveniles begin life as males and most change to females after the first year. Thin lines refer to a minority of fish (males or females) that do not change sex (data from Pollock, 1984).

spermiation approaches, the testis consists mainly of ducts containing sperm, with more prominent interstitial tissue. Clumping of spermatozoa and proliferation of fibrous tissue is seen in some testes after the end of the breeding season (Suparta *et al.*, 1984). The sex ratio of males to females decreases with age, indicating protandrous sex inversion. Most juveniles become functional males by the age of 2 years and protandrous sex inversion commences after the spawning period. Almost all the life cycle of A. *australis* takes place in estuaries and they remain within the estuarine system for most of the year. During the 2 to 3 month spawning season (July–September), adults undertake large-scale migrations and concentrate at the sandbars at the entrances to Moreton Bay, where they spawn. However, a large proportion of the adult population does not participate in the annual seaward spawning migration. Gonads develop to a similar size in migratory and non-migratory adults. Successful reproduction, as indicated by the presence of hyaline-stage oocytes, only occurs in migratory fish. In non-migratory females, oocytes develop to the yolk-stage, but then become atretic and are resorbed. In both groups, stored energy reserves decrease as the gonads increase in size; however, the rate of decrease is greater in migratory fish. Feeding almost ceases in migratory fish during the spawning period. In non-migratory fish, feeding is reduced during the late spawning period, possibly because abundant energy reserves are available due to oocyte resorption. An examination of sex ratios within age classes indicates a decreasing tendency of females to participate in the spawning migration with age. All male fish older than 2 years participate in the spawning migration (Pollock, 1984).

The adaptive significance of the hermaphroditism and spawning strategy of A. *australis* are somewhat obscure, but are possibly related to getting the eggs into the marine environment for dispersal during peak ebb tides, and to responses to predation on both the adults and the eggs. Such factors have been suggested as important in the behaviour of the closely related but more widespread riverbream A. *berda* which is found in estuaries throughout the Indo–West Pacific. The spawning of A. *berda* has been studied in the Kosi system of KwaZulu-Natal, South Africa (Garratt, 1993). Spawning takes place in the Kosi estuary mouth at night and eggs are transported out to sea during peak ebb tides. It is postulated that night spawning may reduce egg predation by planktivores. There is a preponderance of males in the spawning aggregation (sex ratio 8.8:1, m/f) and indirect evidence suggests that males and females are continually recruiting to the aggregation, spawning, and moving back up into the estuary, so that there are no more than 2000 individuals at the mouth at any one time. It appears that this species has developed a spawning strategy that overcomes the necessity to move out into the marine environment, thereby avoiding some of the larger predators, while at the same time ensuring that the eggs are well dispersed. As with *Lates calcar-ifer*, protandry ensures that the larger, fitter fish contribute more to the gene pool, and that the number of eggs spawned is higher, because fecundity is related to fish size.

### *Tenualosa toli* (terubok)

*Tenualosa toli* is a commercially and culturally important tropical shad (Clupeidae) which lives in the fast-flowing, turbid estuaries and adjacent shallow coastal waters of Sarawak. It is protandrous hermaphrodite (Blaber *et al.*, 1996) and its life cycle is discussed in Chapter 3 (Fig. 3.2). At present it is the only known hermaphrodite among the Clupeiformes and is the only member of its genus that is a hermaphrodite: the well-studied *T. ilisha* (hilsa) is gonochoristic and so is *T. macrura*. Ageing based on otoliths indicates that individuals of *T. toli* may not live more than about 2 years, in contrast to the 5–6 years for *T. ilisha* (De and Datta, 1990). Male fish spawn towards the end of their first year, change sex (transitional gonads have been recorded in fish from 14 to 31 cm SL) and spawn as females in their second year. Spawning takes place in the middle reaches of estuaries and females spawn all their eggs at once (semelparity). Fecundity is linearly related to fish length but shows signifi-cant seasonal and site variations.

For *T. toli*, which inhabits a highly 'physically controlled' environment (Dutrieux, 1991), with extremes of salinity, turbidity and current speed (Chapter 2), the size advantage and gene dispersal models of Ghiselin (1969) may be relevant to the evolution of a hermaphroditic reproductive strategy. The size advantage model predicts protandry where the younger stages must hunt for a suitable environment. In addition, where these first-year (smaller) fish must reproduce, as in *T. toli*, which has only a 2 year life span, it is better that they are male rather than female. The gene dispersal model predicts sequential hermaphroditism when a change in ecology reduces gene flow between populations of a species, and the isolated populations could suffer the deleterious effects of genetic drift. This is particularly relevant to *T. toli*, where populations in large Sarawak estuaries appear to be relatively isolated from one another. Although sex reversal has not been recorded in other clupeids, females of a number of tropical clupeids, e.g. *Herklotsichthys quadrimaculatus* (Williams and Clarke, 1983) and *Amblygaster sirm* (Conand, 1991), reach a larger size than males. Fecundity is related to body size in most fishes and hence one advantage of larger body size is increased fecundity. Kawasaki (1980) has postulated that as a larger body size may decrease the rate of intrinsic population increase ($r$) because of greater age at maturity, one strategy to compensate for this would be to increase growth rates and increase the body size and gonad size of females, while keeping males small with small gonads. The sex reversal employed by *T. toli* may be one way of achieving this goal; another would be differential growth rates of males and females.

### Methods of spawning

Most teleosts are iteroparous, that is they spawn more than once during their lives. There are two basic forms of iteroparity, synchronous spawning and serial or batch spawning. In the former, all the eggs are spawned at once, whereas in the latter, the ovaries contain batches of oocytes at different stages of development leading to repeated or multiple spawning. Coastal and estuarine teleosts in the subtropics and tropics are mainly serial spawners with a long spawning season (Longhurst and Pauly, 1987). Houde (1989) commented that spawning in the tropics is characterized by protracted spawning seasons and multiple batch spawning, in contrast to spawning in temperate regions where spawning seasons are short, with one or few batch spawnings. This generalization should, however, be interpreted with caution in relation to estuarine fishes. The lack of uniformity in spawning methods within families, and even within genera, suggests an evolutionary plasticity that can respond relatively quickly to the variable nature of the many different types of estuaries in the subtropics and tropics. Some examples from closely related species serve to illustrate this point: *Gerres filamentosus* spawns in three successive batches from October to February (wet season) in the Cochin estuary in southern India (Kurup and Samuel, 1991a). In contrast, in KwaZulu-Natal, South Africa, *Gerres* species, including *G. filamentosus*, spawn only once but the spawning season is throughout the year and the fish leave the estuary to spawn (Cyrus and Blaber, 1984c). In the Sciaenidae in the Gulf of Mexico, *Cynoscion nebulosus* may spawn up to 41 times between June and September (Tucker and Faulkner, 1987) and *Seriphus politus* up to 24 times (De Martini and Fountain, 1981). On the other hand, individuals of the Indo–West Pacific sciaenids *Johnius carouna* and *J. weberi* spawn only once between July and September in estuaries in Malaysia (Yap Yoon Nian, 1995) and *Nibea albida* spawns in two successive batches, with a short interval, in the Cochin estuary in India (Kurup and Samuel, 1991b).

Multiple spawning increases the total number of eggs produced in one spawning season and spreads the risk of egg and larval predation over a long time period. At the same time, it acts to buffer any short-term adverse fluctuations in the abundance of suitable planktonic larval foods.

Few tropical estuarine teleosts are semelparous, that is, all the oocytes in the ovary are at the same stage of development, and all the eggs are spawned at one time, after which the fish dies – the classical pattern of many temperate salmonids. The alosid shad (Clupeidae) *Tenualosa toli* of Sarawak estuaries is semelparous, but although each individual spawns only once and then dies, the spawning season for the species as a whole lasts from about May to November (Fig. 3.2) (Blaber *et al.*, 1996). It has a reproductive pattern very different from that of its congener, the hilsa, *Tenualosa ilisha*, of India and Bangladesh, which is iteroparous and probably a multiple spawner in some estuaries and a single spawner in others (Raja, 1985).

Those species that complete their entire life cycle in estuaries (Chapter 3) are usually small species and exhibit a wide variety of reproductive strategies. The small clupeid *Gilchristella aestuaria*, which is found from India westward to south-east Africa, is semelparous, grows to maturity and spawns in one year (Day *et al.*, 1981); the burrowing goby *Croilia mossambica* of East Africa is probably iteroparous, but is a synchronous spawner (Blaber and Whitfield, 1977b); and another burrower, *Gobioides rubicundus* (Gobioididae), which is found in the estuaries of the Bay of Bengal, is both iteroparous and a batch spawner (Kader *et al.*, 1988). The cichlid *Etroplus suratensis* of Indian and Sri Lankan estuaries is a multiple spawner (Pathiratne and Costa, 1984).

## Spawning seasons and stimuli

Unlike temperate fishes, those of the tropics experience relatively little seasonal change in temperature or daylength. This has led to speculation that these phenomena may not be important as stimuli to reproductive development or spawning (Wootton, 1990; Jobling, 1995). The stimuli for the onset of reproduction in many tropical freshwater fishes are changes in water level or river flow at the beginning of the wet season (Lowe-McConnell, 1987). The stimuli for spawning in a wide variety of tropical reef species have been reviewed by Johannes (1978). In fringing reef environments, the timing and intensity of spawning of small clupeoids is extremely variable – all species spawn throughout the year with one or two periods of more intense activity – and spawning correlates with particular environmental conditions, especially moon phase and, less importantly, rainfall and temperature (Milton and Blaber, 1991). In tropical estuarine species, the factors that stimulate reproductive activity are less clear, perhaps because of the great variety of abiotic and biotic influences of marine, estuarine and freshwater origin that come together in this environment. Changes in flow rates due to freshwater inflow also cause changes in salinity and turbidity; there are subtle changes in temperature and daylength, and changes in tidal influences and moon phase. Seasonality in patterns of food abundance, particularly in the plankton, may also be significant because of their importance for developing larvae. Spawning usually occurs at a time when environmental conditions are most favourable for larval survival. Longhurst and Pauly (1987) showed that the spawning seasons of several hundreds of coastal fish species of the Indo–West Pacific are protracted over many months, but that there is a tendency for reproductive quiescence during the strong monsoon wind period when the sea is very turbulent.

In subtropical regions, reproductive activity usually coincides with even small rises in water temperature. Gonad development in *Sciaenops ocellata* in the Gulf of Mexico takes place during the summer as water temperatures rise and the fish leave the estuaries for bays and coastal waters. Most spawning is in September and October (Murphy and Taylor, 1990). However, although rising temperature may cause reproductive development and activity, the proximal stimulus to spawning appears to be related to moon phase because peak spawnings always occur on nights of the full or new moon (Peters and McMichael, 1987). As these times also correspond to spring tides, the chances of the eggs or larvae being transported into oceanic waters and widely dispersed are greater (Holt *et al.*, 1981, 1985). Also in the Gulf of Mexico, spawning aggregation size of *Cynoscion nebulosus* was highly correlated with water temperature, but in *Pogonias cromis*, dissolved oxygen concentrations were the most important environmental correlate (Saucier and Baltz, 1993). Similar patterns have been observed in other sub-tropical sciaenids, for example *Stellifer rastrifer* in southern Brazil moves out of estuaries into coastal waters to spawn as temperatures rise in summer (Giannini and Paiva-Filho, 1990).

Rainfall with consequent changes in river flow, salinity and turbidity, exerts a powerful influence on the breeding cycle of many estuarine fishes. In subtropical south-east Africa, west Australia and south-east Queensland, blind estuaries open to the sea at the onset of the wet season in summer. Adult fish with maturing gonads then move out of the estuary to spawn in the sea. Spawning at this time allows the larvae and postlarvae to be transported into, or to migrate into, blind estuaries (when they are still open) (Harrison and Whitfield, 1995) as well as open estuaries, possibly using the strong wet-season salinity and turbidity gradients as cues. Many subtropical species have this type of life cycle, including prominent families such as Mugilidae (van der Horst and Erasmus, 1978; Blaber, 1987), Gerreidae (Cyrus and Blaber, 1984c; Albaret and Desfossez, 1988) and Sparidae (Pollock, 1982).

In equatorial regions, the spawning of some species also coincides with the start of the wet season. For example, in South East Asia, New Guinea and northern Australia, *Lates calcarifer*, *Megalops cyprinoides* and *Selenotoca multifasciata* move out of estuaries into coastal waters to spawn during spring tides after the onset of the rains. If the rainy season is delayed, there is a delay in spawning. The proximal stimulus to spawning is thought to be a combination of rising salinity on the high tides and increased water level (Moore, 1982). Salinities of 30‰ are optimal for the spawning of *Lates calcarifer* and an essential characteristic of their spawning grounds: salinities as high as this are found only in the coastal zone and not within estuaries. It has also been suggested that organic material leaching from freshwater swamp forests in the wet season and being carried into coastal waters may act as a spawning stimulus, especially for *L. calcarifer* (Moore, 1982). As is the case for many subtropical species, the spawning periodicity of *L. calcarifer*, *M. cyprinoides* and *S. multifasciata* allows the postlarvae and juveniles to move into suitable habitats (in this case flooded backswamps and swamp forests) where food is abundant and predator numbers reduced.

The economically important hilsa, *Tenualosa ilisha*, of the large estuaries and rivers of the Indian subcontinent undertakes upstream migrations during the monsoon period, in response, it is believed, to flooding of the rivers. This commences with the start of the monsoon in July and with peak breeding in fresh waters between September and December (Raja, 1985). Nair (1958) showed that the temperature may influence the maturation and ripening of gonads, and that heavy rainfall, high turbidity and strong river flow provide favourable conditions for spawning. Also Quddus (1982) postulated that these conditions are favourable because larval survival is increased by high plankton densities.

The stimuli for reproductive activity in Indo–West Pacific sciaenids such as *Johnius carouna* and *J. weberi* are thought to be the increases in salinity and dissolved oxygen that occur in estuaries in the dry season (June to September) (Yap Yoon Nian, 1995). As with many other essentially marine taxa, there may be a minimum salinity in which the eggs can survive and develop. There is no information on the proximal stimuli for spawning.

Reproduction of the cichlid fishes *Etroplus suratensis* and *E. maculatus* in Indian and Sri Lankan estuaries occurs during the dry seasons when water is clear and current speed insufficient to disturb bottom sediment. This temporal pattern of breeding ensures that these Asian cichlids reproduce at a time when parental behaviour can operate effectively. Any increase in turbidity or current speed inhibits the breeding behaviour of *E. suratensis* (Samarakoon, 1983).

## Spawning sites and behaviour

The paradigm that most estuarine fishes are marine migrants that leave estuaries to spawn in the sea, and that the larvae or juveniles use estuaries as nursery grounds (J.H. Day *et al.*, 1981; J.W. Day *et al.*, 1989), accounts for the situation in many subtropical regions, particularly the Gulf of Mexico, southern Africa and parts of Australia. In these areas, most estuaries are well defined and there is a clear distinction between estuary and sea (type 2, Chapter 2). In southern Africa, therefore, Whitfield (1990) was able to divide estuarine fishes into two groups according to the location of their spawning sites. The marine migrants are mainly larger species that spawn at sea, enter estuaries as juveniles and return to the sea prior to reaching sexual maturity. The estuarine group consists of small species that spawn in the estuary where they spend their whole life. They produce fewer, usually demersal, eggs and may have parental care. Most of the marine migrant group spawn large numbers of small, pelagic eggs. These two broad reproductive strategies must both ensure the maintenance of viable populations and adequate dispersal of larvae to suitable habitats.

In the tropics, there is often no clear physical distinction between sea, estuarine coastal waters and estuaries. Most of the estuarine fishes spawn in estuarine conditions. Species from an array of families that are marine spawners at higher latitudes spawn in estuaries in the tropics. They include many species of Ariidae, Engraulididae, Gerreidae, Haemulidae, Mugilidae, Sciaenidae and Sillaginidae. In addition, many of the species that occur only in tropical estuarine areas are estuarine spawners, particularly among the Clupeidae, Engraulididae and Sciaenidae. Almost all of these groups, which fall into the marine migrant category in the subtropics, lay large numbers of pelagic eggs. One species that requires salinities of 35‰ for optimum survival and development of the eggs is the commercially important milkfish, *Chanos chanos* (Swanson, 1996). This species does not spawn in estuaries, although juveniles and adults are common in estuaries and coastal waters throughout the tropical Indo–West Pacific.

The anadromous hilsa, *Tenualosa ilisha*, of India spawns in freshwater stretches of the large rivers of the subcontinent (Pillay, 1958; Chandra, 1962; Raja, 1985). There are no well-defined or fixed spawning grounds. It is thought that they spawn during the rains when they reach fresh water, and when conditions for egg development and larval survival are suitable. They are highly fecund and each individual may release up to two million eggs into the water column.

In contrast to the hilsa, the spawning of some species is much more site specific. In the sciaenids *Cynoscion nebulosus* and *Pogonias cromis* in the Gulf of Mexico, spawning sites for both species are usually located in deep, moving water between barrier islands as well as in channels in open water where water depth ranges from 3 to 50 m. Spawning site selection depends on a particular range of environmental conditions, mainly temperature and oxygen levels, and spawning locations vary seasonally and yearly depending on variations in these hydrological parameters (Saucier and Baltz, 1993).

Those species that form the estuarine category are similar in both the subtropics and tropics – the smaller species of families such as Cichlidae, Gobiidae, Eleotridae and Syngnathidae. They have similar reproductive strategies in both latitudes, usually laying fewer, larger eggs and in some cases exhibiting parental care of eggs and young.

Parental care is well represented in a number of families in estuaries. Mouthbrooding of eggs or young or both is recorded in at least 10 families (Kuwamura, 1986). A concise review of parental care in cichlids is given by Lowe-McConnell (1987). Most are freshwater species, but a few *Oreochromis* species that are numerous in East and West African estuaries are mouthbrooders. The males guard nests in spawning areas and after spawning, the females pick up the eggs and brood them in the buccal cavity. Newly hatched fry remain close to the female and the female's mouth acts as a refuge when danger threatens. *Cichlasoma urophthalmus* of Central America is recorded spawning in mangrove creeks in salinities between 10‰ and 26‰ (Loftus, 1987; Martinez-Palacios and Ross, 1992), mainly in *Rhizophora* and *Thalassia* habitats, and is one of the few American species that exhibits parental care.

The cichlids *Etroplus maculatus* and *E. suratensis*, studied in Sri Lankan estuaries by Ward and Samarakoon (1981), reproduce twice during the year in the drier premonsoonal and monsoonal seasons. When breeding in isolation, *E. maculatus* pairs select dense vegetation, which camouflages nests. During the peak breeding time in July, *E. maculatus* pairs move to sparse vegetation for nesting as a compromise between survival of young and availability of adult food, because non-breeding cannibalistic conspecifics also forage in vegetated areas. Under these pressures, most *E. maculatus* nest in colonies, and this behaviour helps decrease both actual and attempted cannibalism. The sympatric *E. suratensis* does not forage during nesting, and selection of nest sites is determined by factors that favour larval survival. Both sexes in both species guard the young. One member of an *E. maculatus* pair stands guard over offspring while the other leaves the territory to forage; their roles are reversed every few minutes. This guarding of the young continues until they are almost the size of the parents (Ward and Wyman, 1977). The total parental investment in *E. maculatus* is equivalent to the full-time investment of a single parent. On the other hand, in *E. suratensis*, both parents are vigilant over offspring, and neither forages, so parental investment in juvenile survival is twice as high.

Although relatively few species of cardinalfishes (Apogonidae) occur in estuaries, those that do, such as *Apogon ruppelli*, are mouthbrooders. The male is usually responsible for incubating the eggs (Thresher, 1984).

Many species of pipefish and sea horses (Syngnathidae) are entirely estuarine. In pipefish such as *Doryichthys heterosoma* of South East Asia, *Syngnathus djarong* of East Africa, *Enneacampus kaupi* of West Africa and *Trachyrhamphus bicoarctata* of Australia, the care of the eggs and fry is the responsibility of the male. He carries them in a pouch on the ventral surface of the abdomen (Norman and Greenwood, 1975). The eggs are transferred from the female to the male during mating when the oviduct of the female is inserted into the brood pouch of the male. After hatching, the fry stay near the pouch and hide in it at the least sign of danger. In seahorses such as *Hippocampus kuda*, the male also broods the eggs, but the pouch is under the tail and the young cannot return to it after hatching because of its narrow aperture.

Not all parental carers are small species. The ariid catfishes, some species of which reach more than 1 m in length, produce small numbers of very large eggs (up to 20 mm in diameter) that are carried in the mouth of the male. During incubation, the male does not feed. After hatching, the young remain in and around the male's mouth until they reach a length of about 40 mm (Rimmer and Merrick, 1982). As Rimmer and Merrick (1982) state, "The mode of reproduction found in ariids is undoubtedly an important aspect of their success in fluviatile, estuarine and marine environments. Buccal incubation does not necessitate specific substrata or water quality criteria as the mobile adults are able to select the environmental conditions necessary for developing eggs and larvae. This feature, in addition to apparently wide physiological tolerances, has enabled ariids to colonise a wide range of habitats in tropical and sub-tropical regions." It is worth noting, however, that the absence of a pelagic larval phase in this group has probably restricted their geographic distribution. For example, they do not occur in suitable estuaries of the Solomon Islands, probably because the nearest populations in New Guinea are separated from the Solomon Islands by deep oceanic water that cannot be crossed by benthic ariids. Data from the Gulf of Carpentaria in northern Australia indicate that juvenile ariids, unlike the juveniles of most other estuarine fishes, usually migrate offshore into deeper waters, and only later move into estuaries as large juveniles or subadults.

### Spawning in highly variable or extreme environments

Some small estuarine species exhibit considerable flexibility in their repro-ductive strategy which enables them firstly to cope with the variability of estuarine conditions, and secondly to take advantage of suitable spawning conditions when they occur. *Solea bleekeri* is a small sole (up to 130 mm long) that is common in the turbid and muddy estuaries of south-east Africa. In the St Lucia system when salinities are either well above or well below that of sea water, the species migrates to spawn at sea from September to December, along with other fish species that spend part of their life cycle in estuaries. However, Cyrus (1991) found that when salinity conditions are suitable in St Lucia, part of the population of *Solea bleekeri* does not migrate and spawns within the system. A similar phenomenon has been recorded in St Lucia for the engraulid *Thryssa vitrirostris*, which also spawns at sea (Blaber, 1979).

Unusual or prolonged changes in environmental conditions can also modify the reproductive ecology of estuarine fishes, particularly the timing of the onset of maturation and spawning. In the Casamance estuary in Senegal, West Africa, the amount of fresh water entering the system has declined since about 1920, due, it is thought, to climate change and a reduction in overall rainfall. Salinities in Casamance are now usually higher than that of sea water, with values reaching 91‰ 207 km from the sea (Albaret, 1987). The hypersaline conditions have affected the reproduction of the dominant species in several ways. The clupeid *Ethmalosa fimbriata* has been recorded spawning in salinities as high as 66‰ but maturation of the gonads takes place at a smaller size (130–150 mm) compared with those that spawn in the sea (126–185 mm). Early maturation and spawning at small sizes (100–140 mm) has also been recorded for this species when it is isolated in blind estuaries (Charles-Dominique, 1982). The cichlid *Sarotherodon melanotheron* also spawns at a smaller size in the high salinities of Casamance (140 mm) compared with other West African estuaries (180 mm), Other modifications to its reproductive strategy include a doubling of fecundity and more frequent spawning (Albaret, 1987).

The capability of *Ethmalosa fimbriata* and *Sarotherodon melanotheron* to increase their reproductive effort may be a key adaptive strategy to deal with uncertain conditions, and a major element for ensuring their survival under extreme conditions (Albaret, 1987).

*Gerres nigri* has been recorded spawning at 66‰ in Casamance estuary and the mullets *Liza grandisquamis* and *L. falcipinnis* at 50‰, but no other modifications to their reproduction have been observed. Other Mugilidae living in the high salinities of Casamance, notably *Liza dumerili*, *Mugil bananensis* and *Mugil cephalus*, are marine spawners.

## 6.3  REPRODUCTIVE CHARACTERISTICS OF MAJOR TAXA

Considerable detail on aspects of the reproductive biology of a number of species has been presented in section 6.2, particularly for the Sciaenidae, Gerreidae and Sparidae (pp.192–196). For other important estuarine taxa what is known of the reproduction of selected species is summarized below.

## Elopiformes

Two elopiform genera, *Elops* and *Megalops*, are encountered in most tropical estuaries. Both are predators that reach a large size (pp.172–174), and both are marine spawners in which only juveniles and subadults occur in estuaries. All elopiformes have a leptocephalus larva, similar to that of anguillid eels.

*Elops lacerta* of West Africa spawns in the sea and juveniles spend 1 to 2 years in estuaries before moving to coastal waters in the third year of life (Ugwumba, 1989). The unusual life history of *Megalops atlanticus* of the tropical Atlantic has recently been summarized by Garcia and Solano (1995) for the Colombian coast (Fig. 6.3). The leptocephali of *M. atlanticus* migrate into estuaries where they grow to about 28 mm, they then enter a second larval stage when their length decreases to about 13 mm. After this, growth increases again and the larvae become juveniles. At a length of about 100 mm they move to fresh waters. Prior to sexual maturity the fish leave the estuary (>400 mm) and they reach sexual maturity at a length of about 1 m. Spawning aggregations of 25 to 250 *M. atlanticus* have been recorded up to 25 km from the coast. Fecundity is very high, with a 2 m fish containing about 12 million eggs. The life cycle of *Megalops cyprinoides* of the Indo–West Pacific appears to be similar in most respects to that of *M. atlanticus* (Pandian, 1969; Coates, 1987). Garcia and Solano (1995) comment that *M. atlanticus* exhibits a "non-tropical" life history strategy: growing to a very large size, being long-lived and reproducing late in life. However, in these features it is similar to the tropical centropomid *Lates calcarifer* (pp.187–188), although unlike *L. calcarifer*, *Megalops* does not change sex. Perhaps rather than being 'non-tropical', these species are merely employing the strategy of allowing the largest (fittest) fish to contribute most to the gene pool at the expense of putting very large numbers of eggs into relatively few older fish. For maximum success, however, this strategy requires low adult mortality and a stable environment, especially in relation to spawning grounds and larval habitats.

## Clupeiformes

Most clupeoids are small and relatively short-lived but their life history strategies vary considerably. Lewis (1990) recognized two basic life-cycle strategies in the many clupeoid species used as tuna baitfish in the Indo–West Pacific. These two categories are equally applicable to estuarine species.

Type 1 are species with a short life cycle (less than one year), which are small (70–100 mm), grow rapidly, attain sexual maturity in 3–4 months,

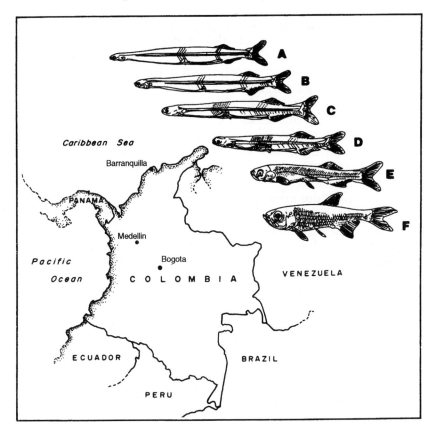

**Fig. 6.3** Sequential development stages (A–F) of *Megalops atlanticus* superimposed on a map of Colombia (modified from García and Solano, 1995).

spawn over an extended period, and have batch fecundities of 500–1500 oocytes per gram of fish. They include the small stolephorid anchovies and sprats of the genus *Spratelloides*.

Type 2 are species with a longer life cycle that are larger (> 100 mm), attain sexual maturity in about a year, may have a more restricted spawning season depending on whether they are subtropical or tropical, and have batch fecundities of at least 500 oocytes per gram of fish.

### Clupeidae

This family contains some of the most economically important fishes of tropical estuaries. Most fall into the type 2 category of Lewis (1990) and undertake spawning migrations that are well known to, and targeted by fishers. Two important genera in the tropical Atlantic are *Brevoortia* in the west and *Ethmalosa* in the east. They have broadly similar reproductive strategies. The fish have seasonal migration patterns and spawn mainly in the sea (pp.169–171, 216 section 7.2). The spawning season is protracted in the tropical *Ethmalosa fimbriata*, but in the subtropical *Brevoortia patronus* it is restricted and its timing depends upon local water temperatures, being earlier in the Gulf of Mexico than on the south-east coast of the USA (Longhurst and Pauly, 1987). Juveniles occur only in estuaries, where they remain for about a year before returning to the sea.

The tropical shads, *Tenualosa*, of the Indo–West Pacific each have different reproductive strategies that are related to their habitats and overall life histories. All are relatively large (> 300 mm) type 2 species. *Temualosa ilisha* of the Bay of Bengal and its estuaries and rivers is gonochoristic, anadromous and long-lived (at least 6 years); it spawns in fresh waters. *T. macrura* of Indonesian waters (Fig. 8.1) is also gonochoristic, lives for only about 2–3 years, inhabits coastal waters, and enters the lower reaches of estuaries to spawn in salinities of about 25‰ (Ahmad *et al.*, 1995). *T. toli* of Sarawak is a protandrous hermaphrodite, lives for only 2 years, lives mainly in estuaries, and spawns in low salinities of 3–13‰ (pp.50–51, 191–192).

Many of the smaller non-commercial clupeids found in tropical estuaries have either a wholly estuarine life history, such as *Gilchristella aestuaria* of south-east Africa, or an estuarine to freshwater life cycle as in *Pellonula afzeliusi* of West Africa. Both these species are usually short-lived and produce relatively few eggs (Ikusemiju *et al.*, 1983; Whitfield, 1990; Baran, 1995), but have multiple batch spawning over a protracted spawning season.

### Engraulididae

In contrast with many of the clupeids, most tropical anchovies are estuarine spawners. In the Indo–West Pacific, the genera *Coilia*, *Lycothrissa*, *Setipinna* and *Thryssa* have mainly estuarine life histories (extending into fresh water for some species). They fall into the type 2 category of Lewis (1990). Several small (< 120 mm) but economically important genera of the tropical West Atlantic are primarily marine spawners and they are probably mainly in the type 1 category. The biology of the approximately 14 species of *Anchoa* in the tropical West Atlantic and the 16 species in the tropical East Pacific is poorly known. Many live in coastal waters and freely enter estuaries. In the Gulf of Nicoya in Pacific Costa Rica, *Anchoa curta* and *A. spinifer* have protracted spawning seasons from January to October (Whitehead, 1985) and spawning takes place at estuary mouths and offshore but not within estuaries (Ramirez *et al.*, 1989). The widespread species *A. hepsetus* of the West Atlantic spawns in harbours, estuaries and sounds (Whitehead, 1985) while *Anchoa mitchilli* of the eastern USA and Gulf of Mexico is a marine spawner (Flores-Coto *et al.*, 1988). Many anchovies such as *Anchoa* and *Cetengraulis* have oval or elliptical eggs.

*Cetengraulis edentulus*, which is widespread in north-eastern South America, spawns in the sea at night and close inshore in Venezuela, with peak spawning in November in the rainy season. The closely related *C. mysticetus* of the tropical East Pacific spawns over shallow mudflats, also at night in the wet season.

Few of the coastally abundant type 1 Indo–West Pacific marine anchovies of the genera *Stolephorus* and *Encrasicholina* have an estuarine phase in their life cycle. For those that do, such as *Stolephorus baganensis* and *S. tri*, unfortunately almost nothing is known of their biology, but assuming that it is similar to that of others in the genus, then they are very short-lived multiple spawners that spawn throughout the year with one or two periods of more intense spawning (Milton and Blaber, 1991).

### Ariidae

The 150 or so species of ariid catfishes have a worldwide subtropical and tropical distribution and are particularly common in coastal waters, estuaries and fresh waters of South America, South East Asia and Australasia. Their reproduction and development has been reviewed by Rimmer and Merrick (1982), who state that all species are gonochoristic, and for all species that have been investigated, the male practises oral (buccal) incubation. Features of the spawning and behaviour are discussed above (p.198) and are listed in Table 6.1.

Anadromous migrations associated with spawning have been observed for a number of species, such as *Osteiogeniosus militaris* in South East Asia and *Arius heudoloti* in West Africa. Others such as *A. thalassinus* of the Indo–West Pacific and *Arius felis* of the Gulf of Mexico are marine spawners. Estuarine spawning takes place in many of those which have a completely estuarine or estuarine/freshwater life cycle, such as *A. leptaspis*, *A. mastersi* and *A. melanopus*. Most ariids have a single annual spawning in the wet season in the tropics or in spring in the subtropics (Kailola, 1990). The spawning season may be protracted but spawning stimuli are largely unknown. *Arius thalassinus* start breeding at temperatures of 25–28 °C and *Arius leptaspis* at temperatures above 26 °C (Rimmer and Merrick, 1982). Fecundities are low and range from 14 to 184 large oocytes of 9.5 to 25.0 mm in diameter. Fertilization is external

**Table 6.1** Summary of reproductive characteristics of a selection of fishes from subtropical and tropical estuaries (ND, no data)

| Family and species | Region* | Sex† | S/I(n)‡ | Spawning site§ | Spawning stimuli | Fecundity (millions) | Egg diameter (mm) | Parental care¶ | Reference |
|---|---|---|---|---|---|---|---|---|---|
| **Ariidae** | | | | | | | | | |
| *Arius thalassinus* | IWP | S | I(s) | M | Temperature | 0.0002 | 15–20 | MB | Rimmer and Merrick (1982) |
| *Arius felis* | TWA | S | I(s) | M and E | ND | 0.00005 | 12–14 | MB | Rimmer and Merrick (1982) |
| *Arius heudeloti* | TEA | S | I(s) | M | ND | 0.00003 | 10–19 | MB | Rimmer and Merrick (1982) |
| *Arius melanopus* | TWA | S | I(s) | E | ND | ND | ND | MB | Yáñez-Arancibia *et al.* (1988) |
| *Arius proximus* | IWP | S | I(s) | E | ND | 0.00002 | 10–15 | MB | Kailola (1990) |
| **Centropomidae** | | | | | | | | | |
| *Lates calcarifer* | IWP | H | I(s + b) | MC | Rainfall, salinity, tides, leaching | 10–40 | 0.45 | None | Moore (1982), Davis (1987), Cheong and Yeng (1987) |
| **Cichlidae** | | | | | | | | | |
| *Etroplus suratensis* | IWP | S | I(b) | E(FW) | Clear water, low flow speeds | 0.002 | ND | Guarder | Samarakoon (1983), Ward and Samarakoon (1981) |
| *Oreochromis mossambicus* | IWP | S | I(b) | E(FW) | Dry season? | 0.0004–0.002 | 1.2–3.6 | MB | De Silva (1986) |
| **Clupeidae** | | | | | | | | | |
| *Gilchristella aestuaria* | IWP | S | I(b) | E | ND | 0.002 | 1.0 | None | Blaber (1979) |
| *Ethmalosa fimbriata* | TEA | S | I(s) | M and E | Salinity? | 0.016–0.2 | 0.5 | None | Blay and Eyeson (1982) |
| *Tenualosa ilisha* | IWP | S | I(b) | FW | River flow, fresh water, turbidity, plankton | 0.1–2.0 | 0.7–0.9 | None | Raja (1985) |
| *Tenualosa toli* | IWP | H | S | E | Salinity, plankton | 0.1–1.2 | 0.9 | None | Blaber *et al.* (1996) |
| **Engraulididae** | | | | | | | | | |
| *Coilia dussumieri* | IWP | S | I(b) | E | Turbidity? | 0.001–0.005 | 0.8–0.9 | None | Fernandez and Devaraj (1989) |
| *Thryssa vitrirostris* | IWP | S | I(b) | M and E | Salinity? | 0.012 | 1.2 | None | Blaber (1979) |

| | | | | | | | | | |
|---|---|---|---|---|---|---|---|---|---|
| Gerreidae | | | | | | | | | |
| *Gerres filamentosus* | IWP | S | I(b) | M and E | ND | 0.06–0.4 | ND | None | Kurup and Samuel (1991a) |
| *Diapterus rhombeus* | TWA | S | i(b) | M | ND | ND | ND | None | Austin and Austin (1971) |
| Mugilidae | | | | | | | | | |
| *Liza falcipinnis* | TEA | S | i(b) | M and E | ND | ND | ND | None | Baran (1995) |
| *Liza macrolepis* | IWP | S | i(b) | M and E | ND | 0.01–9.5 | 0.7 | None | James *et al.* (1983) |
| *Mugil cephalus* | CT | S | i(s + b) | M | ND | 0.45–4.8 | 0.8 | None | Silva and De Silva (1981) |
| Sciaenidae | | | | | | | | | |
| *Cynoscion nebulosus* | TWA | S | i(b) | M | Temperature | 4.0–6.0 | 0.8 | None | Tucker and Faulkner (1987) |
| *Johnius carouna* | IWP | S | I(s) | E | Salinity and oxygen | 0.004–0.09 | ND | None | Yap Yoon Nian (1995) |
| *Johnius weberi* | IWP | S | I(s) | E | Salinity and oxygen | 0.009–0.09 | ND | None | Yap Yoon Nian (1995) |
| *Nibea albida* | IWP | S | I(b) | E | ND | ND | ND | None | Kurup and Samuel (1991b) |
| *Pseudotolithus elongatus* | TEA | S | I(b) | M and E | ND | 0.05–0.4 | ND | None | Fontana and Leguen (1969) |
| *Sciaenops ocellata* | TWA | S | I(b) | M | Temperature and lunar phase | 0.2–3.3 | 0.4–0.6 | None | Peters and McMichael (1987), Wilson and Nieland (1994) |
| Soleidae | | | | | | | | | |
| *Solea bleekeri* | IWP | S | I(b) | M and E | Salinity | ND | ND | None | Cyrus (1991) |
| Sparidae | | | | | | | | | |
| *Acanthopagrus australis* | IWP | H | I(b) | M | ND | ND | 0.8 | None | Pollock (1984, 1985) |
| Syngnathidae | | | | | | | | | |
| *Doryichthys heterosoma* | IWP | S | I(s) | E(FW) | ND | 0.00003 | 1.0–1.5 | BP | Unpublished |
| *Enneacampus kaupi* | TEA | S | I(s) | E(FW) | ND | 0.00004 | 1.0–1.5 | BP | Baran (1995), Dawson (1981) |

* CT, circumtropical; IWP, Indo–West Pacific; TEA, tropical East Atlantic; TWA, tropical West Atlantic.
† H, hermaphrodite; S, single sex.
‡ S/I(m) denotes parity and type of spawning: S, semelparous; I, iteroparous; s, single spawner; b, batch spawner. So, for example, I(s) denotes an iteroparous single spawner.
§ E, estuary; FW, fresh water; M, marine; MC, marine coastal.
‖ BP, brood pouch; MB, mouthbrooder.

and small numbers of eggs are laid on the substratum where they are picked up by the male. Buccal incubation by the male may continue for up to 9 weeks before the fully developed, actively feeding young are released. Males usually fast for the whole of the incubation period (Rimmer and Merrick, 1982; Coates, 1991). The young of some species, such as those of *A. felis, A. melanopus* and *Bagre marinus* of the Gulf of Mexico, move into shallow estuarine nursery areas at a small size (Day *et al.*, 1989). The juveniles of other species, such as *A. proximus* and *A. argyropleuron* of the Gulf of Carpentaria, remain offshore and only move into estuaries as subadults.

### Mugilidae

Most of the 95 or so species of mullet are subtropical or tropical and occur in shallow coastal waters and estuaries, with a few species venturing into fresh water. They are one of the most characteristic and numerous families of fishes in tropical estuaries (see also pp.52–53, 123–124, Fig. 3.1.). Most species migrate to the sea to spawn (Wallace, 1975b; van der Horst and Erasmus, 1978; Blaber, 1987). Spawning has seldom been recorded in estuaries, although the gonads of most species reach an advanced state of development just prior to emigration to the sea. This has led to some confusion and uncertainty with regard to spawning sites. For example, in a series of papers on the Negombo Lagoon in Sri Lanka, Wijeyaratne and Costa (1987a,b,c, 1988) concluded that although *Liza tade, L. macrolepis* and *Valamugil cunnesius* were in ripe condition in the estuary, spawning probably occurred in adjacent marine waters. Similarly in India, Sathyashree *et al.* (1987) found that *Valamugil speigleri* with maturing gonads are only available in the Vellar estuary, Killai Backwaters and Pitchavaram mangrove areas, whereas mature, oozing and spent fishes are found only in the inshore waters, indicating that *V. speigleri* breeds in coastal waters and not in the estuary. Also in India, Kurup and Samuel (1983) found that *Liza parsia* spawned in the Cochin estuary but only in "near marine conditions". The situation in West Africa may be more complex, but 'near marine' conditions still seem to be required for spawning. For example, in the Ébrié Lagoon of Ivory Coast, West Africa, *Liza grandisquamis* is known to spawn in the area near the mouth where it is subject to marine conditions, but probably also spawns in the adjacent sea; similarly *L. falcipinnis* has been found spawning in the lagoon, but marine spawning is also probable; and *Mugil curema* is a marine spawner, as it is in the West Atlantic, and has never been found spawning in the lagoon (Albaret and Legendre, 1985; Day *et al.*, 1989).

The spawning grounds for most species are probably close inshore, although in most cases the precise sites are unknown. There is indirect evidence that *Mugil cephalus* in south-east Africa may spawn at least 3 km from the shore (Bok, 1984). Data from the former USSR and USA suggested that *M. cephalus* spawns at night, offshore over deep water (Dekhnik, 1953; Arnold and Thompson, 1958). Helfrich and Allen (1975) observed *Crenimugil crenilabis* spawning at night in shallow water within atoll lagoons. *Liza vaigiensis* and *Valamugil parmatus* occur in dense schools in coastal waters of the Arabian Sea during the wet season, a phenomenon thought to be connected with spawning (Moazzam and Rizvi, 1980).

*Mugil cephalus* undertakes large pre-spawning aggregations in estuaries and coastal waters (Wallace, 1975b; Whitfield and Blaber, 1978e). Large numbers of fish with ripe gonads school in estuary mouths before moving into the sea, a phenomenon noted throughout the species' worldwide range (Thomson, 1963; Apekin and Vilenskaya, 1978). The spawning run of *M. cephalus* at St Lucia in KwaZulu-Natal, south-east Africa, is very regular and (since records began in 1968) has always taken place between 12 April and 12 May (pp.48–49).

In the subtropics most mullet spawn from winter to early summer and in the tropics for a long period encompassing the wet season. They are multiple spawners with fecundities ranging from about $10^4$ to $10^6$ eggs depending upon size and species. There are few data on spawning stimuli, although the above spawning periods and results of Thomson (1966) suggest that temperature may be important. The eggs are vulnerable to environmental conditions, particularly salinity, which may be the reason that most species spawn in marine conditions. Lee and Menu (1981) demonstrated that the optimum salinity for the hatching of *M. cephalus* eggs is between 30‰ and 40‰ with a peak at 35‰. In addition, other authors (Tang, 1964; Kuo *et al.*, 1973) report that the fertilized eggs of *M. cephalus* must stay in suspension to develop, and that they sink in still water. In the calm waters of estuaries, a combination of lack of much water movement, less depth and lower salinity (lower specific gravity) might cause the eggs to touch bottom, thereby inhibiting development. In a review of oceanic and estuarine transport of fish eggs and larvae, Norcross and Shaw (1984) state "spawning often takes place close to gyral, upwelling or other directional circulations". The spawning grounds of most mullet lie in the vicinity of estuary mouths (Wallace, 1975b); hence the distance between spawning grounds and nursery areas is short. Some of the factors influencing recruitment of juveniles into nursery areas are discussed in Chapter 4 and their degree of estuarine dependence in Chapter 7.

## Polynemidae

The threadfins are common in the more turbid coastal waters and large estuaries of the tropics where several species are of economic importance. Some species are hermaphrodite and others gonochoristic.

*Polydactylus indicus* of the estuaries of India grows to over 1 m in length and is a hermaphrodite throughout its life. It has an ovotestis in which the ovarian part becomes active alternately with the testicular part; because of this, the egg-bearing period for an individual is reduced to half. Visual differentiation of the sexes can be misleading and only histological sections of gonads accurately indicate the active sex. When the ovary is active, the testis in the same individual lies dormant and vice versa. Spawning takes place in inshore marine areas (Kagwade, 1976, 1988).

In marked contrast to the previous species, *Polydactylus octonemus* is a small, inshore, gonochoristic species of the Gulf of Mexico that reaches a length of about 230 mm (Dentzau and Chittenden, 1990). They disappear from inshore waters in November–December and spawn over deep water in a discrete period from mid-December to mid-March when water temperatures are lowest. The pelagic larvae and juveniles presumably use current transport to reach their estuarine and inshore nursery grounds. Dentzau and Chittenden (1990) speculate that longshore components of a cyclonic gyre on the shelf-break off Texas and Louisiana transport the eggs, larvae and juvenile *P. octonemus* from the outer shelf to their shallow nurseries. Further, they suggest that this short-lived (about 1 year) species has apparently evolved a life history strategy that emphasizes small size, early maturity, a short life span and high mortality.

Most polynemids are marine spawners, including *Eleutheronema tetradactylum*, *Polydactylus sheridani*, *Polydactylus multiradiatus* and *Polydactylus sextarius* of the Indo–West Pacific, *Galeoides decadactylus* and *Polydactylus quadrifilis* of West Africa, and *Polydactylus approximans* of the tropical East Pacific. Spawning of some species is known to occur in both estuaries and the sea; *Eleutheronema tetradactylum* breeds both in the sea and the Chilka Lake system of eastern India (Mohanty, 1975), and *Galeoides decadactylus* is occasionally found spawning in West African estuaries (Baran, 1995).

## Haemulidae

Grunters of the genus *Pomadasys* occur in subtropical and tropical estuaries in all regions but little is known of their reproductive biology. All are apparently gonochoristic, most are marine spawners, and juveniles as well as postspawning adults move seasonally into estuaries. The spawning times of 24 species of haemulids are reviewed by Konchina (1978), who states that the spawning period is protracted (3–10 months) and that three species spawn throughout the year.

*Pomadasys commersonni* of the Indo–West Pacific is an economically important species of East and south-east Africa. Maturity is reached at 400 mm and spawning takes place offshore in depths of about 75 m from September to December (late dry season) (van der Elst, 1981). Fry and juveniles as well as postspawning adults migrate into estuaries during spring and early summer. The closely related *Pomadasys kaakan* and *P. argenteus* also occur through much of the Indo–West Pacific. Both are multiple spawners in salinities of > 31‰, with egg diameters of 0.6–0.8 mm and fecundities of 0.6–2.5 × $10^6$ (van der Elst, 1981; Abu Hakima *et al.*, 1983; Zhang-Renzhai, 1987). In eastern Australia, they spawn over an extended period from August to March, but with a major peak in reproductive activity from September to November (Bade, 1994). In other regions the essentially marine reproductive cycle appears to be similar, but there is a paucity of published detail. In West Africa, *Pomadasys jubelini* is a rainy-season marine spawner, with juveniles and adults moving in and out of estuaries (Baran, 1995); *Brachydeuterus auritus* spawns in the sea from February until June with a peak in May and its spawning grounds are between 10 and 30 m deep (Barro, 1979). In the tropical West Atlantic, *Genyatremus luteus* and *Conodon nobilis* are marine spawners in the coastal waters of northern South America where they have a very protracted spawning season (Lowe-McConnell, 1962). In the tropical East Pacific, *Pomadasys leuciscus* and *P. macracanthus* are a significant part of the fauna of coastal lagoons in Central America (Warburton, 1978). They are also marine spawners, and juveniles and postspawners enter the coastal lake systems at the time they are open to the sea (salinities 15–34‰) between August and November.

### Carangidae

Almost all carangids found in tropical and subtropical estuarine waters are juveniles or subadults (Blaber and Cyrus, 1983; Gunn, 1990). Adults are usually marine and the spawning of most species is thought to take place in deeper water, often in the vicinity of coral reefs. Williams (1965) recorded *Caranx ignobilis* spawning from July to March with a peak during the rains of the north-east monsoon (November to March). The juveniles of *Caranx ignobilis* are one of the commonest carangids in Indo–West Pacific estuaries, and in South and East Africa recruit mainly in the wet season. This species and *Caranx melampygus* have a lunar spawning cycle (Johannes, 1978) and spawn mainly over outer reef slopes (Williams, 1965; Johannes, 1978). The finding by Williams (1965) of spawning schools of *C. ignobilis* off Zanzibar composed almost entirely of one sex, suggests segregation in the immediate pre-spawning period. In the subtropical waters of the Gulf of Mexico, *Caranx chrysos* is a summer spawner, with a peak spawning from June to August. Its fecundity varies from $0.041 \times 10^6$ in a 288 g fish to $1.5 \times 10^6$ in a 1076 g fish and spawning occurs in oceanic water deeper than about 200 m (Goodwin and Finucane, 1985). Migrating to spawn over deeper water may be related to a reduction in predation: it may result in eggs and early larvae drifting over deep water, out of range of the benthic and demersal predators of the adult's normal, shallower habitat (Johannes, 1978).

### Gobiidae

Despite the fact that the gobies are the largest family of fishes (at least 1500 species), almost nothing is known of the reproduction of most species. They are a numerically dominant component of many estuarine habitats, particularly mangroves, where mudskippers are a prominent and familiar group. However, in his recent review of the mudskippers, perhaps the best known of all gobies, Clayton (1993) states "It is doubtful if there is any other group of fishes in which so much general interest is based on so little knowledge."

Almost all reproductive styles and strategies can be found among the great diversity of gobies in subtropical and tropical estuaries. In a review of gobies on tropical high islands of the Pacific (e.g. parts of Fiji and Solomon Islands), Ryan (1991) attributes their dominance of estuaries and streams partly to their reproductive strategy, which for most species includes a marine larval stage. In some genera, there is evidence of rapid evolutionary flexibility in reproduction in relation to differing habitat demands such as occur in tropical estuaries. For example, Kishi (1978) showed that in *Tridentiger obscurus*, the eggs of freshwater river and lake populations were similar in size, but those of estuarine fishes were twice as large in volume. No correlation was evident between egg size and female body size, but the observed decline in size throughout the breeding season suggests the influence of changing environmental conditions. The possible effect of external factors, such as salinity, could be ruled out by the identical size of eggs in populations in different salinities and indicates a possible genetic basis for the egg-size difference.

Most gobies are gonochoristic, but hermaphrodites are not uncommon. The species *Lythrypnus dalli* of the tropical East Pacific actually shares characteristics of both successive and simultaneous hermaphroditism, which allows considerable flexibility. Simultaneous hermaphroditism may be retained in this species, to facilitate rapid sex change from female to male, and hence maintain flexibility so that unsuccessful males can revert to reproduction as females (St Mary, 1994).

Some estuarine species, such as *Gobiosoma robustum* in Terminós Lagoon, Mexico, spawn throughout the year (Zavala Garcia *et al.*, 1988), whereas others, such as *Glossogobius giuris* of the Indo–West Pacific, have a well-defined breeding season. The latter species breeds in the rainy season (between November and February in East Africa) and deposits adhesive eggs on underwater plants (Bruton and Kok, 1980). The burrowing goby *Croilia mossambica*, also of East Africa, is likewise a wet-season breeder and lays about 50 sticky eggs, but the spawning site is unknown (Blaber and Whitfield, 1977b). Spawning in estuaries during the wet season ensures that a proportion of the eggs and larvae will be transported out to sea and hence dispersed. The success of this strategy is evidenced by the dominance of gobiid larvae in most tropical estuarine and coastal ichthyoplankton samples (section 6.4).

Information on the reproduction of the mudskippers has been comprehensively reviewed by Clayton (1993), who reports much geographic variation in spawning seasons and fecundity, even within species, and comments that further careful research is required. Fecundities appear to vary from about 1000 to 23 000, depending on location and/or species. The three mudskipper genera, *Boleophthalmus*, *Periophthalmodon* and *Periophthalmus*, as well as the gobiid genus *Acentrogobius*, spawn in the Ganges estuary in the monsoon periods when salinity is low, but when turbidity, temperature and plankton biomasses are high. In this respect they follow the same pattern as many of the other estuarine species of the region by ensuring food for the larvae (plankton) while at the same time reducing predation (turbidity). Most mudskippers lay their eggs in a burrow and stick them to the walls and roof with filamentous threads (Clayton, 1993). Optimum conditions for hatching of the eggs are 28 °C and 15–25‰, and Clayton (1993) suggests that the failure of eggs to hatch at higher salinities may contribute to the absence of mudskippers from otherwise suitable mangrove areas, such as in parts of the Red Sea where salinities can exceed 40‰.

### Chondrichthyes

A detailed account of reproduction in cartilaginous fishes can be found in Dodd (1983), but relatively little is known about most tropical species. Most sharks and rays are marine, some are found in freshwater and very few are estuarine. Although the juveniles and subadults of many tropical species enter estuaries, adults are uncommon, and spawning takes place in the sea.

Exceptions to this include the infamous Zambezi or bull shark, *Carchar-*

*hinus leucas*, which has a circumtropical distribution and penetrates far
into fresh waters for long periods. It has even been recorded breeding in
Lake Nicaragua (Last and Stevens, 1994). It is viviparous with litter sizes
of 1 to 13. In KwaZulu-Natal, South Africa, it gives birth adjacent to
estuaries and the voraciously piscivorous young then move into estuaries
to feed (van der Elst, 1981). The Ganges shark, *Glyphis gangeticus*, of the
Ganges–Hooghly system and *G. glyphis* of Borneo and northern Australia
are large freshwater or estuarine sharks about which almost nothing is
known (Compagno, 1984; Last and Stevens, 1994). It is believed,
however, that they are viviparous. Another smaller shark, *Scoliodon
laticaudis*, is widespread in the coastal waters of the Indo–West Pacific and
is common in the estuaries of Malaysia, Borneo and Sumatra where it
lives for extended periods and probably breeds. It is viviparous with litter
sizes of 1 to 14 and can breed throughout the year (Compagno, 1984).

Like the carcharhinid sharks, a number of stingrays are reasonably
common in the lower reaches of estuaries but few can be considered
estuarine. *Himantura chaophraya* lives and breeds only in fresh and
estuarine waters from northern Australia to Thailand, but little is known
of its biology. Similar poorly known species include *Himantura signifer* and
*H. oxyrhynchus* of the rivers and estuaries of Borneo.

Sawfishes (Pristidae) are ovoviviparous and several species are often
encountered in tropical estuaries. The largest species, *Pristis pectinata*, has
a circumtropical distribution and penetrates far up estuaries. In south-east
Africa it is common in large estuaries such as the St Lucia system where
it breeds. It gives birth to 15 to 20 young, which have soft, sheathed
saws that avoid damaging the mother. *Pristis clavata* of northern
Australia is a smaller species, confined to estuarine coastal waters,
estuaries and fresh water, but nothing is known of its reproduction (Last
and Stevens, 1994).

## 6.4  ICHTHYOPLANKTON

Although much information is available on the ichthyoplankton of
tropical coral reef areas (Leis and Rennis, 1983; Leis, 1991), relatively
little research has been undertaken in subtropical and tropical estuaries,
apart from some work in south-east Africa (Harris and Cyrus, 1995;
Harris *et al.*, 1995), East Africa (Little *et al.*, 1988a) and Thailand
(Janekarn and Boonruang, 1986). Most research on estuarine ichthyoplankton has been conducted in temperate areas (Melville-Smith and
Baird, 1980; Whitfield, 1989a,b; Warlen and Burke, 1990; Neira and
Potter, 1992, 1994) where the faunal affinities and physical conditions
are very different.

The larval fish assemblages in subtropical and tropical estuaries consist of two main components: larvae of species that spawn in the estuary and larvae of species that spawn in the sea. Although it has been widely accepted that estuaries function as nursery grounds (Chapter 7) for juveniles of many marine species that recruit to the estuary as postlarval juveniles, it is only more recently that it has been realized that some species recruit at the larval postflexion developmental stage, via a combination of active and passive processes (Harris, 1996). Harris (1996) concluded that environmental factors, together with ontogenetic behavioural changes and local currents, result in a net shoreward movement of larvae to inshore nursery habitats in KwaZulu-Natal, southeast Africa. The resulting structure of KwaZulu-Natal estuarine ichthyoplankton assemblages apparently depends primarily on environmental conditions, such as salinity, temperature and especially turbidity, at the time of recruitment, together with zoogeographical considerations and the proximity of spawning sites (Harris, 1996). Species having larvae that recruit to KwaZulu-Natal estuaries include *Elops machnata*, *Megalops cyprinoides*, many Gobiidae, *Rhabdosargus* spp., *Ambassis* spp., *Leiognathus equulus*, *Terapon jarbua*, *Johnius dussumieri* and *Thryssa vitrirostris*.

In their large scale study of Thai coastal ichthyoplankton, Janekarn and Kiørboe (1991a) found that larval density was negatively correlated with turbidity. In this respect it is interesting that in Sarawak and Sabah the mean density of ichthyoplankton is likewise higher in most of the small, clearer-water estuaries than in the larger, more turbid estuaries. However, most of the numbers in the smaller estuaries are contributed by Gobiidae and diversity is low (Blaber *et al.*, 1997). In the larger, more turbid estuaries of Sarawak the ichthyoplankton are primarily from taxa the adults of which are found in estuarine or turbid water, whereas those in the smaller clearer-water estuaries are found mainly in marine and clearer waters (Blaber and Blaber, 1980; Blaber, 1981a; Cyrus and Blaber, 1992). Some in the large turbid estuaries, such as *T. toli* (Blaber *et al.*, 1996) and *Coilia* (Whitehead *et al.*, 1988), spawn in the estuaries, while others such as Sciaenidae and Polynemidae spawn both in the lower reaches of estuaries and in adjacent shallow coastal waters. There is no clear cut-off point between these large estuaries and the shallow coastal waters in terms of salinity or turbidity (Blaber, 1981a). However, the fish fauna in these waters is different from that of the offshore shelf (Latiff *et al.*, 1976), with few species in common.

Gobiid larvae are a dominant component of larval fish assemblages of many estuarine and marine areas throughout the world, from temperate South African and west Australian estuaries (Beckley, 1986; Neira *et al.*, 1992), subtropical African estuaries (Harris and Cyrus, 1995; Harris *et al.*, 1995), temperate surf zones (Whitfield, 1989a), tropical mangroves

(Janekarn and Boonruang, 1986; Little *et al.*, 1988a), tropical nearshore waters (Janekarn and Kiørboe, 1991a) to tropical continental shelves (Young *et al.*, 1986). In Sarawak and Sabah they are also a dominant component in most estuaries (Blaber *et al.*, 1997). This widespread occurrence of goby larvae in large numbers is probably a result of the very large number of species in this, the largest family of marine fishes (Nelson, 1984), especially in the tropics, and the relatively long duration of their larval phase (Thresher, 1984).

The numbers of fish larvae in tropical estuaries are highly variable, but the mean densities are similar and similarly variable; for example: 0.03–9.2 m$^{-3}$ in the estuaries of Sabah and Sarawak (Blaber *et al.*, 1997), 0.01–0.52 m$^{-3}$ in the Kosi estuary of northern KwaZulu-Natal (Harris *et al.*, 1995), 0.15–10.3 m$^{-3}$ in the St Lucia system of KwaZulu-Natal (Harris and Cyrus, 1995), 1.2–2.0 m$^{-3}$ in an East African mangrove creek (Little *et al.*, 1988a) and 0.04–0.20 m$^{-3}$ in Thailand (Janekarn and Kiørboe, 1991b). In Southern Hemisphere temperate estuaries, similar densities have been recorded: 3.9 m$^{-3}$ in the Swartkops estuary in South Africa (Beckley, 1985), 0.22–0.47 m$^{-3}$ in the Swartvlei estuary in South Africa (Whitfield, 1989b) and 0.04–3.14 m$^{-3}$ in the Swan estuary in west Australia (Neira *et al.*, 1992). In tropical non-estuarine areas, values are also similar. In the Andaman Sea off western Thailand, mean densities were generally > 0.1 m$^{-3}$ away from shore and < 0.1 m$^{-3}$ nearshore with overall coastal densities of 1.5–1.9 m$^{-3}$ (Janekarn and Kiørboe, 1991a). The mean density in the Arabian Gulf was 3.89 m$^{-3}$ (2.8–10.3 m$^{-3}$) (Houde *et al.*, 1986) while on the north-west shelf of Australia mean densities were lower (0.01–0.04 m$^{-3}$) (Young *et al.*, 1986).

The diversity of the ichthyoplankton assemblage in Sarawak and Sabah estuaries (56 taxa, 26 families) is lower than that of other tropical estuaries of the Indo–West Pacific, with the exception of a mangrove creek in Kenya (at least 25 families; Little *et al.*, 1988a). In Thailand, 44 families were recorded (Janekarn and Boonruang, 1986); in the Kosi and St Lucia estuaries, both in KwaZulu-Natal, 61 families (153 taxa) and 44 families (85 taxa) respectively were collected (Harris and Cyrus, 1995; Harris *et al.*, 1995). The tropical shelf waters of north-west Australia yielded 104 taxa (Young *et al.*, 1986) and the Arabian Gulf 84 taxa (Houde *et al.*, 1986). Even the temperate waters of the Swan estuary in west Australia contained 74 species (Neira *et al.*, 1992). The relatively low diversity in Sarawak and Sabah estuaries is possibly the result of their rigorous physical nature, particularly the very high turbidities and current speeds, or in small estuaries where these factors are less, the low biomass of zooplankton available as food for the larvae (Blaber *et al.*, 1995b).

# Chapter seven

# Estuarine dependence

## 7.1 INTRODUCTION

An understanding of the degree to which tropical fishes are dependent on estuaries, and the reasons for this, is not only necessary for studying their ecology, but vital if fisheries management and conservation are to be effective. This is especially the case in the tropics where the ecological functioning of many estuaries is being extensively modified by humans. 'Estuarine-dependent' species have been variously defined and subdivided (Lenanton and Potter, 1987; Day *et al.*, 1989; Potter *et al.*, 1990; Whitfield, 1994a), but for present purposes they are defined as those for which estuaries, or similar habitats, are the principal environment for at least one part of the life cycle, and without which a viable population would cease to exist (Blaber *et al.*, 1989).

The 'estuarine' category of fishes (section 3.1) that live only in estuaries are obviously estuarine dependent. However, this group seldom forms more than about 10–15% of the fish fauna of subtropical and tropical estuaries (Longhurst and Pauly, 1987). In this chapter we are largely concerned with whether or not the concept of estuarine dependence has validity among the majority of fishes that spend only part of their life history in estuaries. In their case it is important to ask the question, what are the advantages (or disadvantages) and survival value in having an estuarine phase in the life cycle and is it obligatory? The answer must vary from species to species. Following from this, we must examine the factors that may make estuaries desirable places to live and how these may vary geographically and according to estuary or habitat type.

## 7.2 REGIONAL CONSIDERATIONS

Whether or not particular species, populations, communities or certain life history phases of fishes that occur in estuaries are dependent on the estuarine environment, as opposed to the sea or even fresh water, has given rise to much discussion and speculation (e.g. Hedgpeth, 1982). Unfortunately there is little agreement in general terms about the validity of estuarine dependence, due mainly to parochialism and the all-too-ready acceptance by workers everywhere of the paradigm developed for fishes of the estuaries of the south-eastern USA (Gunter, 1967; McHugh, 1967; Day and Yáñez-Arancibia, 1985). The concept of an estuarine-dependent phase in the life cycle of coastal fishes arose primarily from studies of temperate and warm temperate estuaries in the USA and in southern Africa. In these non-tropical areas, many marine species have been shown to be dependent on the estuarine environment during their juvenile phase (Day *et al.*, 1981; Deegan and Thompson, 1985). For example, in the estuaries of the Louisiana deltaic plain, species utilizing the estuary as a nursery make up 35–72% of all species, and usually over 90% of individuals (Deegan and Thompson, 1985). Species such as *Anchoa mitchilli*, *Brevoortia patronus*, *Cynoscion arenarius* and *Micropogonias undulatus* are largely estuarine dependent and use the estuaries as nursery grounds (Gunter, 1967). In these warm temperate estuaries of the south-eastern USA, Gunter (1967) states that the water temperature cycle (4–40 °C) is probably chiefly responsible for the seasonal movements and recurrent cyclic activities of fishes. Such extreme seasonal changes in temperature do not occur in the tropics.

The degree of estuarine dependence among fishes of the subtropics and tropics has been questioned by Longhurst and Pauly (1987), who analysed studies from 20 subtropical and tropical areas, and concluded that in only two areas could estuarine dependence really be demonstrated. These were in southern Brazil and South Africa, in systems that could only be considered marginally subtropical. Blaber (1981a) and Blaber and Blaber (1980) postulated that most of the estuarine fish fauna of the subtropics and tropics is not 'estuarine' *per se*, but is a fauna characteristic of shallow, turbid areas, often with variable salinity. Areas with such characteristics occur over very large areas of the sea in South and South East Asia, West Africa and northern South America, but are confined mainly to estuaries in southern Africa and in parts of the USA and Australia. Many of the taxa considered largely estuarine dependent in, for example, South Africa, such as *Gerres*, *Thryssa* and *Elops*, are common in estuarine coastal waters throughout South and South East Asia.

The situation in the tropics is, however, often complex, and the very diversity of the estuarine environments and the flexibility of many of the species (together also with difficulties of defining the limits of estuarine waters) confounds any attempt to generalize or package species into those that are estuarine dependent and those that are not. For example, in the Gulf of Carpentaria in northern Australia, offshore sampling in Albatross Bay and concurrent sampling in the Embley estuary (Blaber *et al.*, 1989,

1990b) showed that 27 species (17 of which are Gobiidae) were truly estuarine, juveniles of 14 offshore species lived only in the estuary, and juveniles of 24 estuarine and shallow marine species were exclusively found in the estuary. Thus of a total of 197 species in the Embley estuary, 15% are estuarine while another 19% live as juveniles only in estuaries. These two groups, which form approximately one-third of the species, can be considered estuarine dependent. Juveniles of 24 species that live in both the estuary and sea may be termed 'estuarine opportunists' (Lenanton and Potter, 1987). A subsequent study of the shallow inshore waters of Albatross Bay (Blaber *et al.*, 1995a) revealed further complexity, and showed that the inshore zone also acts as a nursery for some species. It is the first nursery ground for some species the adults of which live mainly offshore, such as the carangids *Caranx sexfasciatus* and *Scomberoides tol*, the sciaenids *Johnius amblycephalus* and *Otolithes ruber*, and the ephippid *Drepane punctata*. It is also inhabited by larger juveniles of many species the fry of which live in the estuary, such as *Ambassis nalua*, the hemiramphids *Arrhamphus sclerolepis* and *Hyporhamphus quoyi*, and the mugilids *Liza subviridis*, *Valamugil buchanani* and *V. cunnesius*. The adults of these species are generally abundant in both inshore and estuarine waters. And finally, the inshore zone also supports juveniles of leiognathids and clupeids that are also common as juveniles in the estuary and offshore. So the overall juvenile component of the inshore coastal waters fauna consists of fishes apparently restricted to the inshore habitat, larger juveniles that have emigrated from the estuary, and fish that occur in all three habitats. Hence we return again to the problem of defining the limits of 'estuarine' and these limits are to some extent subjective (Blaber, 1991).

In southern Africa, Whitfield (1994a) divided the 142 estuarine fishes into five categories in relation to estuarine dependence: estuarine species (species that breed in estuaries, 28%), euryhaline marine species (species that breed in the sea and their juveniles occur in estuaries, 43%), marine species (species that occur in estuaries in small numbers as occasional visitors, 21%), euryhaline freshwater species (these also may breed in estuaries, 5%), and obligate catadromous species (species that use estuaries in transit between the sea and fresh water, 3%). Therefore according to this classification, 71% of the species are completely or partially dependent on estuaries. It is important to note, however, that this classification includes a mixture of temperate and warm temperate species as well as those of subtropical south-east Africa – many of the subtropical species included in this classification are at the southern limits of their distribution, and as discussed earlier, are not strictly estuarine dependent in the tropics.

There is no doubt that the degree of estuarine dependence in some families is greater than in others. For example, the juveniles of most species of Mugilidae are found only in sheltered estuaries and few occur in coastal waters (Blaber, 1987), and most members of the family must be considered estuarine dependent. A number of the tropical clupeoids are also primarily estuarine dependent as juveniles: *Ethmalosa* in West Africa and *Brevoortia*, *Anchoa* and *Cetengraulis* in the West Atlantic. Others such as *Tenualosa* in Asia are dependent on estuaries as spawning sites. In contrast, juvenile Sciaenidae are widespread throughout coastal waters and estuaries in both the tropical East Atlantic and Indo–West Pacific, and although often abundant in estuaries, few species are strictly estuarine dependent (Longhurst and Pauly, 1987; Sasekumar *et al.*, 1994a). It is possible, as suggested by Longhurst and Pauly (1987), that some of the tropical carangids are estuarine dependent. However, although the juveniles of species such as *Caranx sexfasciatus* and *C. ignobilis* are found mainly in estuaries (Blaber and Cyrus, 1983), others, such as *Caranx bucculentus* (Brewer *et al.*, 1994) and most of the genus *Carangoides*, are not.

The majority of species found in tropical estuaries are probably not estuarine dependent, but the question of total estuarine dependence may be relevant only in estuaries outside the 'core areas' (section 7.3). It has less relevance where there is extensive estuarization (*sensu* Longhurst and Pauly, 1987) of the continental shelf, where most species remain in similar (large and widespread) habitats for their whole life. Many more species spawn in tropical estuarine areas, including particularly the Sciaenidae, Engraulididae and Clupeidae, than they do in temperate or warm temperate estuaries.

Notwithstanding this, however, the *shallow* estuarine waters found either in the sheltered waters of open estuaries or in the mangrove habitats of estuaries and estuarine coastal waters of the subtropics and tropics, are vital nursery grounds and important habitats for the feeding or reproduction of a number of species.

## 7.3   EVOLUTIONARY PERSPECTIVES

The region with the greatest diversity of coastal fish species is the tropical Indo–West Pacific (Nelson, 1984), where vast areas of estuarine coastal waters in South and South East Asia, south to parts of northern Australia, form what can be termed the 'core area' for the majority of Indo–West Pacific estuarine species. Briggs (1974) proposed that the tropical Indo–Malayan region was the centre of speciation among Indo–West Pacific shallow-water fishes, with its unparalleled variety of habitats, from coral reefs to seagrasses to mangroves and estuaries. Whether the greater number of species in this region is due to increased speciation rates or reduced extinction rates remains a subject of controversy (Briggs, 1974; Woodland, 1990). Nevertheless, the similarity of tropical estuarine fish faunas across the great breadth of the Indo–West Pacific is striking. For example, at a species level, two-thirds of the species in the St Lucia coastal lake system of KwaZulu-Natal, South Africa, and 71% of species in Madagascar estuaries occur in South East Asia. Almost all species in the estuaries of northern Australia are also common throughout South and South East Asia. For most families, there is also a gradual reduction in species diversity away from the core area, particularly where physical conditions change and mangrove habitats are not dominant. Such marginal areas that are restricted to the relatively small areas of estuaries occur mainly west of India and east and south of New Guinea. This is especially the case along south-east African, east Australian and South Pacific shores where shallow, turbid and variable salinity conditions give way to coral reefs and clearer waters.

The restriction of this habitat to estuaries is particularly marked along the south-east African coast where the shelf is narrow and there is a high-wave-energy coastline. Here, although the Moçambique Current transports many tropical species southwards, the species diversity declines rapidly as winter water temperatures fall below 15 °C south of KwaZulu-Natal (Whitfield, 1994a), and the only shallow, quiet water habitats are within sheltered estuaries. The decline southwards of this habitat is also reflected in the lack of extensive mangrove forests, so characteristic a part of the estuaries and estuarine coastal waters of the core area. The estuaries of south-east Africa are dominated by juveniles that do not occur in any numbers in the sea, leading to the supposition that the populations may be estuarine dependent. Spawning of most of these species takes place in the sea (Whitfield, 1994a) and ichthyoplankton research shows that large numbers of larvae occur in the coastal zone prior to migrating into the estuaries (Harris and Cyrus, 1995; Harris *et al.*, 1995). What is not known, however, is the extent to which recruitment is dependent upon larvae that have been advected southwards by the Moçambique Current, and the quantitative significance of this recruitment in relation to local spawning. Some of the most abundant families have long larval durations (e.g. about 40 days for gobiids; Thresher, 1984); sufficient to allow transport from regions far to the north. Hence, although the juveniles in the sub-tropical estuaries of south-east Africa may be locally estuarine dependent, the overall maintenance and fluctuations of population size may not be dependent on their survival to spawning age. A somewhat analogous situation exists on the east Australian coast. Here the contrasts between sea and estuary are less well defined due to the larger amounts of shallow sheltered waters protected from the ocean by the Great Barrier Reef, or offshore islands in southern Queensland, and the greater size of many of the estuaries and the larger areas of mangroves. The main contrasts are in terms of variable salinity and higher turbidities in the estuaries relative to the coastal waters. Nevertheless, as in the south-east African situation, there is a southward attenuation of tropical estuarine species and characteristic 'core area' habitat.

Whitfield (1994b) has postulated that few new species have evolved in south-east African estuaries because these systems are unpredictable environments that lack any degree of permanence. This, together with the seasonal and interannual variability of their abiotic environment, and because the lifetime of individual estuaries is only of the order of a few thousand years, has prevented much speciation and led to a dominance of estuaries by ecologically flexible marine migrant species with a wide geographic distribution. Whitfield (1994b) further suggested that estuarine fish communities on a global basis may mirror the southern African situation, with domination by ecologically flexible taxa with a low speciation potential. While this may be true in the Indo–West Pacific for southern African, and perhaps southern Australian, estuaries that are marginal habitats for much of the tropical Indo–West Pacific estuarine fish fauna, it is hardly the case across the vast tropical sweep of northern Australia, South East and South Asia (the 'core area'), where extensive species radiation has occurred in many estuarine families, notably the Ariidae, Clupeidae, Engraulididae and Sciaenidae. What is apparent is the reduction in diversity as the estuarine habitats become marginal and perhaps less predictable. Hence only the most flexible of the tropical core area species survive in southern African and east Australian estuaries. The degree of endemism also increases south of the subtropics in both Africa and Australia (Ekman, 1953; Poore *et al.*, 1994; Whitfield, 1994a).

The species-rich families that show the greatest reduction in diversity away from the Indo–West Pacific core area include the Ariidae, Sciaenidae and Engraulididae (Table 7.1). Significantly, some of these are families without a pelagic larval phase (Ariidae) or are ones with short larval durations (Sciaenidae), suggesting that these limitations to dispersal may prevent their colonizing estuaries far from their tropical centre of distribution.

In the tropical East Atlantic, the tropical estuarine environment is restricted to the West African region and the species composition across the region is relatively uniform. Here there is a complex interrelationship between depth, estuaries and species, with a gradation of closely related species. For example, in the Sciaenidae, *Pseudotolithus typus* and *P. elongatus* are mainly estuarine while *P. senegalensis* and *P. brachygnatus* are coastal (Baran, 1995).

**Table 7.1** Approximate numbers of species in estuaries of three regions of the Indo–West Pacific, showing attenuation of subtropical and tropical estuarine species away from the estuarine waters of the Indo–Malayan 'core' area (from northern Australia to India). Data from Munro (1967); Fischer and Whitehead (1974); Menon (1977); Trewavas (1977); Fischer and Bianchi (1984); Nelson (1984); Smith and Heemstra (1986); Whitehead *et al.* (1988) and Kailola (1990)

| Family | East Australia | Indo–Malayan 'core' area | East Africa |
|---|---|---|---|
| Ariidae | 6 | > 50 | 2 |
| Carangidae | 19 | 26 | 17 |
| Clupeidae | 11 | 49 | 10 |
| Cynoglossidae | 9 | 22 | 13 |
| Engraulididae | 10 | 49 | 5 |
| Leiognathidae | 12 | 21 | 5 |
| Sciaenidae | 12 | 60 | 5 |

## 7.4   ADVANTAGES AND DISADVANTAGES OF AN ESTUARINE PHASE IN THE LIFE CYCLE

### Costs and benefits

Estuarine dependence among tropical marine teleosts is not confined to particular families or even genera. In many closely related species, the adults are sympatric offshore but only some of their juveniles are estuarine dependent. For example, juvenile *Gerres filamentosus* and *G. oyena* are estuarine-dependent whereas *Gerres subfasciatus* is not (Cyrus and Blaber, 1983; Blaber *et al.*, 1989); juveniles of *Leiognathus equulus* and *L. splendens* are found only in estuaries, whereas juveniles of many other *Leiognathus* species are abundant offshore. The reasons for these differences frequently relate to feeding requirements and physical preferences (Cyrus and Blaber, 1983, 1987a). Miller *et al.* (1985) viewed the movement of juveniles into estuaries in the south-eastern USA in terms of costs and benefits, with unpredictable environmental conditions as costs and with reduction in predation (but see pages 223–228), more food and shelter as benefits. However, they caution that until the predator-avoidance part of this theory is properly quantified, the overall advantages of this life history strategy are difficult to determine. Nevertheless, because a large number of species in the south-eastern USA follow the 'juveniles in estuaries and adults in sea' model, it must be assumed that the benefits outweigh the costs, and that survival is enhanced through this strategy.

In regions where there is a very definite boundary between estuary and sea, and where estuaries are unpredictable and often isolated, such as in erratic rainfall areas of South Africa, west Australia, north-west Africa and western Central America, spawning in the sea, as opposed to within estuaries, may be advantageous for dispersal and may ensure that at least some larvae reach favourable habitats.

However, why some species are dependent on estuarine habitats remains to be clarified. Three broad hypotheses have been advanced to explain the high densities of fish and the dependence of certain species on tropical estuaries. They relate to (1) reduced predation linked with turbidity and depth, (2) increased food supply for postlarvae and juveniles, and (3) shelter for postlarvae and juveniles. They are not mutually exclusive and in many instances are inextricably connected.

### Predation

It has been suggested that predation on juvenile fishes is less in estuaries, firstly because, their turbid waters reduce the effectiveness of large visual fish predators (Blaber and Blaber, 1980; Cyrus and Blaber, 1987a), secondly, because shallow waters exclude large fishes, and thirdly, because structure such as seagrass or mangroves enables small fishes to hide from predators. Evidence supporting the turbidity hypothesis comes from comparisons of fish densities across a variety of coastal habitats which show that the abundances of certain species respond positively to increases in turbidity. Further evidence comes from observations of greater abundances of piscivorous species in the creeks of the Dampier mangroves of north-west Australia that receive no run-off from the land, and thus have much clearer waters (Blaber *et al.*, 1985). Evidence supporting this hypothesis also comes from estuarine systems in which there is little or no mangrove habitat (e.g. the St Lucia system in South Africa) (Cyrus and Blaber, 1987a). Thus, if fish are responding solely to turbidity, it is possible that mangrove vegetation has little effect on the dependence of fish on estuaries.

Habitat structure may frustrate the predatory ambitions of some larger fishes, but others enter mangroves and seagrassbeds in search of food, especially at night. In the large seagrass areas around Groote Eylandt in the Gulf of Carpentaria, Australia, an array of larger species (e.g. sharks and carangids) move into shallower waters, particularly at night. The majority of these are more abundant in tall seagrass than in short seagrass, and more abundant in short seagrass than in open areas. These movements are related to feeding on the smaller resident and juvenile fishes that show the same patterns of abundance (Blaber *et al.*, 1992). Studying the fishes of small tropical seagrass beds adjacent to coral reefs in the Caribbean, several workers (Weinstein and Heck, 1979; Robblee and Zieman, 1984; Baelde, 1990) also noted diel movements related to patterns of feeding activity and suggest that such seagrass beds are important feeding grounds for predatory fishes. However, there is much less evidence for large numbers of piscivores entering mangrove forests at high tide. In the Embley estuary in the Gulf of Carpentaria, of the more than 1000 fish, weighing over 600 kg, trapped in an area of only 9000 m$^2$ over one high tide period, very few were piscivores; the majority consisted of benthic invertebrate feeders such as dasyatids, gerreids and ariids (Blaber *et al.*, 1989). On the tropical east coast of Australia, low numbers of large piscivores were also recorded in the mangroves of the Alligator and McIvor estuaries by Robertson and Duke (1987) and in the Trinity estuary by Blaber (1980). There is thus evidence that the structure provided by the prop roots, pneumatophores and tree debris, as well as invertebrate burrows, in mangrove forests gives small fish some measure of protection from predation by larger fish. However, in the deeper channels of larger mangrove estuaries, especially in the lower reaches, large piscivorous fishes such as carangids and sharks may be abundant (Blaber *et al.*, 1989). Juvenile and subadult sharks are very abundant piscivores of the coastal waters and lower reaches of estuaries of the Gulf of Carpentaria. Here nine species were found to feed in a density-dependent manner, mainly on small fishes: in the estuaries, Clupeidae and Mugilidae were the most common prey, while in coastal waters, Clupeidae, Leiognathidae and Haemulidae constituted much of the diet (Salini *et al.*, 1992). There seems little doubt that the large numbers of juvenile fishes in larger estuaries and coastal waters are attractive prey for such piscivores. Indeed it may not be only the density of the juveniles that is attractive, but their size and inexperience: it has recently been postulated for freshwater fishes that because prey fish rapidly develop effective escape behaviours, piscivore capture success is often a strong function of prey size and development stage (Juanes, 1994).

Juvenile fish can certainly avoid many of the larger fish predators by moving into very shallow waters, as described for juvenile *Mugil cephalus* in Negombo Lagoon in Sri Lanka (De Silva and Silva, 1979) and for some species in St Lucia in South Africa (Whitfield and Blaber, 1978e). Reduced predation on species that frequent shallow littoral areas, especially where there are beds of aquatic macrophytes, is illustrated by examining the prey selectivity of the suite of piscivorous fishes in St Lucia. Most species are eaten in a density-dependent manner, but juvenile *Oreochromis mossambicus*, *Rhabdosargus holubi*, *Leiognathus equulus* and *Terapon jarbua*, which live mainly in shallow, or shallow and vegetated, areas, suffer relatively little mortality from fish predation (Fig. 7.1).

However, the strategy of moving into shallow water may make fishes

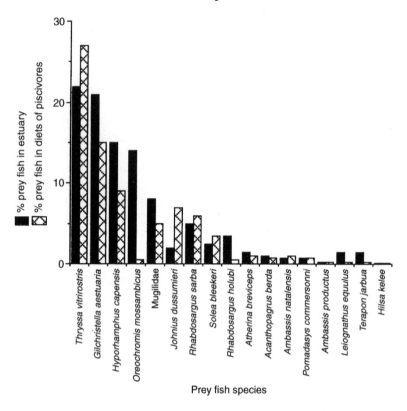

**Fig. 7.1** Relative proportions of prey fish species in the diets of a suite of pisci-vorous fishes in Lake St Lucia, KwaZulu-Natal, South Africa, compared with their relative abundance in the lake (modified from Whitfield and Blaber, 1978a).

vulnerable to predation by piscivorous birds. Other reptile and mammalian (non-human) predators are also a danger in most tropical estuaries. Levels of bird predation on estuarine fishes vary both regionally and in relation to the physical characteristics of individual estuaries. In a detailed study in the St Lucia system of South Africa Whitfield and Blaber (1978a, 1979b,c) divided the avian predators of fish into three groups according to their way of hunting: diving birds, wading birds and swimming birds.

### Diving birds

The depth at which various species of fish swim (Fig. 7.2) can be correlated at least partly with their vulnerability to predation by birds that

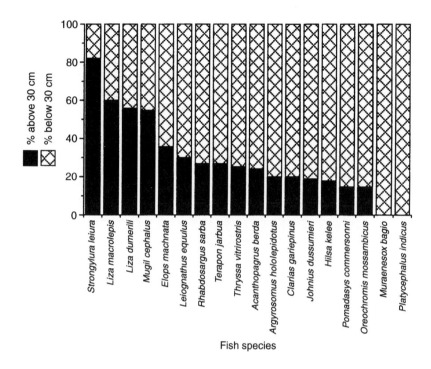

**Fig. 7.2** Swimming depths of 18 species of fish in Lake St Lucia, KwaZulu-Natal, South Africa (modified from Whitfield and Blaber, 1978c).

dive onto their prey. The fish eagle, *Haliaeetus vocifer*, is common at St Lucia (approximately 100 birds) and preys predominantly on fishes that are pelagic or enter shallow littoral areas. Mugilidae are the most important prey in St Lucia because of their abundance and habit of swimming in surface waters. Other pelagic species taken include the belonid *Strongylura leiura*. Not all pelagic species, however, are equally vulnerable. The abundant engraulid *Thryssa vitrirostris* is not caught, perhaps because it is too small to detect and difficult for the eagle to grasp in its talons. Other surface dwellers such as *Elops machnata* are also not taken, probably because most individuals of this species are too large. Generally, fishes less than 1 kg are selected because they are easier to transport. Benthic species that form part of the diet of fish eagles include *Acanthopagrus berda*, *Oreochromis mossambicus*, *Clarias gariepinus* and *Pomadasys commersonni* – all species that frequent shallow littoral areas, at which time they become visible and vulnerable to fish eagles. The breeding season of the fish eagle (when food requirements are greater) occurs in the dry season when lake levels are lower, wind velocities less and turbidities low, all factors combining to make fishes more easily detected and caught. The Caspian tern, *Hydroprogne caspia*, also dives for its food, which consists of smaller pelagic species such as *Thryssa vitrirostris*, *Hyporhamphus capensis*, *Gilchristella aestuaria* and smaller Mugilidae. Like the fish eagle it also takes some benthic species that move into shallow waters during the day. The third common diving predator at St Lucia is the pied kingfisher, *Ceryle rudis*. This species has a similar diet to that of the Caspian tern but the prey fishes are usually smaller. Much of its food consists of juvenile *Oreochromis mossambicus* that frequent the shallows during the day. Therefore most sizes of fishes, except the very largest, are vulnerable to predation by these three diving birds.

## Wading birds

The fishing behaviour of four species of herons at St Lucia is related to the depth at which they are able to forage and the presence or absence of aquatic macrophytes. Thus their leg (tarsometatarsal) length determines the maximum depth at which they can fish (Table 7.2). All species select fish that swim in surface waters and are abundant in the littoral zone. The prey of all the herons is similar although the little egret, *Egretta garzetta*, feeds on invertebrates as well as considerable numbers of the goby *Glossogobius giuris* and small juveniles of *Oreochromis mossambicus*. The great white egret, *Egretta alba*, the grey heron, *Ardea cinerea*, and the Goliath heron, *Ardea goliath*, all prey predominantly on *Oreochromis mossambicus* and Mugilidae but select different size groups (Table 7.2).

## Swimming birds

The abundant swimming bird predators at St Lucia are the reed cormorant, *Phalacrocorax africanus*, white-breasted cormorant, *P. carbo* and the white pelican, *Pelecanus onocrotalus*. Reed cormorants usually feed in littoral waters less than 1 m deep, remaining submerged for up to 14 s. They catch mainly small fishes (1–15 g, with a peak in the 3–4 g size class) such as *Solea bleekeri*, *Oreochromis mossambicus*, *Acanthopagrus berda*, *Johnius dussumieri* and *Terapon jarbua*. The larger white-breasted cormorant eats larger fishes further offshore and can reach any part of the bottom of St Lucia (2.5 m), where it preys mainly on *Thryssa vitrirostris*, *Johnius dussumieri*, *Rhabdosargus sarba* and *Pomadasys commersonni* ranging in weight from 1 to 214 g with a peak in the 10–20 g size class. White pelicans eat a wide variety of fish species with weights up to 2 kg. These pelicans congregate in thousands near the mouth of the estuary

**Table 7.2** Foraging zonation of the little egret, great white egret, grey heron and goliath heron at St Lucia and the main prey species, prey weight range and modal weight

| Species | Tarso meta tarsal length (mm) | Mean wading depth (mm) | Fish prey species (% frequency) | Weight range (g) | Modal weight (g) |
|---|---|---|---|---|---|
| Little egret | 113 | 100 | *Glossogobius giuris* (18)<br>*Oreochromis mossambicus* (14)<br>Other Gobiidae (7) | 1–14 | 1 |
| Great white egret | 149 | 160 | *Oreochromis mossambicus* (56)<br>Mugilidae (19)<br>*Acanthopagrus berda* (13)<br>*Terapon jarbua* (13) | 1–45 | 8 |
| Grey heron | 152 | 190 | *Oreochromis mossambicus* (37)<br>Mugilidae (31)<br>*Acanthopagrus berda* (16)<br>*Thryssa vitrirostris* (11) | 1–110 | 20 |
| Goliath heron | 231 | 325 | Mugilidae (33)<br>*Oreochromis mossambicus* (29)<br>*Acanthopagrus berda* (24)<br>*Pomadasys commersonni* (5) | 11–297 | 55 |

during the annual spawning migration of *Mugil cephalus*, at which time vast numbers of these mullet are consumed.

From the above it is evident that bird predation has a significant impact on fish numbers in the St Lucia coastal lake system. However, the extent to which this phenomenon can be extrapolated varies. As with piscivorous predation, the depth and turbidity of the water affect which species of birds can prey on fishes in estuaries. The more turbid the water, the less effective are all the visual hunting methods of piscivorous birds. In small, blind estuaries, birds may have a significant impact on fish numbers (Blaber, 1973), but in large and highly turbid open estuaries with high flow rates, such as those of Borneo and Sumatra, avian predators are relatively few in number. Their effects are probably greatest in the coastal lake environments of Africa, India and South America and in the clearer-water estuaries of parts of Australia. Large deltaic mangrove areas also usually have high numbers of piscivorous birds, for example the Sundar-bans of India (Mukherjee, 1971; Sanyal, 1992) and the Orinoco delta of Venezuela.

Other significant vertebrate predators of fish in tropical estuaries are crocodiles and monitor lizards. The Nile crocodile, *Crocodylus niloticus*, is common in many African estuaries with the St Lucia system having probably the highest number (500–620 individuals of > 1–5 m, Pooley, 1982) in any one estuarine area. The average daily intake of fish for a 2.5 m crocodile is 1.1 kg (Whitfield and Blaber, 1979d), hence the amount of fish eaten in St Lucia is obviously very great. This predation covers a wide range of fish sizes, with small crocodiles eating mainly small fish and larger crocodiles taking mostly adult fish. Their impact upon migrating schools of *Mugil cephalus* near the mouth of St Lucia was estimated to be 3.9 tonnes over one month in 1979 (Whitfield and Blaber, 1979d). The estuarine crocodile, *Crocodylus porosus*, of Asia and Australia and the American crocodile, *Crocodylus acutus*, which ranges from Mexico to South America, are both mainly estuarine in distribution (Guggisberg, 1972; Johnson, 1973). Their diets are probably mainly fish but comprehensive data are lacking.

The monitor lizards *Varanus indicus* and *V. salvator* range from northern Australia through New Guinea and South East Asia and live mainly in mangrove and estuarine areas. Their diet is thought to be largely fish and Crustacea (Cogger, 1988).

A number of tropical mammals are known to include estuarine fishes in their diets but their overall impact on fish populations is probably negligible. In Africa the water mongoose, *Atilax paludinosus*, has a diet consisting of 16% fish at St Lucia (Whitfield and Blaber, 1979e). In South East Asia several otter species live in coastal regions and estuaries, and include fish in their diet. The flat-headed cat, *Felis planiceps*, a fish-eater (Medway, 1978), occurs occasionally in mangroves, and in the Sundarbans of India even tigers include fish in a very mixed diet (Sanyal, 1992).

## Food supply

Both the quantity and types of food may differ between estuaries and adjacent waters. Some of the various specializations and ontogenetic changes in feeding ecology shown by estuarine fishes have been described in Chapter 5. The fact that many of the foods available in sheltered estuarine waters are rare or absent in offshore waters may enhance the importance of estuaries to juvenile fishes. This is particularly the case with regard to detritus and the microfauna and flora associated with it, as well as aquatic macrophytes and their epifauna and flora. The presence of mangroves in tropical estuaries increases the diversity and quantity of food available to juvenile fishes (Blaber, 1980, 1987), and Robertson and Duke (1987, 1990) have shown highly significant differences in the densities of juvenile fishes between mangrove and other nearshore habitats that are immediately adjacent to each other. Furthermore, densities of zooplankton in mangrove habitats are greater (by an order of magnitude) during the late-dry-to-mid-wet-season recruitment period of fishes, than in the middle of the dry season. Most newly recruited fish in estuarine or mangrove habitats are zooplanktivores. Robertson *et al.* (1992) showed that during the major recruitment period (wet season) of juvenile fishes into north-east Australian estuaries, crab zoae were dominant in the diets of many juvenile fishes, and that these zoae are two orders of magnitude more abundant in the mangrove-lined estuaries than in adjacent coastal waters. In addition, these zoae are the larvae of the mangrove-leaf-eating crabs (Sesarminae) abundant in the mangrove forests. Hence there is a strong trophic connection between the feeding ecology of the juvenile fishes and the mangroves of the estuary. Chong *et al.*, (1990) postulate that in West Malaysia, coastal mangroves function more importantly as feeding grounds than as nursery grounds for juvenile fishes, and showed that mangroves are used during high tides by at least 34 species of fish. Adults and juveniles of numerous species of both the Ariidae and Sciaenidae feed on the very abundant prawns, crabs and polychaetes in the Matang system of West Malaysia (Singh and Sasekumar, 1994; Yap *et al.*, 1994). Subadults of fishes in northern Australian estuaries, such as *Lates calcarifer* which feed mainly on penaeid prawns, and *Lutjanus argentimaculatus* which feed mainly on sesarmid and grapsid crabs, gastropods and alpheid shrimps (Salini *et al.*, 1990), are consuming taxa that have direct trophic links with the mangrove habitat. The biomass and production of invertebrate benthos in tropical estuaries, both infauna in the substrata and epifauna on the mangroves, is highly variable (Alongi and Sasekumar, 1992). However, the numbers and biomass of mangrove epifauna – barnacles, gastropods, bivalves, hydroids, sponges, bryozoans, tunicates and particularly brachyurans – may be an order of magnitude greater than that in the adjacent sea (Sutherland, 1980; Day and Grindley, 1981; Day *et al.*, 1989; Alongi and Sasekumar, 1992). In West Malaysian estuaries, Sasekumar *et al.* (1984) found that the major prey of the suite of fishes within the mangrove forests consists of crabs, encrusting epifauna, meiofauna, sipunculids and detrital aggregate; but in open-water areas, natant prawns, polychaetes and fish are the most common prey. In a comprehensive study of the foods of fishes of the estuarine area of West Africa, Longhurst (1957, 1960) found that although most benthic feeding fishes were relatively unspecialized and ate a wide variety of invertebrates, those with empty stomachs were much commoner offshore than in estuaries. The density of suitable foods, such as polychaetes, small Crustacea and molluscs, was thought to be higher in the estuaries.

There is therefore considerable evidence for this second hypothesis that there is a greater supply of food for both adult and juvenile fish in estuaries than in adjacent habitats. Many of the prey types are found only within estuaries and may have obligate links with specific habitats such as mangroves, seagrasses or detritus-rich substrata. Also, any estuarine food dependence by the fishes may relate not only to the larger quantity of food but to its greater diversity and the availability of wider size ranges than occur in adjacent waters. The wider size range is particularly important because during their relatively rapid growth, the postlarvae and juveniles of most species must go through sequential ontogenetic changes in the size of food taken.

### Shelter

The third hypothesis concerns the structural complexity provided by the prop roots, pneumatophores and fallen logs and branches in mangrove forests, and the dense habitats provided by some seagrasses and fringing reed beds. These estuarine habitats are also likely to play a role in determining the dependency of some juvenile fish on the estuarine habitat (Thayer *et al.*, 1987; Robertson and Duke, 1990). The effects of differing structure are illustrated by reference again to the Solomon Islands mangroves (Chapter 4). The physical characteristics of the Solomon Islands estuaries are similar to those of other tropical Indo–West Pacific estuaries in terms of their salinity, temperature and turbidity regimes (Blaber, 1980; Blaber *et al.*, 1989), but there are two different types of mangrove estuary in Solomon Islands, based on substrata and mangrove tree species, each with different fish faunas. The hard-substratum estuary with an abundance of mangrove tree debris is inhabited mainly by species that apparently need the cover or structure provided by the debris. Pomacentridae and some species of Apogonidae and Gobiidae, together with juvenile lutjanids and serranids, predominate. These species are largely absent from the soft-substratum estuaries that have little debris cover. Data from elsewhere on debris-strewn mangrove estuaries with different substrata are not available, so direct comparison with the Solomon Islands situation is not possible.

### 7.5   CONCLUSIONS

The above three hypotheses which refer basically to predator avoidance, food and shelter, are each probably important to some degree. However, their relative significance, in terms of dependence and use of tropical and subtropical estuaries by fishes, will vary depending upon the fishes in question and the nature of each estuary. There is no worldwide generalization applicable to the question of estuarine dependence. There are perhaps specific regions, mainly on the margins of the tropics, where a general estuarine dependence among the euryhaline marine fish community has validity, but throughout the tropics, the question should be examined critically on an estuary-by-estuary, and species-by-species basis, if it is to be of any value for conservation or fisheries management.

# Chapter eight

# Estuarine fisheries

## 8.1 INTRODUCTION

The fishes of most subtropical and tropical estuaries are a valuable and usually readily accessible human food resource and have formed the basis for the development of a great variety of fisheries. Most tropical estuaries are in developing countries and such fisheries often constitute the main source of both food and income for people living along their shores. In this chapter we examine the productivity of major estuarine fisheries and how their catch levels compare with those of other areas and latitudes. We look also at the types of fisheries found in tropical estuaries, how they may interact with one another, the state of the fish stocks and the extent to which people are dependent on the fisheries. Finally we look at how and whether various tropical fisheries are managed. Because this book is about fishes, the emphasis in this chapter is on those fisheries that take mainly fish, and only passing references can be made to those involving other living resources such as crustaceans and molluscs. This chapter is also not about fisheries science – although much of the content has been generated by this science – and the reader is referred to the many excellent fisheries science texts such as Pitcher and Hart (1983), Beverton and Holt (1993), Cushing (1975) and Ricker (1958). See also Pauly (1994) for some of the important differences between tropical and temperate fisheries.

Although it is impossible to be too precise due to the problems delimiting estuaries (sections 2.1, 2.6.), the geographical coverage is restricted to those fisheries within estuaries and in immediately adjacent estuarine coastal waters. It does not include some of the larger offshore shelf fisheries such as those of the Gulf of Thailand, Gulf of Mexico and West Africa, although it must be recognized that many of these are contiguous with more inshore fisheries and may involve the same fishers undertaking seasonal, usually climate-driven, changes in their activity. Details of most of the important tropical shelf fisheries can be found in Longhurst and Pauly (1987), Lowe-McConnell (1987) and Pauly and Murphy (1982).

As most subtropical and tropical estuarine fisheries are in developing countries, the types of fisheries and methods of fishing may be markedly different from those in more developed or industrialized countries. Also, in developing countries the range of species retained by fishers is much greater. For example, most of the small species that are discarded as unsaleable bycatch by trawlers in northern Australia are commercially valuable in similar fisheries in the different cultures of most of South East Asia.

The types of fisheries can be divided into three main sectors (Harden Jones, 1994; Rawlinson *et al.*, 1995):

- subsistence – where the fishers predominantly consume all of their catch or give it away, but do not sell it;
- artisanal – where the fishers sell part of their catch, but also retain part for their own consumption;
- commercial – where the fishers sell all of their catch.

Within the subsistence and artisanal sectors are included 'traditional fisheries'. Many of these have a very long history and form part of the culture of human estuarine communities. They may also have a long-standing and complex interrelationship with the environment, and are increasingly coming to be regarded as part of the overall ecology of the tropical estuarine environment ('The Kuala Lumpur Statement', Asian Wetland News, 1995).

An additional category exists in the few developed countries that have subtropical and tropical estuaries, such as Australia, South Africa and the USA:

- recreational – where fishing is carried out as a sport or leisure pastime and not primarily for producing food or income. Nevertheless, the service infrastructure associated with recreational fishing usually encompasses economically important income-generating activities.

In developing countries, there are often resource conflicts between subsistence/artisanal fisheries and commercial fisheries, and in developed countries between recreational and commercial fisheries. Both situations require important resource allocation decisions from whoever manages the resources.

## 8.2   PRODUCTION AND PRODUCTIVITY

The fish yields in terms of tonnes landed per $km^2$ per year of a selection of subtropical and tropical estuaries are shown in Table 8.1. The reported values are mainly from larger systems because these are the ones that

**Table 8.1** Production (tonnes km$^{-2}$ year$^{-1}$) of fish in selected tropical estuaries (modified partly from Marten and Polovina, 1982)

| Country | Estuary | Type | Production (t km$^{-2}$ year$^{-1}$) | Reference |
|---|---|---|---|---|
| India | Lake Chilka | Coastal lake | 3.7 | Jhingran and Natarajan (1969) |
| | Lake Pulicat | Coastal lake | 2.6 | Jhingran and Gopalakrishnan (1973) |
| | Mandapam Lagoon | Coastal lake | 5.6 | Tampi (1959) |
| | Hooghly–Matlah | Open estuary system | 11.4 | Jhingran (1991) |
| | Vellar–Coloroon | Open estuary system | 11.1 | Venkatesan (1969) |
| Malaysia | Larut–Matang | Open estuary system | 38.6* | Choy (1993) |
| Philippines | San Miguel Bay | Estuarine coastal waters | 23.8† | Mines et al. (1986) |
| Ivory Coast | Ébrié Lagoon | Coastal lake | 16.0 | Durand et al. (1978) |
| South Africa | Kosi system | Coastal lakes, open estuary | 1.0 | Kyle (1988) |
| Ghana | Sakumo Lagoon | Coastal lake | 15.0 | Pauly (1976) |
| Malagasy | Pangalanes Lagoon | Coastal lake | 3.7 | Laserre (1979) |
| USA | Texas bays | Estuarine coastal waters | 12.1 | Jones et al. (1963) |
| Colombia | Cienaga Grande | Coastal lake | 12.0 | INDERENA (1974) |
| Mexico | Caimanero Lagoon | Coastal lake | 34.5 | Warburton (1979) |
| | Terminós Lagoon | Coastal lake | 20.0 | Yáñez-Arancibia and Lara Dominguez (1983) |
| Venezuela | Tamiahua Lagoon | Coastal lake | 4.7 | Garcia (1975) |
| | Lake Maracaibo | Coastal lake | 1.9 | Nemoto (1971) |
| | Tacarigua Lagoon | Coastal lake | 11.0 | Gamboa et al. (1971) |
| El Salvador | Jiquilisco | Estuarine coastal waters | 1.7 | Hernandez and Calderon (1974), Philips (1981) |

* Includes penaeid prawn catches and not coastal waters.
† Probably an overestimate as trawlable biomass is only 2.13 tonnes km$^{-2}$.

support significant fisheries and for which data exist. The tonnages are based on total catches in relation to estuarine area and in most cases do not reflect sustainable yields. The values range from 1 to 38 tonnes $km^{-2}$ $year^{-1}$ and are generally higher than those for tropical rivers and lakes, but the range is similar to that reported for tropical continental shelves and coral reefs (Marten and Polovina, 1982; Lowe-McConnell, 1987). For temperate waters as a whole, Lowe-McConnell (1975) quotes a range of 7.3 to 71.3 tonnes $km^{-2}$ $year^{-1}$ and Saila (1975) provides figures to show that annual yields of fishes in estuaries and coastal lagoons, produced without supplementary feeding, are reasonably similar over a wide geographic range. However, production in most temperate tidal estuaries, as opposed to shallower coastal lagoons, is relatively low. For example, production in the Forth estuary, Scotland, was estimated at 4 tonnes $km^{-2}$ $year^{-1}$ (Elliott and Taylor, 1989).

Much of the variability in reported yields may be due to different accuracies in the measuring of catches, and the often poor quality of original statistics caused by inadequate catch sampling, frequently exacerbated by over-extrapolation. However, the biomass of fish in different tropical estuaries also varies greatly, ranging from 0.4 to at least 70 tonnes $km^{-2}$, with most falling in the range 7 to 29 tonnes $km^{-2}$ (Chapter 3; Robertson and Blaber, 1992). Therefore at least some of the variability is due to the different productivities of estuarine systems. The relatively high yield of 34.5 tonnes $km^{-2}$ $year^{-1}$ for Caimanero Lagoon in Mexico consists primarily of mullet species, of which *Mugil curema* contributes about 10 tonnes $km^{-2}$ $year^{-1}$. These figures compare favourably with typical aquaculture-based fish yields from tidally stocked enclosures (Warburton, 1979). Table 8.1 does not include yields from aquaculture in estuaries, which may far exceed 100 tonnes $km^{-2}$ $year^{-1}$, depending on methods and whether fertilization and feeding take place. The highest fish yields of all are for intensive aquaculture facilities located in tropical estuaries, where yields of 80 to 130 tonnes $km^{-2}$ $year^{-1}$ have been recorded (Warburton, 1979).

There is little doubt that many tropical estuaries are zones of high productivity due to a combination of shallowness and high nutrient input from rivers. In addition, the vegetation in and adjacent to tropical estuaries, particularly mangroves, contributes to this productivity. In Terminós Lagoon and the adjacent coastal waters of Campêche Sound in the Gulf of Mexico, fish yields are controlled by climatic and meteorological conditions, the amount of river discharge and the tidal amplitude; these factors play a major role in affecting the movement patterns of fish between the lagoon and the sea (Yáñez-Arancibia *et al.*, 1985).

The importance of fish production from tropical estuaries can be illustrated with reference to South East Asia. Here estuarine capture fisheries are estimated to contribute approximately 1.4 million tonnes or 21% of the total marine capture fisheries. Finfish constitute approximately 1 million tonnes, prawns 400 000 tonnes, and crabs about 12 000 tonnes (Chong and Sasekumar, 1994).

## 8.3 TYPES OF FISHERIES

Fisheries can be variously defined in terms of the method used to catch fish (e.g. the Indian bag net fisheries), in terms of the species sought (e.g. the Australian barramundi fishery) or with regard to a particular location (e.g. the Lake Chilka fishery). In this section, examples of different fisheries, some single species and some specific to particular estuaries, are reviewed and compared with particular regard to the state of the resources and how the resource is used and allocated between different sectors.

### Problems and issues in tropical estuarine fisheries

Most tropical developing countries face the same problems of increasing pressures on their natural resources, and their fish stocks are no exception. The main problem in determining the status and viability of most tropical estuarine fisheries is the lack of accurate catch data. In developed countries, fisheries landings are well documented, and long-term records of fisheries statistics are carefully maintained. Hence it is possible to obtain at least some information on various changes in catches and to attempt to estimate the state of the stocks. In most tropical countries, such statistics are either not collected or unreliable, due to inadequate infrastructure and expertise, often exacerbated by political instability.

Notwithstanding the above, there is strong evidence that most subtropical and tropical estuarine fisheries are either fully exploited or overexploited. The reasons for this can include any or all of the following.

1. Overfishing – used here in the sense of exceeding the maximum sustainable yield and including all three kinds defined by Pauly (1994): growth overfishing, recruitment overfishing and ecosystem overfishing. In most developing countries this has been as a result of four factors.
   (a) More fishers – high population growth rates and increased food requirements; the breakdown of traditional practices and controls and the irresistible introduction of commercial fishing to meet national food and economic needs. In developed countries, the

increase in numbers of recreational fishers is noteworthy, particularly in the south-eastern USA and in Australia.

(b) More efficient fishing gear – of most importance has been the almost universal introduction of monofilament gill nets.

(c) Mechanisation of boats – the introduction of both inboard diesels and outboard engines for small, often traditional craft has enabled fishers to travel greater distances more quickly and to deploy more complex gear. The introductions of more efficient gears and of engines were often encouraged, facilitated and financed by aid projects from developed countries, particularly during the 1960s and 1970s. Developed countries have sometimes also assisted with the commercialization of tropical estuarine fisheries, such as in parts of West Africa and South America. Fortunately fisheries aid is now more enlightened and generally in tune with local cultural as well as economic needs.

(d) Lack of accurate scientific data – any contemplated management controls can only be effective if the stock status, biology and socio-economics of the fishery are understood, and the relevant authority has the power and the will to enforce conservation measures. This is a problem not only in developing countries: the imposition of restrictions on recreational fishers in Australia and the USA has frequently met with opposition.

2. Habitat loss or degradation – Most larger tropical estuaries have become a focus of human activity leading to the following problems for fisheries.

(a) Loss of nursery habitat – this involves mainly mangrove destruction for industrial or port developments, or, particularly in South East Asia, for aquaculture developments. Large areas of mangrove forest have also been cleared for woodchips or charcoal production.

(b) Industrial and sewage pollution – this takes a great variety of forms but almost all subtropical and tropical systems now suffer some form of anthropogenic input.

### The broad spectrum of fisheries

In this section, widely different fisheries in subtropical and tropical estuaries are briefly reviewed, both to illustrate their fascinating diversity and to provide examples of some of the issues listed above. We examine: the hilsa fishery of the Bay of Bengal, a huge single-species fishery of several developing countries; the barramundi fishery of northern Australia, a small, well-managed, single-species fishery of a developed country; the Kosi fishery of south-east Africa, a multispecies fishery with traditional and modern components; the Lake Chilka fishery of India, a large multispecies fishery in a developing country; the Larut–Matang fishery of Malaysia, an intensive multispecies artisanal and commercial fishery typical of most of the smaller systems of South East Asia; the Pichavaram fishery of south-east India, a multispecies subsistence and artisanal mangrove fishery; the Ébrié Lagoon fishery of Ivory Coast, a multispecies fishery with traditional and commercial sectors; the Valenca delta fishery of north-eastern Brazil, a disrupted, traditional, canoe-based fishery; and the Gulf of Nicoya fishery of Costa Rica, an artisanal and commercial fishery. The main features of each of these fisheries are shown in Table 8.2.

### The hilsa fishery, South Asia

Various aspects of the biology of the hilsa, *Tenualosa ilisha*, have been described earlier (pp.171, 195, 202). This anadromous clupeid is probably the basis of the largest tropical estuarine fishery. The species extends from the Arabian Gulf to at least Burma (Fig. 8.1), but the largest fisheries are in the Bay of Bengal and its estuaries in India and Bangladesh. The fishery has been extensively reviewed by Pillay and Rosa (1963), Dunn (1982), Melvin (1984) and Raja (1985). The last-named author states that the popularity, socio-religious significance and traditional public knowledge of hilsa are reflected in the proverbs and historical records of the Bengal area and that no other fish is as highly prized. The largest

**Table 8.2** A selection of subtropical and tropical estuarine fisheries and their characteristics

| Country | Fishery | Type* | Tonnes year$^{-1}$ | Management |
|---|---|---|---|---|
| Bangladesh | Hilsa | S, A, C | >100 000 | None |
| Australia | Barramundi | R, C | >1 500 | Legislative, enforced |
| South Africa | Kosi system | S, R | 45 | Traditional and legislative |
| India | Lake Chilka | A | ~6 000 | Traditional |
| | Pichavaram | A | 200 | Traditional |
| Malaysia | Larut–Matang | S, A, C | 60 000 | Legislative |
| Ivory Coast | Ébrié Lagoon | A, C | 7 000 | Traditional and legislative |
| Brazil | Valenca delta | S, A | Unknown | Traditional |
| Costa Rica | Gulf of Nicoya | A, C | 6 300 | Traditional and legislative |

*Fishery types: A, artisanal; C, commercial; R, recreational; S, subsistence.

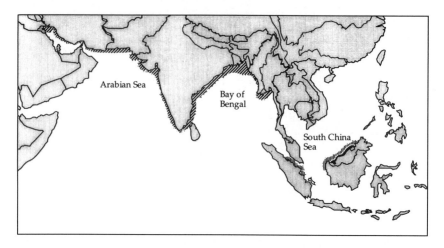

**Fig. 8.1** The distribution of the hilsa, *Tenualosa ilisha* (diagonal hatching). The distributions of the more restricted species *T. macrura* (vertical hatching) and *T. toli* (black) are also shown.

catches of hilsa come from the Ganges delta and upper Bay of Bengal region, with Bangladesh probably taking the largest share (well over 100 000 tonnes p.a.), followed by India (about 25 000 tonnes) and Burma (about 5000 tonnes). Unfortunately, the nature of the fishery has precluded very accurate records of annual catches (Dunn, 1982), as is the case with many of the fisheries of the developing world, and estimates of total yields vary widely. However, there seems little doubt that the overall catch in the Bay of Bengal region is now at least of the order of 200 000 tonnes (FAO, 1995). It is the most important fishery of the region and for example, currently makes up about 25% of the total fish landings in Bangladesh.

The hilsa fishery has subsistence, artisanal and commercial sectors although there is considerable overlap between all three, and very large numbers of people are dependent on the fishery. The subsistence sector comprises mainly the fishing activities of women and children who catch juveniles in the estuaries and rivers; the artisanal consists of smaller, mainly non-mechanized boats working the estuaries; while the commercial sector consists of larger, mechanized vessels working in the Bay of Bengal. There are conflicts of interest between the sectors, with the commercial fishers believing that the catching of juveniles by the subsistence sector adversely affects adult stocks, and the artisanal riverine fishers contending that the expansion in the marine fishery has reduced the number of hilsa available in the estuaries. These issues and the paucity of accurate catch data emphasize the importance of instituting detailed fisheries data collection. This is an expensive task for such a complex fishery, but until it is achieved, the present situation, where there is no management of the hilsa fishery, cannot be rectified.

*Fishing methods*
A wide variety of methods are used to capture hilsa, but today, gill nets are responsible for the bulk of the catch. Most of the 50 or so gears listed by Pillay and Rosa (1963) fall into seven basic types.

- Clap nets – these are purse-shaped nets with a bamboo frame that are suspended from the prow of the boat which drifts downstream. The mouth of the net can be closed when fish enter it.
- Gill nets – almost all are now monofilament and a wide range of mesh sizes are used. Lengths of up to 1 km are common in the estuary (Fig. 8.2), with fleets of many km in the coastal and offshore areas.
- Fixed bag nets – these are anchored where there are strong currents in the estuary and are used mainly for juvenile hilsa.
- Lift nets – these usually have a V-shaped frame and are either laid on the bottom near the bank or deployed from boats in slower-flowing regions.
- Barrier or stake nets – these are deployed at high tide from poles set in the intertidal region and encircle large areas. The fish are retrieved from the net as the tide recedes.
- Cast nets – these are used either from small boats or, in Orissa, eastern India, from tiny, specially built platforms in the estuary.
- Traps – these take a variety of forms and are mainly passive, relying on fish swimming into them. Many consist of bamboo fencing in the form of a V that directs the fish to an area where they can be scooped out with hand nets.

*Types of exploitation and changes in the fishery*
The number of fishers actively engaged in the fishery cannot be estimated with any certainty. In Bangladesh there are probably about 200 000 hilsa fishers scattered in fishing villages along the major rivers and marine coasts. These fishers work as units that range from a single boat with two or three men using a lift net, to two boats working together with a small fleet of gill nets, to larger boats with a crew of 10 to 15 using larger fleets of gill nets, and to motorized sea-going boats (Figs 8.2 and 8.3) (Nuruzzaman, 1987). Historically the fishery operated from small (6–15 m) boats powered by sail and oars. These boats were used to catch hilsa that ascended the rivers to spawn in the monsoon season. The fishery was restricted to the estuaries and rivers because of the distance that could safely be travelled by small boats and the proximity to

**Fig. 8.2**   Artisanal gill netting for hilsa on the River Meghna in Bangladesh.

markets. From about 1960 the fishery began a long transition from traditional fishing methods to more modern equipment. The two most
important changes were the mechanization of vessels using diesel engines
and the introduction of monofilament nylon gill nets (Melvin, 1984). The
mechanization made it possible to fish large new areas in the Bay of
Bengal and increased the distances that could be travelled from markets.
The newer and larger types of fishing gear (e.g. gill nets of up to 5 km
long) that could not be employed in estuaries could be used in the sea.
The situation in the Orissa and West Bengal Provinces of India was
similar to that described for Bangladesh, with a change from a completely
traditional fishery to a more mechanized and more commercial one.

The number of mechanized vessels operating in the Bangladesh sector of
the Bay of Bengal increased from 1400 to 2800 between 1979 and 1983
(Melvin, 1984) and there are now believed to be about 4000. They are in
addition to the estimated 12 000 non-mechanized traditional boats
involved in the estuarine and riverine part of the hilsa fishery (Nuruzzaman, 1987). In India in the mid 1980s there were approximately 1000
mechanized boats in the hilsa fishery of the Bay of Bengal and at least
7000 non-mechanized traditional riverine boats (Raja, 1985).

**Fig. 8.3** Motorized sea-going gill-netter (background) landing its catch of hilsa (foreground) from the Bay of Bengal.

The changes that have occurred in the fishery are reflected in the catch records of hilsa for Bangladesh. There is apparently a clear trend, with a gradual increase in catches from the coastal and marine sector (commercial) and with the artisanal and subsistence riverine sector remaining static. From the early 1960s catches from the estuarine/riverine hilsa fishery were of the order of 90 000 to 100 000 tonnes (Ahsanullah, 1964) and these have apparently been maintained (FAO, 1995) in spite of a massive increase in landings from the marine sector (up to 130 000 tonnes in 1993) over the same time period (FAO, 1995). Unfortunately, the accuracy of the figures may be questionable as no comprehensive catch data recording systems exist. There is no doubt, however, that the hilsa stocks of Bangladesh, and probably India as well, must be under much pressure from the increased levels of exploitation, but in the absence of reliable data their actual status remains unknown.

### The barramundi fishery, Australia

The barramundi or sea bass, *Lates calcarifer*, is sparingly distributed throughout the estuaries and coastal waters of northern Australia, Papua

New Guinea and South and South East Asia. Aspects of its biology have
been described in earlier chapters (pp.151–153, 187–188). In northern
Australia it is harvested by three different sectors: commercial, recreational
and traditional (subsistence) aboriginal fishers. In contrast to the hilsa
fishery, it is relatively small in terms of tonnages landed (Fig. 8.4), but
it is a well-documented and carefully managed fishery. It extends from
the Ashburton River in west Australia to the Noosa River in Queensland
(Fig. 8.5) and is described in detail by Lea *et al.* (1987), Quinn (1987)
and Kailola *et al.* (1993).

Commercial fishing is carried out almost entirely using large-mesh
monofilament gill nets (150–200 mm stretch mesh size) in coastal waters
and the tidal portions of estuaries across northern Australia. Most of the
catch comes from the large rivers of the Northern Territory and the Gulf
of Carpentaria, such as the Alligator, Mary, Daly, Roper, McArthur and
Norman Rivers. The fishers use outboard-powered dinghies to set, check
and clear the nets, and fish are taken to a mother vessel or land-based
camp for filleting, packaging and freezing (Kailola *et al.*, 1993).

Recreational fishing for barramundi is an increasingly popular leisure
activity in Australia, particularly near population centres such as Darwin
or Cairns. Anglers fish for barramundi in freshwater billabongs, creeks,

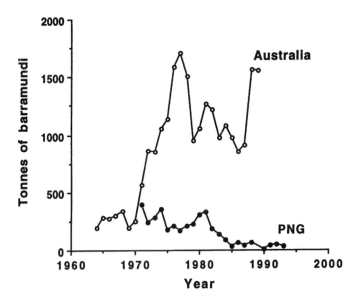

**Fig. 8.4** Commercial landings of *Lates calcarifer* in Australia and Papua New
Guinea (PNG) (data compiled from various sources).

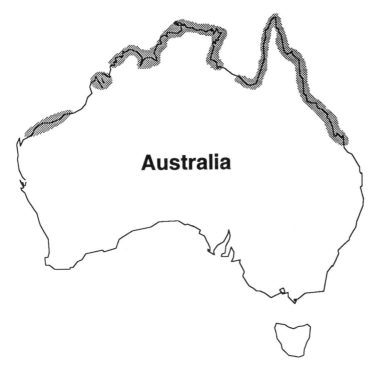

**Fig. 8.5** Distribution of the *Lates calcarifer* fishery in northern Australia (modified from Kailola *et al.*, 1993).

the upper tidal reaches of rivers and throughout the estuarine and coastal habitats of the species (Kailola *et al.*, 1993). The increasing number of recreational fishers, together with improved access to remote areas, has increased the fishing effort in recent years and this sector may now take as much barramundi as the commercial sector. The species forms the basis of a lucrative tourist industry in the Northern Territory, with many charter fishers catering for large numbers of both Australian and foreign fishing tourists. The total monetary value of this industry plus the recreational catch probably now exceeds that of the commercial fishery (Grey, 1987).

The traditional fishery for barramundi carried out by aboriginal people is small compared with the commercial and recreational sectors, and many of the traditional fishers have now been absorbed into one or other of these sectors. The traditional method of catching barramundi was using spears and fish traps, but they are now usually taken using handlines or short gill nets set from dinghies. Most of the fish caught by aborigines are consumed within the fishers' own community (Lea *et al.*, 1987).

The barramundi fishery of Australia is now heavily regulated. The need for careful management was reinforced by the discovery that genetically discrete stocks exist in different groups of estuaries, with more than 16 such stocks in the Northern Territory alone (Salini and Shaklee, 1988). Commercial fishing effort has been reduced by closing the freshwater areas of rivers in Queensland, Northern Territory and west Australia to commercial fishing and restricting it to the lower reaches and coastal waters. There is a closed season from 1 October to 31 January to protect spawning populations, and certain estuaries are completely closed to commercial fishing (e.g. the Embley in the north-east Gulf of Carpentaria). The Northern Territory Government also has a buy-back scheme for commercial licences to try and increase the availability of fish for the recreational industry (Griffin, 1995). In the Gulf of Carpentaria, the commercial fishery is limited entry and each licence holder is required to have a full-time commitment to the gillnet fishery for a minimum of 20 weeks a year. Each must also meet a minimum catch quota value of A$10 000 derived solely from the fishery and must complete detailed catch logbooks. In 1981 there were 191 fishers but these had declined to 113 by 1986 (Quinn, 1987). However, although the number of fishers has declined, the fishing effort has remained the same. Catch statistics indicate a relatively stable fishery (Fig. 8.4), with catch and distribution of effort remaining constant (Quinn, 1987).

Recreational fishers are also subject to regulation, with minimum fish sizes and bag limits enforced in Queensland and Northern Territory. Closed seasons also apply to recreational fishers.

Although interactions between the commercial and recreational sectors of the fishery have often been vocal, the views of both have largely been accommodated into successful management strategies. The value of these strategies is evident when the fishery is compared with the barramundi fishery of adjacent Papua New Guinea. Here, no management has been enforced and large numbers of juveniles are retained, possibly contributing to the drastic declines in catches at Daru (Kare, 1995). A viable commercial fishery has ceased to exist in some areas (Fig. 8.4).

### The Kosi fishery, South Africa

The Kosi system of coastal lakes opening to the sea through an estuary close to the border between South Africa and Moçambique has been described in Chapter 2 (section 2.6.) and its fishes in Chapter 3 (pp.54–57). The system has been described as one of the least spoilt estuaries of the region (Begg, 1978), and paradoxically, supports a thriving and important traditional subsistence and artisanal fishery, together with a recreational tourist fishery (Kyle, 1988).

The traditional inhabitants of the area are the Tembe people, who have fished the Kosi system in a similar way for hundreds of years (Bulpin, 1966) and are very much part of the ecology of the system. The main means of fishing is by the use of fence-and-barrier fish traps (Fig. 8.6), the design and function of which has not altered in living memory. The fence of the traps usually begins at the edge of a channel, curving towards the middle and then usually turning upstream, forming a figure '6'. At the end of the guide fence is a fish-catching basket with a valve that permits the fish to enter freely, but inhibits escape. From the basket, another fence continues a short distance back towards the bank, forming a funnel to guide fish into the basket. Fish moving through the channels encounter the fence and are guided and funnelled into the basket during the night. The trap owner spears them and takes them away the following day (Kyle, 1988). Similar fish traps exist in many of the estuaries of East Africa, notably the Rio Inharrime of Moçambique, but many have fallen into disuse due to the unrestricted use and availability of gill nets. The traps are made of plant materials, particularly mangrove wood. Approximately 1100 kg of wood is used for each trap and they require constant maintenance to retain their catching efficiency.

**Fig 8.6** Traditional fish trap made from mangrove wood and *Phragmites* in the Kosi estuary of KwaZulu-Natal, South Africa.

**Fig. 8.7** View of the network of fish traps in the lower reaches of the Kosi estuary of KwaZulu-Natal, South Africa. The sea is visible in the centre left of the picture.

The complexity and extent of the fish trap system in Kosi is illustrated in Figs 8.6 and 8.7. In the last 50 years the number of operational traps has ranged between 110 and 67. They are owned by families (about 56) and are passed down from generation to generation. The catches from the fish traps average about 40 tonnes of fish per year, with *Pomadasys commersonni* (37%), *Mugil cephalus* (33%), *Acanthopagrus berda* (5%), *Valamugil buchanani* (4%), *Rhabdosargus sarba* (3%), *Liza macrolepis* (3%), *Gerres rappi* (3%), and *Caranx* spp. (2%) the most important species taken (Kyle, 1988).

The fish traps generate income as well as providing sufficient food for most of the local people. Fish trap owners sell surplus fish to other local people on site, many of whom transport them to market for resale at a higher price.

Other traditional methods of catching fish in the Kosi system include spearing fish, which is a highly skilled activity used to obtain free-swimming fish. Spearing is done by individuals or in small groups, or occasionally in groups of up to 200 in large 'fish drives'. Other methods include small amounts of gill netting and line fishing.

Recreational fishing by tourists using rod and line is confined mainly to the Kosi Lakes where there are few fish traps and tourists are accommodated in camps next to Lake Nhlange. Species taken are mainly *Pomadasys commersonni* (50%), *Rhabdosargus sarba* (30%), *Acanthopagrus berda* (7%), *Caranx* spp. (3%) and *Sphyraena* spp. (2%). None of the Mugilidae or Gerreidae taken by the traps are caught by recreational fishers, whose total catch is of the order of only about 5 tonnes per annum. Tourist fishing is strictly controlled and its value lies in the income generated by the infrastructure activities, much of which flows to the local people.

The Kosi fishery is now carefully managed to avoid conflicts between traditional fishers and recreational fishers on the one hand, and to avoid overexploitation on the other. Neither the traditional fishery nor the recreational fishery is significantly affecting fish populations, and current catches (45 to 50 tonnes p.a.) appear sustainable. The use of illegal gill nets and a rapidly increasing population probably pose the greatest threats to the continued viability of the fishery.

### The Lake Chilka fishery, India

The physical characteristics and fish fauna of Lake Chilka on the north-east coast of India have been described in Chapters 2 and 3 (pp.35–36, 61). Its fishery can be broadly described as artisanal with a mixture of traditional and modern methods. The lake supports an extremely complex array of fishing gears that can be divided into three main types: (i) nets, (ii) large impoundments constructed from bamboos in shallow areas, known locally as 'janos', and (iii) traps. Net fishing is practised throughout the year but jano and trap fisheries are seasonal (Jhingran, 1991).

Net operations account for 50–66% of annual landings, with various types of seine nets and gill nets the main gears. Other nets include cast nets and lift nets. The netting is carried out in leased grounds (exclusive fishing rights) and in other unleased grounds on a fee basis. The most important species taken by the netters are *Pseudosciaena coibor* (10%), *Mugil cephalus* (7.5%), *Eleutheronema tetradactylum* (7%), *Nematalosa nasus* (5%), *Mystus gulio* (4.5%), *Tenualosa ilisha* (4.5%), *Lates calcarifer* (2%) and *Osteiogeniosus militaris* (1.5%), together with penaeid prawns (3.5%).

In 1989 there were 112 exclusively leased-out janos in the lake and these accounted for between 13% and 22% of annual landings (Jhingran, 1991). The mullets *Mugil cephalus* and *Liza macrolepis* form 80% of the catches with *Mystus gulio*, *Lates calcarifer*, *Eleutheronema tetradactylum* and penaeid prawns also taken. There were 67 leased-out trap fisheries catching mainly penaeid prawns in 1989.

The annual landings of fish increased slowly from the 1930s onwards

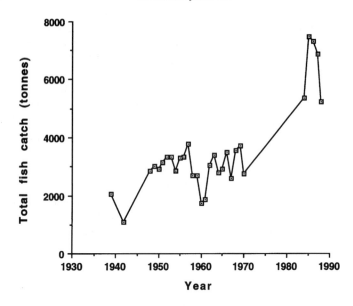

**Fig. 8.8**   Total fish catch in Lake Chilka, India, over 50 years (data compiled from various sources).

and were fairly stable at around 3000 tonnes from 1950 to 1970. Thereafter catches increased to a high of nearly 8000 tonnes in 1985; this increase coincided with the introduction of monofilament gill nets and greater use of outboard engines. The catch has since declined and the resources are now probably overexploited (Fig. 8.8). There has been much discussion and concern about how to regulate fishing effort and conserve the fishery (Jhingran, 1991), and detailed conservation measures have been proposed (Jhingran and Natarajan, 1969). However, the nature of the various artisanal methods and the socio-economic necessity for retaining fishes of all sizes has precluded effective management. For example, the diversity of gear types has made minimum mesh size regulations impractical and unenforceable. Physical changes to the system, including silting and shallowing of large areas, and the mobile nature of the lake mouth, have caused a reduction of tidal influences and may have hampered recruitment of some marine migrant species. This is particularly important because about three-quarters of the fisheries production of Lake Chilka depends on recruitment of fishes from the sea.

### The Larut-Matang fishery, Malaysia

The mangrove estuaries of the Matang area and their fishes have been described in Chapters 2 and 3 (pp.17, 61–62). They are very heavily fished by both commercial and artisanal sectors and as such, are similar to many such areas in South East Asia. The forests of the Matang Mangrove Forest Reserve have been managed on the basis of sustainable yields, with a 30 year timber-harvesting rotation, since the early part of the 20th century. The people of the villages in the area are involved in either the forestry or fishing industries. The forest industry provides direct employment for about 1400 and indirect employment for 1000 and the total annual value of the forest products is about US$9 million. In contrast to this, the fishing industry provides direct employment for 2500 and indirect employment for 7500, with an estimated annual value of about US$30 million. The fisheries thus provide four times as much employment and over three times the economic returns of the forestry, clearly illustrating the value of fishing in these mangrove estuaries and coastal waters (Ong, 1982; Khoo, 1989).

All fishers and boats are licensed and the coast is divided into three zones: zone A, from the estuaries to 8 km from shore; zone B, from 8 to 19 km offshore; and zone C, from 19 km offshore. Trawlers and purse seiners are only licensed to operate in zones B and C, whereas traditional fishing gears are free to operate anywhere. During 1990 the following numbers of each type of gear were licensed in the Larut–Matang District (Choy, 1993): 585 trawls, 540 gill nets, 47 hooks and lines, 10 portable traps, 9 bag nets, 3 seine nets, 1 purse seines.

In 1990 there were 2542 fishers; of these 1250 were involved in trawling, and the remainder operated traditional gears – 970 used mainly gill nets. Most of the boats are small, locally constructed and suitable for shallow seas, narrow river mouths and creeks, and 79% are less than 20 GRT (gross registered tonnage; able to operate in all zones). The above figures exclude about 200 unlicensed mechanized push-net operators who fish the very shallow coastal areas and river mouths (Fig. 8.9).

In 1990 a total of 58 265 tonnes were landed (73% from trawling, 7% from gill nets) plus another 2300 tonnes or so from the push-net operators. The main species landed are sciaenids, *Rastrelliger* spp., ariids, *Sardinella* spp., *Lates calcarifer*, *Megalops cyprinoides* and large numbers of juveniles of a wide variety of inshore and estuarine species. The small modal size of most species caught (Fig. 8.10), together with the large number of juveniles, indicates that the fishery is stressed, but there are at present no catch restrictions. The fishery also catches large numbers of commercially important penaeid prawns. The estuaries are also fast becoming important sites for cage culture of *Lates calcarifer* (Fig. 8.11),

**Fig. 8.9** A push-net fishing boat in the Larut-Matang estuary system of West Malaysia.

and in 1990, 72 fishers operated about 2500 cages that produced 26 tonnes with a value of about US$100 000.

### The Pichavaram fishery, India

The Pichavaram mangroves form part of the Vellar–Coloroon estuary system of east Tamil Nadu, India and their fish fauna has been summarized in Chapter 3 (pp.60–61). The fishery is typical of many such mangrove areas of India (Chandrasekaran and Natarajan, 1992) and is artisanal. The fishers use a variety of traditional gears such as cast nets, drag nets, stake nets, gill nets, traps and hook and line. The dugout canoe is almost universally used for most of the types of fishing.

Stake nets have the highest catch per unit effort and are operated in the main channels and catch fish on the rising tide. Drag nets of about 10 m long are used in shallow water, and intermittently dragged by two men for about 50 m. The catch is placed in earthen pots tied to the waists of the fishers. Most fishing, except cast netting and gill netting, is undertaken at night. Gill nets are about 100 m long and tied from poles, while cast nets are operated either standing waist deep in water, or from canoes.

**Fig. 8.10** Sorting catches of large numbers of small fishes from the Larut-Matang estuary system of West Malaysia.

Another method of fishing is by bund construction, in which a fish fence trap is constructed to mid-tide height and lined with a mat of bamboo cane. When the water level increases at high tide the bund is submerged, but as the tide ebbs, the fish, notably mullet, become trapped and can be collected. These bunds, which build up over time due to sediment deposition, become permanent structures and have led to shallowing of many areas and alterations in tidal flow patterns (Krishnamurthy and Jeyaseelan, 1981).

The total annual catch in Pichavaram is about 200 tonnes and this apparently varies little from year to year. The bulk of the catch is penaeid prawns (80%) while crabs form about 4%. The remainder consists of fish of more than 50 species. Almost half these fish are mullet (five species). The rest consists mainly of ariid catfishes, *Pomadasys kaakan*, *Chanos chanos*, *Leiognathus* spp., *Ambassis* spp., *Etroplus suratensis* and *Cynoglossus* spp., but about another 100 species have also been recorded in catches.

According to Chandrasekaran and Natarajan (1992), most of the fish are juveniles that are caught with small-mesh gear, and the incomes derived from their sale are very low. The only restrictions on fishing are

**Fig. 8.11** Cage culture of *Lates calcarifer* in the Larut-Matang estuary system of West Malaysia.

in relation to a traditional 'Padu' system of allocating exclusive fishing rights to fishers of each locality. This serves to reduce competition between fishers and may help to control effort. However, any management of this sort of fishery is extremely difficult, and none has been attempted because fishing is the sole means of livelihood and main source of food for most of the local people.

### The Ébrié Lagoon fishery, Ivory Coast

The lagoon network of the Ivory Coast consists of three different lagoon complexes stretching along the coast for about 300 km. The nature of the fish fauna of the Ébrié Lagoon (a coastal lake of 566 km$^2$) has been extensively studied (Albaret and Écoutin, 1989, 1990) and has well-developed traditional (subsistence and artisanal) and commercial fisheries.

The commercial fisheries, using mainly ring nets and beach seine nets (up to 2 km in length) operated by a salaried workforce, began in about 1960 and increased fourfold between 1964 and 1975, to become responsible for at least 70% of the catch (5000 tonnes p.a.) The permanent opening of the lagoon to the sea in 1950 brought a more marine influence and increased the catches of marine migrant species. The traditional fishery takes about 2000 tonnes per annum.

Six species make up the bulk of landings in both fisheries: *Ethmalosa fimbriata* (61%), *Tilapia* spp. (6%), *Elops lacerta* (6%), *Chrysichthys* spp. (5%), *Tylochromis jentinki* (4%) and *Sardinella maderensis* (4%).

The history of the Ébrié Lagoon fishery has been described by Verdeaux (1979) and summarized by Kapetsky (1981). They state that it provides a good example of the interactions of complex socio-economic factors that have to be accounted for when drawing up management strategies for allocating resources between commercial and traditional fisheries.

Verdeaux defines three phases of development from pre-colonial times to 1979.

- The pre-colonial period, which existed until the end of the 19th century, saw individual and group fishing existing side by side. The 'community' fishery used fish traps constructed from palm fronds and wood, and seine netting made from natural fibre. These blocking traps were built and operated by either family groups or the villagers and the catch was for preservation and sale by family members. Individual fishing was done using smaller traps, cast nets or bottom longlines with the catch going towards daily subsistence. There were no organized fish markets at the time.

- The colonial period (*c.* 1910–1960) saw a move away from community fishing for a number of reasons. Cotton was introduced by Europeans and was used to make a variety of enmeshing nets which could be used by one person. Deforestation made wood more difficult to come by and the opening of an artificial channel to establish the port of Abidjan saw salinities alter and introduced the wood-boring mollusc *Teredo*, which destroyed the wooden fish traps.

- The period after 1960 saw marked economic expansion of the fishery with the reintroduction of large, non-selective fishing gear – purse and beach seines – requiring a large labour force. These were run by non-indigenous people (primarily Ghanaians) with locals being paid for their labour. The situation led to increasing social and economic conflict between the small group of commercial operators, who were growing wealthy from large catches, and local Ivorian fishermen who continued to fish fixed areas using individually operated gear with selective mesh sizes. The situation was exacerbated by the encroachment of commercial operations into areas of the lagoon traditionally reserved for Ivorian fishermen. Both groups saw fisheries as a means to raise capital quickly for investment elsewhere – for example, farmland – with neither group committed to conservation of the fishery resource in the long-term. Verdeaux (1979) refers to this as a "dynamique de pillage" resulting in overexploitation which is borne out by recent declining catch trends (FAO, 1995). Seine mesh sizes have been reduced as low as 13 mm and the modal size of fishes continues to decline. Industrialization of the lagoon shores and various forms of pollution have also had detrimental effects on the fisheries (Écoutin and Delahaye, 1989).

In about 1977, brush-park fisheries were introduced to Ébrié Lagoon by immigrant fishers from Benin (Kapetsky, 1981). Brush-park fisheries or 'acadjas' are a traditional form of low-technology aquaculture practised in many lagoonal areas of West Africa and exist in their most sophisticated form in Benin. They consist of an inner core, or concentric circles, of densely packed tree branches surrounded by an outer, more substantial, wooden framework. They are placed in quiet waters of not more than 1.5 m depth and act as attractants to fish, particularly cichlids and catfish, both because of their luxuriant growth of periphyton (food for cichlids) and for the shelter they provide. They are fished by surrounding the acadja with a net, after which the tree branches are removed and the net pursed, or the fish concentrated and removed with baskets or hand nets (Welcomme, 1972). Yields in Ébrié average 15 kg per fishing of an acadja of about 3 m diameter, with one fishing every 2 months. In Benin, yields as high as 800 tonnes $km^{-2}$ $year^{-1}$ are claimed for acadjas fished twice a year (Welcomme, 1972). The main species taken in this way in Ébrié Lagoon are *Lutjanus goreensis, Tilapia* spp., *Epinephelus* spp., *Pomadasys jubelini, Chrysichthys walkeri* and *Gerres nigri*. This form of traditional fishing offers some advantages in developing countries: acadjas achieve high yields, use simple technology, are labour intensive and are low in cost. Their disadvantages, however, are considerable: they need large quantities of branches and hence have adverse environmental impacts, particularly on mangroves, they obstruct navigation and are perceived by commercial fishers as competing directly for resources.

### The Valenca delta fishery, Brazil

This is a traditional subsistence and artisanal fishery in a mangrove delta in Bahia in north-eastern Brazil. In this estuarine fishery, a complex of factors evolved over four centuries to match traditional canoe-based fishing, using shallow-water purse seines, to the resources available (Cordell, 1974, 1978). Fish catches consisted mainly of estuarine spawning species, such as ariids, centropomids and eleotrids, as well as mugilids and sciaenids. The canoe fishing relies on the fisher's knowledge of spawning behaviour and his ability to find the fish. This is accomplished by organizing environmental knowledge of moons and tides into a system for choosing fishing spots that also seems to be a fairly accurate index of the migration of estuarine species (Cordell, 1989). This knowledge is confined to tight-knit family groups that are limited in number. The establishment of proprietary fishing rights to certain fishing areas that were passed from one generation to the next was probably the most important development that rationalized exploitation. Other factors included cooperation between families and community pressure (Kapetsky, 1981). Unfortunately this fishery nearly collapsed in the 1970s with the introduction of nylon nets. This innovation was intended to increase fish catches in order to supply urban areas, and hence economically benefit the fishers. Because of the high cost of the new nets, however, the traditional fishers could not afford them, and they were purchased by businessmen who employed local fishers to operate them. This led to a struggle for control of the traditional fishing grounds and often physical conflict. Territorial fishing rights were gradually lost and some of the resource became one of common property. Although the new nets briefly increased yields, these soon declined and both the traditional fishers and businessmen were affected. Traditional fishers then began to use individually operated seine and gill nets and expanded their fishing grounds into new areas in the mangroves, taking smaller and smaller fish. In effect, in less than a decade, a four-century-old traditional fishery disintegrated, with drastic consequences for the fish resources (Cordell, 1978).

### The Gulf of Nicoya fishery, Costa Rica

This large mangrove-lined estuarine embayment on the Pacific coast of Costa Rica (Vargas, 1995) supports a substantial artisanal fishery on various stocks in the inner Gulf, as well as a commercial prawn (shrimp) trawl fishery in the outer Gulf. The artisanal fishery lands about 6300 tonnes per year of which 43% is sciaenids, such as *Cynoscion albus, C. squamipinnis, Stellifer* spp. and *Bairdiella* spp. (Szelistowski and Garita, 1989; Herrera and Charles, 1994). The artisanal fishery is barely managed and the fishery is regarded as a common property resource. Legislation has restricted the commercial trawlers to the outer Gulf to protect spawning and nursery areas in the inner Gulf. However, their activities result in the destruction of large numbers of juveniles taken as bycatch, and hence they damage the stocks fished by the artisanal sector (Campos *et al.*, 1984). The artisanal sector uses a great variety of gears including handlines, seines, gill nets and traps. All the stocks have suffered severe declines under the high fishing pressure from both sectors. Hence the government is in the process of instituting tough new management programmes to try and protect the dwindling stocks (Weidner, 1992).

## 8.4  MANAGEMENT

### Definitions and goals

Before any discussion of the advantages, disadvantages and consequences of fisheries management it is necessary to decide the goal of management. The most obvious objectives of fisheries management are the biological objectives of resource conservation and physical yield maximization, the economic objective of profit maximization, and the socio-political objectives concerned with employment and equity. In practice it is not possible to satisfy all of these objectives simultaneously and trade-offs have to be made between them (Harden Jones, 1994).

Most fisheries management has been dominated by the mainly mathematical population dynamics approach, usually based on information about the number of fish caught each year and the amount of effort expended to catch them. There has been considerable discussion about how well this method, involving for example the logistic equation, has served fisheries management in any practical sense. These sorts of questions may, however, be largely irrelevant in most tropical estuaries, where fisheries are usually based on multispecies stocks composed mainly of short-lived species with very complex biological and environmental interactions (Pauly and Murphy, 1982; Longhurst and Pauly, 1987; Day *et al.*, 1989). The problems in managing such fisheries have been summarized by Day *et al.* (1989), who conclude that whatever management rationale, or lack thereof, exists, it is obvious that fish yields per unit of fossil energy used are declining, and furthermore, that most of the world's fisheries are becoming industrialized. Most tropical estuarine fisheries are effectively unmanaged, a few are still traditionally managed and fewer still are managed.

### Developing country issues

Against the above background, what then is the situation in tropical estuaries – most of them in developing countries that are rich in resources and people, but poor in finance, scientific expertise and trained manpower?

The position in South East Asian estuaries and adjacent coastal waters has been summed up by Chong and Sasekumar (1994) as follows:

Open access fishing, whether by large-scale industrial trawlers or by small-scale traditional fishermen, will eventually lead to an overcapacity problem. The problem is that nobody owns the sea; anybody can fish as long as he owns a boat. From their meagre numbers in the late 1960's, the number of fishing boats in the ASEAN region has swelled to 800,000 boats in the late 1980's.

Areas that have large coastal populations, such as the Malacca Straits, the Gulf of Thailand, the Java Sea, south Kalimantan, the Arafura Sea and Manila Bay, have all recorded a decline in the catch-per-unit-effort, a decline in captured fish lengths, an increase in the trash fish component and the virtual disappearance of certain commercial fishes. These characteristics are all symptomatic of over-capacity, overfishing and stock depletion.

Overcapacity implies economic inefficiency in the allocation of fishing effort. To put it simply, it means that fishermen are not getting adequate returns from what they have invested. But fishing in the ASEAN region has always been very complicated, often defying basic economic principles. Many fishermen often remain in the fishery even when there are few gains to be made. Years after years of fathers passing down to sons their fishing skills have created a static fishing population that has no alternative jobs to turn to. Although this has changed in some countries, especially in Malaysia, it is still ingrained in many fishing communities.

Options for management of tropical estuarine fisheries include both regulatory and non-regulatory strategies (Kapetsky, 1981). Strictly enforced legislation has been impossible in most estuarine fisheries in developing countries. This has been for a variety of socio-economic and political reasons, a lack of appropriate legal and constitutional frameworks, a lack of appropriate technical data, and a lack of scientific and economic expertise, as well as financial constraints.

Despite the theoretical value of regulatory options that could be taken up by developing countries, such as the elimination of the most damaging fishing methods (e.g. the use of explosives or cyanide), very little has been achieved. There is frequently a gulf between what is illegal and what happens in practice (e.g. monofilament gill nets are illegal in many countries but are used almost everywhere), because there may be neither the political nor social will to enforce 'in name only' legislation, much of which may have been enacted in colonial times, or more recently to appease developed countries or aid donors. Additionally, most developing countries do not have the infrastructure to effectively enforce fisheries legislation (Marr, 1982).

Non-regulatory strategies such as the re-introduction of traditional or 'appropriate' technologies are attractive from a resource conservation point of view, but have not worked well. In most spheres of human endeavour, the use of old technology is rejected when newer, more efficient methods exist. This is true in fisheries because there is often a pressing economic or even food imperative (hunger) for the immediate maximization of catches. The adverse ecological and stock size effects of gill nets may be profound, but their lure in terms of increased catch are irresistible to the poor. It is now increasingly realized that effective management of tropical estuarine fisheries depends just as much on an understanding of the motivations and behaviour of fishers, fish consumers and the socio-economics of the whole activity, as on scientific knowledge about the fish (Aguero and Lockwood, 1986). Generally, local fisheries staff responsible for working out management strategies, and the expert staff (often from developed countries) providing scientific advice, have biological training and are equipped to remedy perceived problems such as overfishing, catching of juveniles, or inappropriate gears, rather than to recognize or tackle the underlying causes that emanate from the cultural, social or political environment. That this point is now recognized is evident from the recently changed emphases of fisheries aid projects, many of which integrate social and cultural components into their programmes.

### Community-based management

In a recent review of community-based management options and systems in South East Asia, Pomeroy (1995) recognized that the underlying causes of overexploitation are often social, economic or political, and that the main focus of management should be people, not fish *per se*. Fisheries management in many tropical countries has followed the temperate model of working out maximum sustainable yields and having centralized administration (Pomeroy, 1995). The problems with this approach for the tropics are the multispecies nature of their fisheries and the lack of consultation with the fishers who may have much of the knowledge necessary for managing the resource. The growing realization of the need for increased participation by resource users in fisheries management can be seen in a wide range of policies and programmes now emerging in the South East Asian region. Unfortunately only a few localized, long-standing, community-based management systems still exist in the region. Their functions have largely been taken over by national governments that in many cases have failed to develop adequate substitutes, and in some cases there is now no real fisheries management at all (Pomeroy, 1995). Pomeroy (1995) goes on to show that the revitalization of community-based management in South East Asia will not be easy for a whole variety of cultural, economic and political reasons. The often essential move towards integrated management (p.264), involving sectors and areas outside fisheries, while not inimical to local fisheries management, relies on solutions outside the traditional realm of fishing communities. Despite these problems, however, the Philippines, Thailand, Malaysia, Indonesia and Vietnam are taking active steps to revitalize community management of some of their fisheries (Majid, 1992; Piumsombun, 1993; Nikijuluw and Naamin, 1994; Sulaiman, 1994).

To some extent at least, the revitalization of traditional practices or the introduction of ecologically sustainable methods has been overtaken by the great expansion in aquaculture in tropical estuaries, especially in South East Asia. Discussion of aquaculture is beyond the scope of this book, but it is worth noting that its development, from the point of view of the estuary-based fisherman, while providing considerable employment and often a market for trash fish, has often come at the cost of massive mangrove habitat loss (e.g. Thailand lost 50% of its mangroves in the 10 years after 1985), to the ultimate detriment of capture fisheries production.

With careful planning, however, aquaculture developments that avoid habitat destruction and over-eutrophication of the estuary can play an important role in supporting capture fisheries management. For example, in recent years it has become necessary to introduce a 2 month closed season for the terubok, *Tenualosa toli*, fishery in Sarawak in order to protect spawners (Blaber *et al.*, 1995b). The terubok fishers particularly target the spawning females because of the high value of the eggs (at least US$100 per kg), so it was necessary to find them an alternative source of employment and income during the closed season. This was done by constructing aquaculture ponds adjacent to the estuary and fishing villages, and stocking them with *Lates calcarifer* that could be fed on trash fish, grow extremely rapidly and command a high price (Bin Hasman, 1996).

### Resource allocation

Allocation of fisheries resources between competing users is a problem common to both the developing and developed world. In developing countries this usually translates to gear conflicts, where different fishers are using different capture methods, often in different places. In South East Asia, for example, the conflicts usually involve inshore–estuarine fishers, who use traditional gears such as seines, traps and gill nets, versus offshore fishers who use trawls (Marr, 1982). As Marr (1982) states, this is a major problem because there are very large numbers of artisanal and subsistence fishers. After years of continuing conflicts between artisanal and trawl fishers, ASEAN countries finally imposed limited entry or a total ban on trawling in inshore waters in the early 1980s. These restrictions have, however, proven difficult to enforce (Chong and Sasekumar, 1994). In Malaysia the problem has been tackled by requiring trawlers to fish beyond a certain distance from the shore (p.251). Whether this measure has any management value in terms of conservation of stocks is questionable, because for many species of fish, the juveniles are caught in the estuaries and inshore waters by the artisanal sector and the sexually mature adults offshore by the commercial fishers. Most ASEAN governments do not regulate artisanal fishing, and intense exploitation of juveniles can adversely affect recruitment to the adult offshore stocks. In this way the sympathy of governments towards the artisanal sector may have done more harm than good, and has encouraged overfishing in estuarine and inshore areas (Chong and Sasekumar, 1994). The system of separating the fisheries does, however, possibly have value in minimizing physical confrontation between artisanal and commercial fishers (Marr, 1982).

Traditional resource allocation based on ownership or stewardship of specific areas or methods is becoming less common in estuarine fisheries as cultural values and systems change, and populations increase. It still exists in parts of Africa, for example in some West African lagoons and in a few estuaries on the east coast, such as Kosi (pp.246–249), as well as in parts of India such as Lake Chilka (pp.249–250). It is not well suited to dealing with change, particularly where increasing numbers of people need access to dwindling resources.

In developed countries, fisheries resource allocation involves commercial and recreational fishing. The many management difficulties associated with this problem have recently been summarized by Rogers and Gould (1995), who recognized that the difficulty arises from the long-held ideal in countries such as Australia and USA, that fish resources are owned in common by the community and that access should be free and open to all. This scenario would allow fishing pressure to increase to such an extent that stocks would be depleted, and hence has justified government intervention in the form of management arrangements for commercial, recreational and even traditional fishers. Such government intervention has tried to balance the needs of various user groups with the ultimate objective of ensuring that fish stocks are sustained and used efficiently. In the Australian context it is now a requirement that all fisheries be managed on an ESD (ecologically sustainable development) basis. ESD has been defined as follows: "... means using, conserving and enhancing the community's resources so that the ecological processes on which life depends, are maintained, and the total quality of life, now and in the future, can be increased" (Commonwealth of Australia, 1990). In this regard, however, it must be recognized that most decisions everywhere are also influenced by political and economic pressures.

In all states of Australia, fisheries are now very heavily regulated using a number of methods, such as limited entry, quotas, gear restrictions, buy-back schemes and so forth for the commercial sector, and bag limits, size limits and licensing for the recreational fisher. The effectiveness of these measures is increased by the active involvement of the fishers themselves in the management planning process through their representation on relevant decision-making bodies. Such systems are also used in South Africa, but there, one of the main methods has been to designate some species as commercial and others as recreational. Negotiations over the large number of species that were of both commercial and recreational significance involved a lot of bartering and trade-offs. However, the end result was a list of genuinely de-commercialized species which recreational fishers can manage, and a large number of other species where recreational access is limited and where the commercial sector can promote itself. This has gone a long way towards solving resource allocation problems in South Africa (van der Elst and Penney, 1995). In South Africa, almost all estuarine fish stocks are now reserved for recreational fishing and the situation in Australia is moving in that direction, although the commercial barramundi fishery remains important. In the USA, recreational catches of red drum exceed commercial catches in most Texas bays, and management strategies have been devised to provide an equitable and sustainable harvest for both groups, using quotas on both recreational and commercial catches, prohibiting certain gears at weekends, as well as licensing both groups. Management controls are based on research and inputs from monitoring the activities of Texas recreational and commercial fishers.

Rogers and Gould (1995) posed the following key policy questions regarding resource allocation for Australian fisheries, but they are also generally applicable to both developing and developed countries.

- What legal access rights do commercial, recreational, traditional and other user groups have to the resource?
- If a variety of user groups have rights to the resource, are these equal or do some groups have priority?
- What decision criteria should be used to allocate fish stocks to competing users?
- What management strategies or process of consideration should be used to achieve the desired allocation?
- Is an economic rationalist approach to decision making appropriate?

Added to these questions should be the need for clearly defined goals for the management of species and the fisheries based on them (van der Elst and Penney, 1995), together with adequate research and monitoring to allow for any future changes in the fishery.

Because fish stocks themselves change and human usage also changes, the re-allocation of resources is an important issue. In the Australian context at least, and probably in most countries, it is at this time that the situation becomes very complex. There is an increasing body of law supporting the view that commercial fishing access is a quasi property right with an economic and financial value. Against this, recreational, or in the developing-country context, artisanal fishers, often have no legally defined rights other than the freedom of access (Rogers and Gould, 1995).

### Integrated management

Estuaries are boundary areas and are influenced by a variety of marine, freshwater and terrestrial inputs. For this reason it is impractical that they should be managed in isolation, particularly with regard to fisheries. From a biological perspective, many of the fishes have part of their life cycle outside the estuary; from a physical point of view, the turbidity and salinity may be affected by human activities (e.g. agriculture, logging, dams) in river catchments far from the estuary; and economically, estuarine fisheries are often integrated with both marine and freshwater fisheries and increasingly, with aquaculture.

Integrated management plans for coastal area development as a whole are now being formulated and used in both developed and developing countries. Whether or not the plans can be translated into practical and beneficial effects, at least in most developing countries, remains an important question. A good example of how such a plan can work in a developed country concerns the strategic management plan for Moreton Bay, Queensland, Australia (Anonymous, 1991), a large subtropical estuarine embayment near Brisbane. It is an important recreational, economic and educational resource for approximately 1.8 million people. Many of the activities are in conflict with each other and with environmental sustainability. Recreational and commercial fishing depend on maintaining the appropriate qualities of the environment such as water quality and habitat protection, while others, such as extractive industries, remove and modify the resource. The management plan has developed arrangements for integrating all the activities in the bay (Quinn, 1992).

## Chapter nine

# *Human impacts*

## 9.1 INTRODUCTION

Almost all subtropical and tropical estuaries have been affected in some way by humans. Fishing is only one activity in and adjacent to estuaries; an extremely diverse range of other industries and environmental modifications also have a wide variety of impacts on the fishes. The amount of change may be almost total, as in some harbours such as Richard's Bay in KwaZulu-Natal, South Africa, where the shape, hydrology and functioning of the system have been completely altered. Other estuaries are severely degraded by industrial or agricultural impacts. Relatively few remain in anywhere near a 'natural state'. For example, in a major review of KwaZulu-Natal estuaries, Begg (1978) found that only three out of 73 estuaries were problem free, and that the impoverished fauna in many of them is principally due to indirect impacts (through habitat degradation) rather than direct impacts such as overfishing.

Most port cities are built around large estuaries, hence they become the focus of development and shipping activities. Industries involved in or dependent on exports and imports have become established near harbours. Many of the factories are for processing raw materials prior to export or after import, and produce effluents that may be discharged into the estuary. The large cities that have formed around the ports have high populations and sewage is frequently disposed of directly into the estuary. In developed subtropical and tropical countries, some estuaries have been greatly modified to build residential canal estates or marinas. Such developments usually entail canalization and removal of mangrove forests. In both developed and developing countries, the aquaculture industry has likewise been responsible for the removal of mangroves for pond construction.

Upstream from port cities, the estuary often passes through a fertile floodplain that is largely given over to agriculture. Farming practices can affect the ecology of the estuary through modifications to bank vegetation, abstraction of fresh water for irrigation and run-off from irrigation. Such run-off may contain fertilizers and pesticides. In many tropical countries, logging is an important industry and the rivers and estuaries often provide the transport route for shipping the logs. Removal of forest cover in the catchment may also change run-off patterns and increase water turbidity.

The construction of dams for supplying water to cities and for agricultural irrigation may change the amount of fresh water entering the estuary, affect turbidity and salinity, and hence influence the distribution of fishes.

In this chapter we examine how a number of these activities, changes and developments affect the fish and fisheries of subtropical and tropical estuaries. In addition, the importance of conservation, amelioration of pollution, and restoration of ecological functions are related to human dependence on the effective functioning of estuaries, particularly with regard to fish and fisheries. The preservation of at least some of the world's great estuaries in as pristine a state as possible may be difficult but can be shown to have important benefits.

## 9.2   EFFECTS OF INDUSTRY

Most of the research into the effects of industrial activity on estuaries and fish has taken place in developed countries in temperate regions. Recent examples include the work by Elliott and Taylor (1989) and Costa and Elliott (1991) on the Forth and Tagus estuaries in Scotland and Portugal. Many of the effects of such anthropogenic activities on fish (Table 9.1) are probably similar in the tropics, but much less research has been undertaken until relatively recently. The much higher species diversity, the greater complexity of interrelationships, and different chemical reactions and reaction times at higher temperatures in the tropics, mean that caution is needed in extrapolating from temperate data. The increasing industrialization of developing countries has brought with it many of the pollution and effluent problems, but not usually the means, knowledge or scientific expertise to solve them.

Much of the research on the effects of industries stems from human health concerns, either direct (e.g. swimming or washing in polluted estuaries) or indirect (e.g. eating fish containing a high level of pollutants). Other phenomena that excite human concern include mass fish kills (section 9.7), large reductions in commercial, recreational or artisanal fish catches, loss of visual and recreational amenity and bad odours emanating from eutrophication, raw sewage or industrial effluents.

The degree to which industrial effluents affect estuaries and their fishes

**Table 9.1** A summary of the effects of industrial activities on fish in estuaries (modified from Costa and Elliott, 1991)

| Activity | Effects |
| --- | --- |
| 1. Organic enrichment from domestic sewage, pulp mills, sugar mills etc. | Poor water quality, effects on fish migrations, especially in upper reaches |
| 2. Persistent contaminant inputs from industry | Bioaccumulation, presence of detoxifying mechanisms |
| 3. Port activities, dredging and spoil disposal | Damage to feeding and nursery grounds, increased turbidity |
| 4. Power generation and water abstraction | Thermal pollution effects and mortalities on fish screens |
| 5. Reclamation, engineering works | Loss of feeding areas, especially the rich intertidal areas |
| 6. Naval activities, underwater testing | Mortalities of pelagic species |

depends not only on the volume discharged into the estuary, but also on the type of estuary. Small blind estuaries, such as many of those on the south coast of KwaZulu-Natal in South Africa, can become severely degraded (Begg, 1978), whereas larger open systems, with high tidal exchange and hence flushing, are less affected. The time is past when estuaries were regarded as a bottomless sink, and with existing knowledge, careful planning and regulation there is little excuse now for poor industrial planning. The problem is exacerbated in many tropical developing countries by the urgent needs, often at almost any environmental cost, to provide employment and to generate income from industrialization, without any of the mandatory safeguards now in place in most developed countries. Sadly, this is often with the help of entrepreneurs from the developed world.

It is not only modern industries that can severely affect estuaries. For example, the coast of Kerala in India has 31 small, often blind estuaries lying roughly parallel to the Lakshadweep Sea and separated from it by a narrow strip of land 0.4 to 12 km wide. The abundant availability of coconut husk and the generally shallow nature of the estuaries have led to the emergence of the coir industry as a massive cottage industry in the state. Fresh coconut husks, steeped in the shallow regions of estuaries, are allowed to remain soaked in water for periods of 4 to 12 months. Retting is brought about by the pectinolytic activity of microorganisms, liberating large amounts of organic matter and chemicals into the environment. The retting zones in the estuaries are thus exposed to prolonged periods of total oxygen depletion and remarkably high concentrations of hydrogen sulphide, thus causing extensive damage to the living, aquatic resources in the region (Nandan and Abdul-Azis, 1995).

## Trace metals

Trace metals are of great significance as pollutants in estuaries and coastal waters because many are highly toxic to a variety of organisms. Fishes are particularly affected, as are humans, by the way in which these metals can either affect, change, or accumulate through, the food chain. For example, copper, cadmium, mercury and silver are all particularly toxic to phytoplankton and may influence overall primary productivity or alter phytoplankton species composition. Higher organisms are exposed to trace metals both in solution and in food; uptake from solution occurs mainly through the gills and from food. Many invertebrates in particular, accumulate large amounts of trace metals which may then be taken up and accumulated by predatory fish (Phillips, 1994). Fish feeding near the bottom of the food chain, particularly substratum feeders such as mullet, can also accumulate trace metals.

There are a variety of public health standards relating to trace metals in human foods, including fishes. Examples of some of those for Australia are shown in Table 9.2. There remains much debate concerning the applicability of such standards, largely because the medical and epidemiological data on direct links between fish-eating and trace metal poisoning are poor (Phillips, 1994).

The natural levels of mercury in sharks and other top predators in tropical Australian waters have been shown to be sufficiently high to prohibit their use as human food (Lyle, 1984, 1986). Hence mercury accumulation cannot always be attributed to anthropogenic sources.

**Table 9.2** Examples of Australian National Health and Medical Research Council standards for trace metals in fish used as human food (modified from Phillips, 1994)

| Metal | Maximum permitted concentration ($\mu g\ g^{-1}$ wet weight) |
|---|---|
| Arsenic | 1.0 |
| Cadmium | 0.2 |
| Copper | 10.0 |
| Lead | 1.5 |
| Mercury | 0.5 |
| Zinc | 150.0 |

Studies in a number of industrially polluted estuaries in India have, however, revealed trace metal accumulation in fishes derived from industrial effluent. For example, *Chanos chanos* and *Liza macrolepis* in the Ennore estuary, which receives effluent containing mercury from a chloralkali plant, contain mercury. The concentration apparently does not yet pose an imminent threat to human consumers or to the biota themselves, but mercury can accumulate by three orders of magnitude in *Mystus gulio*, a catfish that has a detritus–invertebrate food chain pathway. Total mercury in 29 species of fish sampled from commercial landings varied from 0.01 to $0.48 \mu g\,g^{-1}$ dry tissue (Joseph and Srivastava, 1993). Similarly, industrial effluent discharged into the Rushikulya estuary from a chlor-alkali plant has been identified as the source of mercury which has polluted the estuary since the early 1970s. The mercury concentrations in sediment samples ranged from 1.6 to $192 \mu g\,g^{-1}$. The results indicated that pH was an important factor regulating the availability of mercury; the lower the pH the higher the availability of mercury. The overall assessment indicated that, although localized, the sediment of the Rushikulya estuary is contaminated with high concentrations of mercury with respect to both bioavailability to fishes such as mullet, and toxicity (Panda *et al.*, 1990). That high mercury levels in fish cause mercury accumulation in human fish eaters was demonstrated by Srinivasan and Mahajan (1989) working on sediments and organisms from Thana Creek and Ulhas estuary. Higher concentrations of mercury were recorded in surface sediments at stations situated near effluent discharge points of major industries. Mercury concentrations in organisms were found to be highest in benthic filter feeders and detritus feeders. The blood mercury level in individual people in the region was strongly dependent on how much fish they ate. There was a three-to-four-fold increase in blood mercury content of fish-eaters over non-fish-eaters.

Studies by Law and Singh (1987, 1988) in the polluted Kelang estuary of Malaysia showed that mercury and lead levels in *Arius thalassinus*, *Plotosus anguillaris*, *Dasyatis zugei*, *Lagocephalus lunaris*, *Setipinna taty* and *Johnius carutta* (overall means of $0.22 \mu g\,g^{-1}$ and $0.27 \mu g\,g^{-1}$) were five times greater for mercury and two times more for lead than in an adjacent unpolluted estuary, but still below the maximum permitted levels for human consumption of $0.5 \mu g\,g^{-1}$ and $1.5 \mu g\,g^{-1}$ respectively. Levels of zinc and copper were similar in fish from both estuaries.

The effects of heavy metal accumulation on the health or fitness of tropical estuarine fishes may be considerable. Some of the mechanisms and biochemical pathways are summarized by Jobling (1995). Five species of fish (*Mugil cephalus, Liza macrolepis, Liza parsia, Mystus gulio* and *Rastrelliger kanagurta*) from two polluted stations in Visakhapatnam harbour (India) were compared haematologically with the same species from two control stations. All species from polluted waters showed significantly lower erythrocyte numbers, haematocrit, haemoglobin and thrombocyte percentages; and significantly higher mean cell volumes, leucocyte numbers and lymphocyte percentages, compared with the controls. The pathological haematological characteristics of fish from polluted waters were ascribed to the synergistic effects of the toxicants (Pb, Cd, Cu, Fe, Zn, Mn) rather than to the effects of each toxicant separately (Rao *et al.*, 1990). In addition, the condition factor of both the mullet species (*Mugil cephalus* and *Liza macrolepis*) from Visakhapatnam inner harbour (polluted site) was significantly lower than that from control sites (Bhaskar *et al.*, 1989). Studies by Lin and Dunson (1993) in the USA indicate that the acute toxicity of cadmium to both tropical and temperate estuarine fishes is similar, and may be due to their having the same physiological and biochemical responses to such abiotic stresses. Developmental defects such as pugheadedness, jaw deformities and stunted or missing dorsal spines or rays in three *Haemulon* species, *Diplodus argenteus*, *Kyphosus sectatrix*, *Lagodon rhomboides*, *Archosargus rhomboidalis*, *Sphoeroides testudineus* and *Lutjanus griseus* have been linked to heavy metal contamination in Florida coastal waters (Browder *et al.*, 1993).

### Polychlorinated biphenyls (PCBs)

This group of compounds are primarily used as dielectric fluids in electrical equipment but are also used in hydraulic fluids, oils, plasticizers and paints. They were first produced in 1929 in the USA and since then over 1 million tonnes have been manufactured worldwide. It is estimated that 30% of these have entered the global environment, while the remainder are still in use in equipment or have been dumped in landfills (Phillips, 1994). They are highly toxic to most organisms, persistent in the environment and resistant to degradation. They have been implicated in bird population decreases through egg-shell thinning (Risebrough and Anderson, 1975; Spitzer *et al.*, 1977). They are hydrophobic and lipophilic, and hence accumulate in fatty tissue, so it is to be expected that those fishes with a high fat content such as mullet will be the best accumulators of PCBs. This has been shown to be the case with *Mugil cephalus* in the Brisbane River estuary, Australia, where PCBs measured $500 \, \mu g \, g^{-1}$ (Shaw and Connell, 1980). The same researchers also found that they accumulate in piscivorous birds, with *Pelecanus conspicillatus* having particularly high levels. Concentrations in fishes from non-industrialized tropical areas of Australia, such as the Gulf of Carpentaria, were much lower, with *Sillago* species containing $37 \, \mu g \, g^{-1}$ and *Arrhamphus sclerolepis* $40 \, \mu g \, g^{-1}$.

The way in which PCBs accumulate in fishes in temperate estuaries was found to be via selective transport from sediments and water to fish food organisms, and thence to fish (Samuelian and O'Connor, 1985). A similar pathway is assumed for lower latitudes but few data are available. The uptake of PCBs by *Rhabdosargus holubi*, a warm temperate to subtropical South African endemic, was studied experimentally (De Kock and Lord, 1988). Bioconcentration factors for whole fish were $2.4 \times 10^4$ and the biological half-life of the chemical during depuration was estimated at 50 days.

## Oil and hydrocarbons

Oil refineries and petroleum-based industries are usually located on large estuaries and their effluents have a variety of effects on the estuarine fauna. High levels of contamination seem only to occur in and around industrial complexes and do not spread significantly to non-industrialized areas. In Australia, high levels of petroleum hydrocarbons have been recorded in both birds and fish in the Brisbane River estuary, particularly in the mullet *Mugil cephalus*. The mullet fishery of south-east Queensland has been adversely affected by a kerosene-like taint to the flavour of the fish, caused by accumulation of hydrocarbons from feeding on contaminated sediment and food items from the Brisbane River (Connell, 1974). The concentration of polyaromatic hydrocarbons (PAH) in fishes is apparently related to their tissue lipid content; in a recent study of three fishes (*Mugil cephalus*, *Nematalosa come* and *Arius graeffei*) in the Brisbane River estuary, Kayal and Connell (1995) found a significant relationship between fat content and PAH concentration.

Most of the effects of hydrocarbons on fish populations are indirect and are caused by alterations to the food chains leading to the fish. These may be considerable: for example, in a study of a small estuary at New Bayou, Texas, which receives water discharge from oil and gas fields that is contaminated with petroleum products, Nance (1991) found that a sediment hydrocarbon concentration of $2.5\,\mathrm{mg\,g}^{-1}$ dry sediment depressed the population abundances of 96 macrobenthic species. Both abundance and diversity values were lowest at central sites near the discharge point.

In India the Hooghly–Matlah estuarine system near Calcutta receives vast quantities of pollutants including those from the Haldia oil refinery complex. The estuary receives about $4800–8400\,\mathrm{m}^3\,\mathrm{day}^{-1}$ of oil-borne wastes. This oil refinery effluent is acidic to alkaline, has a low oxygen content, and occasional high BOD values, and large amounts of phenol, oil and chromium. The effluent, although treated, is toxic to fish, fish food organisms and shrimps, as evident from toxicity bioassay studies. The impact of organo- and chemotoxic components of the effluent affected the primary productivity of the Hooghly estuarine water around the waste discharge point of the refineries (Bagchi et al., 1987). Panigrahi and Konar (1989) found significant reductions of zooplankton, phytoplankton and bottom organism densities near effluent discharge points together with severe depletions of dissolved oxygen. Commercial fish catches have also declined in the area.

## Sewage

The overenrichment of estuaries by untreated sewage effluents has the following consequences (Begg, 1978; Phillips, 1994):

- Public health hazards and toxicological problems through seafood consumption;
- oxygen depletion and the development of sludge communities;
- eutrophication;
- modification of species composition and abundances, particularly with regard to fishes;
- aesthetic offence and reduced amenity and recreational values.

According to Phillips (1994), managing water quality with respect to sewage effluent is a function of attaining the appropriate level of enrichment of the receiving waters. This can be achieved only by matching the amount of sewage discharged with the capacity of the estuary to dilute and disperse the effluent. Whereas the treatment and discharge of sewage effluent in developed countries is becoming much better controlled, the situation in subtropical and tropical developing countries remains serious. Here most sewage is discharged in a raw state into estuaries with little concern for its effects.

A study of the Adyar and Cooum estuaries around Madras in India (Nammalwar, 1992) found that the major threat to fishes is the effect on water quality of the discharge of untreated domestic sewage. The sewage output into the Adyar and Cooum estuaries has been estimated at about 8.1 million litres per day and 69.3 million litres per day respectively. The sewage causes anoxic conditions and many fish species have disappeared. Attempts to utilize some of the bacterial and algal growth caused by eutrophication by culturing mullet species have not been very successful due to the low oxygen concentrations. Efforts have been made in other parts of India to use sewage to increase fish production in estuaries. In the vast low-lying areas around the Kulti estuary which receives Calcutta city sewage, the wetlands are being profitably used for recycling of waste water through fish culture. Appropriate management of such systems in non-saline and saline areas, otherwise unsuitable for agriculture, is opening up immense possibilities for increasing fish production (Ghosh *et al.*, 1987). In some Indian estuaries, the incidence of sewage-borne pathogens such as the bacterium *Escherichia coli* is a problem. The seasonal incidence and levels of *E. coli* in environmental samples (water, plankton and sediment) during 1981 and 1982 and in seafoods during 1982 and 1983 was monitored at the Vellar estuary by Ramesh (1987). High *E. coli* counts in water, plankton and sediments were recorded during the monsoon and post-monsoon periods, while low populations coincided with summer. Among seafoods, *E. coli* was detected from 30% of fish, 30% of crustaceans and 51% of bivalves examined. Elsewhere in Asia, severe sewage pollution of estuaries has been reported from the Segara Anakan estuary in Java where coliform bacteria counts have been very high (Thayib *et al.*, 1991).

In Brunei on the island of Borneo, the Brunei River estuary receives major portions of the capital city's stormwater run-off along with treated and untreated sewage. It is treated like a sink to assimilate all nature- and human-related solids, liquids and gases without regard for the risks to its ecosystem. A 1987 study concluded that the existing level of sewage discharged into the river did not have any apparent effect on its ecosystem. Findings indicated that the river's water quality has deteriorated in terms of bacteriological quality and suspended solids and that high levels of bacteriological contamination near the sewage outfall will be a health hazard if not controlled (Yau, 1991).

On Viti Levu, the main island of Fiji, effluents from sewage treatment plants are disposed of in mangroves or discharged into mangrove creeks. Mangrove land is used because of its relatively low value and because the clay soils are suitable for building treatment ponds. The actual mangrove area used for sewage disposal is small but the effects of the sewage can be significant as evidenced by the lush growth of mangroves adjacent to sewage discharge points (Lal, 1984). Circumstantial evidence suggests that there has been a decline in species diversity and catch per unit effort of fish in the Vatuwaqa estuary where much of Suva's sewage (including septic tank effluent) is discharged. Crabs and oysters have disappeared from the estuary (Lal, 1984).

The effect on an estuary of eutrophication caused largely by discharge of treated sewage was demonstrated in a comparison of two degraded estuaries in KwaZulu-Natal in South Africa (Blaber *et al.*, 1984). The composition, interrelationships and fluctuations of the biota of Tongati and Mdloti estuaries were studied in relation to the physical environment and various forms of human interference. Results were compared with previous studies on an adjacent, relatively undamaged estuary. The two estuaries are impoverished in different ways for largely different reasons. Tongati receives treated sewage effluent, is rarely closed from the sea, has low salinities and low oxygen tensions, but is rich in P and N nutrients. Energy values of benthic floc from Tongati were high, and large quantities of the floating exotic water hyacinth, *Eichornia crassipes* occurred. The zooplankton and zoobenthos were impoverished and dominated by freshwater species. Mdloti frequently closed from the sea but was often artificially opened to prevent flooding of sugar cane farmland. It exhibited typical estuarine salinity patterns, was well oxygenated but relatively poor in P and N nutrients. Energy values of benthic floc were low and primary production low. Zooplankton and zoobenthos were impoverished, probably due to the flushing effect of unseasonal artificial mouth opening or pesticide pollution (dieldrin). The fish faunas of both estuaries were similar, impoverished, and dominated by Mugilidae. Evidently the food chain from benthic floc (detritus) to iliophagous fish can remain viable under great environmental stresses in these degraded estuaries.

In the tropical Western Hemisphere, similar sewage contamination problems have occurred in estuaries. A survey of benthic invertebrates of the Nicoya Gulf in Costa Rica, an important artisanal and commercial fishing area, revealed the effects of raw sewage discharge (Maurer *et al.*, 1984). The water quality of a small tropical estuary in Barbados, West Indies, was investigated by Turnbull and Lewis (1981). Raw and partially treated sewage were evident throughout the estuary. Marked concentration gradients were observed along a transect from the inner estuary to the open ocean. Concentrations of micronutrients and biochemical oxygen demand (BOD) decreased along the transect, while water transparency, dissolved oxygen and salinity increased. Suspended particulate material was abundant in surface water throughout the estuary and persisted beyond the mouth into the ocean.

## Dredging

Dredging of estuaries may be undertaken for harbour maintenance and for keeping shipping channels open, to maintain the estuary in a particular state or to alter its characteristics, or for recreational reasons. Whatever the purpose, it involves massive disruption of the substratum together with disposal of the spoil either on the banks or out at sea. It therefore has major impacts on the ecology of the system.

The effects on fishes of frequent harbour and navigation channel dredging in tropical estuaries have not been well documented, but they include changes to the species composition as well as effects on movement patterns. For example, deeper channels encourage the movement of larger fishes, particularly sharks, into the lower reaches of the estuary. The dredging often causes the bottom to consist of a layer of fluid mud, as in the Orinoco estuary in South America and the Chittagong estuary in

Bangladesh (Eisma *et al.*, 1978; Burren *et al.*, 1981). The effects of such a layer on fishes are unknown.

Dredging is carried out at the mouth of St Lucia estuary in KwaZulu-Natal to try and maintain the ecological health of the coastal lake system by maintaining a connection to the sea through the narrow estuary. This is mainly to prevent the build-up of hypersaline conditions. The effects of dredging at St Lucia have been reviewed by Cyrus and Blaber (1988), who have documented the increased and record high turbidities caused by dredging. Studies of the relationship between fish distribution and turbidity emphasize the potential significance of the turbidity changes caused by dredging (Cyrus and Blaber, 1987a,b). Work in temperate areas suggests: (a) that increasing turbidity and siltation together with dredging activities are an inevitable consequence of human activity in estuaries (McCluney, 1975); (b) that the increased potential for entrapment of sediments makes the dredging process partially self-perpetuating (Palmer and Goss, 1979); and (c) that the alteration of the estuarine system beyond natural limits will cause changes that, in many ways, are still unpredictable (Dyer, 1972).

Much of the impact of dredging on estuarine fishes is indirect through its effects on the occurrence and density of benthic prey organisms. At St Lucia, Hay (1985) found that areas dredged 10 years previously were still essentially devoid of benthic animals while adjacent substrata were densely populated. For example, the density of benthic invertebrates in dredged areas was $<200$ animals $m^{-2}$ (only 1 species), while in undredged areas it was 3000–5300 animals $m^{-2}$ (7 to 10 species). Similar reductions in density and diversity have been reported in other subtropical estuaries in Texas (Gilmore and Trent, 1974; Johnson, 1981; Harrel and Hall, 1991).

Therefore despite the ecological advantages of dredging at St Lucia, its overall effects in the long term may be detrimental to the fishes of the estuary and possibly the connected coastal lake system. In other areas in South Africa, dredging has been employed to try and restore some of the estuaries that have been severely affected by siltation resulting from soil erosion in the catchment (Wiseman and Sowman, 1992). This restoration has largely been for recreational purposes and was successful in the blind Mhlangankulu estuary, where dredging was confined to the centre of the lagoon so as not to disturb the fauna and flora in shallower water (Begg, 1978).

The dredging of sand from tropical estuaries for construction purposes is another activity that has not been well documented. The mouth of the Rewa estuary in Viti Levu, Fiji, has been dredged for sand for at least 25 years, but the effects are thought to be negligible compared with the vast sediment loads brought down by the river (Lal, 1984).

## 9.3    EFFECTS OF AGRICULTURE

Like those of industry, the effects of agricultural activities on estuarine
fishes are largely indirect, but nonetheless significant. The impacts are
caused mainly by physical and hydrological changes to estuaries as well
as by contamination with pesticides and fertilizers. The main effects
involve:

- siltation, especially in lower reaches and lagoon areas, caused by deposition of material originating in run-off from agricultural lands and
  resulting in shallowing and increased turbidity;
- encroachment of agriculture onto floodplains and destruction of riparian
  vegetation;
- chemical pollutants entering the estuary, mainly in run-off from farm
  lands.

In most subtropical and tropical lowlands, extensive monocultures have
developed in and around estuaries. The most important of these are sugar
cane and oil palms.

### Sugar cane

Man's insatiably sweet tooth has led to an ever-increasing demand for
sugar, and ever more land has been planted with sugar cane, with higher
yields encouraged by the use of chemicals. The adverse effects of sugar
cane on estuaries and their fishes can be divided into two categories:
effects of farming and effects of mills.

#### *Effects of sugar cane farming*

In the South Johnstone River estuary in North Queensland, the sediment
delivery from highly erosion-prone sugar cane cultivations in the tropical
catchment has increased dramatically. This has subsequently given rise to
elevated flood levels in both the lower and upper catchment areas, as well
as significant modifications to the river bed morphology (Arakel *et al.*,
1989). In South Africa too, at any one time there are about 35 000 ha of
ploughed land in the sugar belt of KwaZulu-Natal that are vulnerable to
erosion through direct exposure to rain (Begg, 1978). Begg (1978) considered siltation derived largely from such agriculture as having a greater
detrimental effect on KwaZulu-Natal estuaries than any other single
factor. Although local regulations in KwaZulu-Natal prohibit planting of
cane right to the bank of the river or estuary, this law is seldom enforced,
resulting in erosion of banks, loss of riverbank vegetation and more
siltation in the lower reaches. In Fiji large areas of mangroves have been

lost to sugar cane cultivation, with consequent loss of juvenile fish habitat (Lal, 1984).

The chemicals used on sugar cane include nematicides, insecticides, fungicides, ripeners, fertilizers and herbicides. The effects, both direct and indirect, of many of these on estuarine fishes are drastic. Nematicides may kill a large proportion of the benthic meiofauna while insecticides such as dieldrin (still in use in many developing countries) actually kill fish. Most of the fertilizers cause eutrophication with subsequent changes to the fauna (Blaber *et al.*, 1984).

### Effects of discharges from sugar cane mills

Begg (1978) quotes the following passage from a KwaZulu-Natal newspaper of 1909: "For years past it has been the custom of sugar mills to discharge 'dunder' into the nearest stream. Most, if not all, of the coast streams end in a lagoon which is landlocked for part of the year. The 'dunder' (organic waste from sugar cane processing) is fatal to fish and when the first rains come the stream becomes so polluted that the fish in the river and lagoon die in tens of thousands." He goes on to state that the situation today is largely unchanged. Serious pollution problems resulting at least partially from sugar cane processing have also been reported from the Mundau–Manguaba estuary complex in north-eastern Brazil (a 79 km$^2$, shallow, tropical, coastal lagoon system consisting of two interconnected water bodies and a channel to the ocean; Oliveira and Kjerfve, 1993), the Paraiba do Sul estuary in south-east Brazil (Lacerda *et al.*, 1993), and Cuba (Areces-Mallea and Gil-Vades, 1980), and direct fish kills have been reported after large volumes of untreated or partially treated waste have been discharged from sugar mills in Lautoka and Labasa in Fiji (Lal, 1984). Mass mortality of fish in the River Mathab-hanga–Churni in West Bengal, India, is a common event, occurring several times a year following the discharge of sugar mill complex effluent (SMCE) from Darshana, Bangladesh (Ghosh and Konar, 1993).

Various ways of trying to avoid discharging the water-borne waste products of milling into estuaries have been attempted, such as using the liquid to irrigate adjacent cane lands, but the disposal of the very high volume of liquids produced is a problem. Treatment of the wastes to render them less harmful is possible but an expensive option.

### Oil palm plantations

Oil palms have been planted over tens of thousands of hectares in many lowland areas of South East Asia and are the basis of a highly valuable agro-industry. Unfortunately they are also the source of some of the worst pollution causing the death of fishes in mangroves (Sasekumar, 1980). The use of arsenical herbicides in the plantations has led to pollution of estuaries with this trace metal (Sarmani *et al.*, 1992). The discharges of raw or partially treated palm oil waste water from over 150 mills in Malaysia have contributed an estimated BOD load which is equivalent to that of nearly 6 million persons (Hum *et al.*, 1981). The problems lie in the organic pollutants in the sludge from oil palm mills. Various methods to try and reduce the BOD of the waste have been tried, including an irrigation biodegradation canal method, but this failed to significantly reduce BOD values. The waste's adverse toxicity effects on fish such as the carp, *Cyprinus carpio* and mullet, *Liza subviridis* demonstrated the seriousness of the problem (Sivalingam and Thavaraj, 1978).

Fortunately, at least in Malaysia, since 1977, stringent environmental regulation has been progressively imposed on the palm oil mills, Malaysia's most polluting industry. The impact of the regulation on international trade and producer welfare has been quite small compared with the relative benefits to society in terms of changes in the levels of dissolved oxygen in the industrial effluents (Khalid and Wan-Mustafa, 1992). This improvement has been achieved by using a variety of techniques including the use of anaerobic pond digestion followed by aerobic unit processes. In this way up to 95% of the BOD can be removed (Hum *et al.*, 1981).

Oil palm is one of a group of tropical agro-industries that cause pollution in estuaries. Others include rubber, cocoa and sago as well as pulp and paper mills. Pulp and paper mill wastes have been a serious problem in the Hooghly–Matlah estuary complex in the Calcutta region of India. This system is extremely important for fisheries production and aquaculture. The degree of pollution at any one time depends upon the tide, but overall is believed to be partly responsible for declining fish catches (Bagchi and Ghosh, 1986).

## 9.4   EFFECTS OF DAMS AND WEIRS

Dams are seldom built across estuaries, but on the rivers above estuaries so as to retain fresh water. The enormous demand for fresh water to supply people and industries in the cities, and to irrigate agriculture, has led to the construction of dams on very many tropical and subtropical rivers. The effects of dams on estuarine fishes are twofold: firstly, they present a physical obstruction to fish movement, of particular importance to diadromous species; and secondly, they alter the flow characteristics of the river, its sediment loads, and most importantly, the amount of fresh water entering the estuary.

## Obstruction

The very existence of a barrier across a river will have a substantial effect on fish migrations and may change the species composition both above and below the dam. Despite the successful use of devices such as fish ladders to allow migratory fish such as salmonids and anguillids to cross the barrier in temperate regions, such techniques have not proved very successful in the tropics.

In many cases, governments have decided that the benefits derived from dam construction outweigh the adverse effects on estuaries. For example, an investigation of the Feni River estuary in Bangladesh (Halder *et al.*, 1991) showed that the estuary was used by 34 species of upstream fresh-water fishes, 11 species of estuarine-dependent freshwater fishes and 9 species of marine fishes at different stages of their lives. A proposed weir constructed in the lower reaches of the estuary would inevitably change the present brackish-water estuary into a freshwater lake during the dry season, and destroy the nursery grounds of fish and prawns. However, it was judged that the overall economic and food benefit to be derived from increased agricultural output in the project area, and fish production from the reservoir, would outweigh the reduction in fisheries and loss of nursery grounds in the estuary.

In Senegal, West Africa, the Guidel Dam is the first of a series constructed on the Casamance estuary (see also section 4.3, and p.126). This system in southern Senegal suffers from hypersaline conditions, due mainly to low rainfall. To combat this problem, the government of Senegal is developing 70 000 ha of rice lands above anti-salt dams. The Guidel Dam, of 1150 ha, was the first one finished and therefore considered a pilot project. The salinity range increased above the dam, but primary production and fishing yields were not greatly affected. However, export of nutrients towards the estuary is prevented and some of the nursery grounds for the commercially important clupeid *Ethmalosa fimbriata* and prawn *Penaeus notialis* have now been lost (Barry and Posner, 1986; Le Reste, 1986).

Changes in fish species above and below a dam have been documented for the lower Tocantins River, a tributary of the Amazon estuary. This study analysed the effects of a dam in a tropical region on the whole fish community. The results indicated an immediate and long-term reduction of fish numbers and species diversity (de Merona, 1987).

Perhaps one of the best examples of the impact of a dam on the movement of fish concerns the Farakka Barrage on the Ganges in India. This dam, built on the Indian side of the border with Bangladesh, controls the water flow of the Ganges system into Bangladesh and south to West Bengal in India. The hilsa, *Tenualosa ilisha*, is an anadromous clupeid that migrates long distances from estuaries to fresh water (p.195) and supports a very important fishery (pp.239–243) in India and Bangladesh. Prior to the construction of the Farakka Barrage, the Ganges River supported a prosperous hilsa fishery that stretched far into the interior of India. Now that movement beyond the barrage is restricted, the hilsa cannot reach some of their historical spawning grounds, and catches above the dam have fallen from $176 \, \text{kg km}^{-1}$ to almost nothing (Melvin, 1984). A similar dam, the Ghulam Mohammed Barrage on the Indus River in Pakistan, also had drastic effects on the migration of hilsa and led to its disappearance above the dam (Husain and Sufi, 1962a), despite the incorporation of fish ladders in its construction. The fishery below the dam did not apparently suffer unduly (Husain and Sufi, 1962b).

The Farakka Barrage is also thought to have influenced the fisheries productivity of the Hooghly estuary through changes in water quality and flow patterns; (Lal (1987) found that turbidity, dissolved oxygen, nitrates, phosphates, temperature, salinity and alkalinity, all of which affect the rate of primary production, became altered. Not all the fisheries effects were negative. For example, the reproductive biology and fisheries of *Polynemus paradiseus* in the Hooghly–Matlah estuarine system changed after the commissioning of the Farakka Barrage. Conspicuous reductions in size at first maturity and an increase in the spawning period of the species were recorded with the alteration in hydrological conditions. This has resulted in a steady increase in annual fisheries yield of the species (Mukhopadhyay *et al.*, 1995).

Other human-made obstacles that affect the movement of estuarine fishes include embankments built across estuaries for roads and railways. If these are solid or with insufficient culverts they reduce tidal exchange and cause damming of water. Many such embankments exist along the coast of KwaZulu-Natal although they are gradually being replaced by more open bridges that allow passage of water (Begg, 1978). Even culverts under roads can inhibit the upstream migration of diadromous fishes unless suitably engineered (McDowall, 1988).

### Freshwater retention and changes in flow patterns

The retention of fresh water in dams and hence the reduction of freshwater input into estuaries is most significant in subtropical areas with highly seasonal rainfall, such as South Africa, Australia and parts of South America. It may be less significant in equatorial high-rainfall areas but information on this point is scarce. The problem has been recently reviewed by Whitfield and Wooldridge (1994). They state, "River inflow is one of the most critical factors influencing southern African estuaries (and by inference those in other areas of similar climate) because of its role in sedimentary and dynamic processes, which in turn determine the biotic characteristics of individual systems." Large dams not only decrease the number of floods reaching an estuary, but reduce the volume of fresh water per flood reaching the estuary. Smaller floods may be completely captured by the dam, and in some parts of South Africa the capacity of existing dams actually exceeds the mean annual run-off. The abstraction of large quantities of fresh water has had the effect of forcing some estuaries into extreme ecological states. For example, the high irrigation demand of sugar cane near St Lucia has contributed to the cyclical hyper-saline conditions. In areas of freshwater shortage, there is now a very real planning problem if sufficient water is to be released to permit the viable functioning of estuaries. A comparison of estuarine responses to differing inputs of fresh water is illustrated in Fig. 9.1.

## 9.5   HABITAT LOSS

The major loss of estuarine habitat in the subtropics and tropics has involved the destruction of mangroves. The total worldwide mangrove area, which is estimated at about $170\,000\,km^2$ with some 60 species of trees and shrubs exclusive to the habitat, dominates approximately 75% of the world's coastline between latitudes 25°N and 25°S (Chapman, 1976; Wong and Tam, 1995). To quote Ong (1995), "Despite the recent better understanding and awareness of the role of mangroves, these coastal forest communities continue to be destroyed or degraded (or euphemistically reclaimed) at an alarming rate. The figure of 1% per year given by Ong (1982) for Malaysia can be taken as a conservative estimate of destruction of mangroves in the Asia–Pacific region. Whilst the Japanese-based mangrove wood-chips industry continues in its destructive path through the larger mangrove ecosystems of the region, the focus of mangrove destruction has shifted to the conversion of mangrove areas into aquaculture ponds, and the consequences of the unprecedented massive addition of carbon dioxide to the atmosphere by post industrial man."

The situation in West Africa is no less serious. Mangroves in West Africa cover an area of over $27\,000\,km^2$ of deltas, estuaries and lagoons, and are biologically linked to the offshore coastal ecosystem which has high fisheries productivity. Their excess organic production is exploited by many marine species, especially fishes and crustaceans that enter the mangrove environment as juveniles and return to the sea as adults. The beneficial effects on marine fisheries of this net energy outflow are at risk from anthropogenic influences causing pollution or destruction of the mangroves (John and Lawson, 1990).

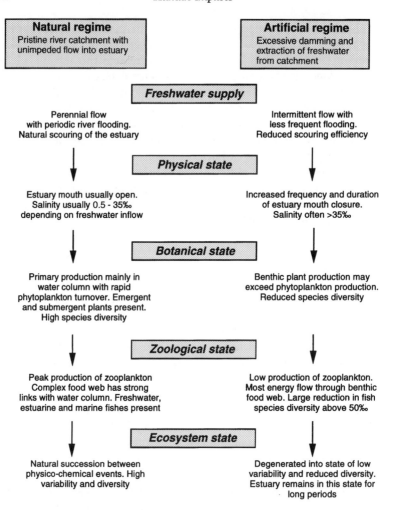

**Fig. 9.1** A diagrammatic representation of estuarine responses to differing inputs of fresh water (adapted from Whitfield and Wooldridge, 1994).

One of the problems in relation to mangrove conservation in relation to fisheries is that hard data usable for quantifying the role of mangroves in fisheries-related food chains are scarce (Pauly, 1985). However, recent research has indicated that there is a positive correlation between offshore fisheries production and mangrove area (Chong and Sasekumar, 1994). Evidence from parts of Indonesia now shows that destruction of mangroves has severely affected the fishery resources of the south coast of

Bali, west and east coasts of Lombok, and in Jakarta Bay (Subani and Wahyono, 1987). Similarly, the filling-in of mangrove swamps and coastal land reclamation in Singapore have affected the productivity of the seas around Singapore, and led to a decline of inshore and brackish-water fish yields (Sien, 1979). Singapore now has only 0.5% of its original mangrove forests (Sasekumar and Wilkinson, 1994).

Unfortunately there are three main competing interests for mangroves or mangrove land (Atmadja and Man, 1994):

- those focused on the value of intact forests, e.g. timber, fisheries and shoreline protection;
- those wanting to convert the land into other uses and clearfell the forests for wood chips;
- those emphasizing the conservation value of mangroves, e.g. for bio-diversity and ecosystem maintenance, and as carbon dioxide traps.

In South East Asia, the clearing of mangroves for aquaculture has been a major cause of habitat loss. More than 50% of the mangrove loss in the Philippines can be traced to brackish-water pond construction. The decrease in mangroves from 450 000 ha in 1920 to 132 500 ha in 1990 has been accompanied by expansion of culture ponds to 223 000 ha in 1990 (Primavera, 1995). In Ban Don Bay, a biologically productive area in Thailand, there has been rapid development. Since the mid 1980s more than 60% of the total mangrove forests along the coast of the bay have been converted for other land uses: 5331 and 2723 ha have been converted to prawn farms and agricultural lands, respectively, leaving only 4160 ha of mangrove forests (Panvisavas *et al.*, 1991). Sasekumar and Wilkinson (1994) state there are spectacular examples of failed aquaculture projects in Thailand involving massive mangrove destruction. For example, in the upper Gulf of Thailand, 365 000 ha of mangroves have been reduced to 600 ha (a reduction of > 99%) through development of prawn ponds. Most ponds have now been abandoned due to low production, diseases, falling prices and rising production costs associated with poor water quality. For a whole variety of reasons, prawn pond ventures usually fail in cleared mangrove areas and any gain is very short term, whereas the long-term loss to fisheries is incalculable. Sasekumar and Wilkinson (1994) conclude that mangrove land is unsuitable for prawn pond and agricultural development and that costs exceed income after 5 years. Mangroves are, however, highly productive if left or managed, and can return high sustainable yields of both fish and timber, as has been demonstrated by the carefully managed Larut–Matang forests of Malaysia. Nevertheless, in countries as far apart as Bangladesh and Costa Rica, prawn-farming areas are still expanding at the cost of large-scale destruction of mangrove forests (Abdus-Shahid *et al.*, 1992; Herrera and Charles, 1994).

In West Africa it is not only aquaculture that poses a threat to the mangroves and their nursery and productivity functions for fisheries. In the Niger delta, the massive destruction of $200\,km^2$ of the vegetation around Tsekelewu Oil Field is attributed to high salinity caused by salt water incursion from the sea caused by oil-related activities. The $200\,km^2$ area suffered varying degrees of damage, but in the transitional saline–fresh water zone in which the electrical conductivity of ground water was greater than $30\,dS\,m^{-1}$, all the vegetation including the relatively salt-tolerant white mangrove, *Avicennia africana*, was killed (Fagbami *et al.*, 1988). Other factors causing mangrove destruction in West Africa include crop production, fish breeding, sugar manufacture and recreation as well as solid and liquid fuel production (Ossom and Rhykerd, 1985).

## 9.6  EFFECTS OF RESIDENTIAL CANAL ESTATES AND MARINAS

In the subtropical regions of both the USA and Australia, as well as in some areas of South Africa and Brazil, unprecedented urban development is taking place in estuarine and coastal areas. Fringing the estuaries and mangroves, there are usually extensive tracts of agricultural lands that have little ecological value and these are now being used for residential development (Morton, 1993). Most such developments involve the clearing of mangroves and the creation of canals and lakes as well as marinas (Fig. 9.2). Quite apart from the loss of natural fish habitat and especially nursery areas, little is known about how and which fish use the newly created habitats.

From an engineering and development perspective, the only economical way of making the optimum use of low-lying estuarine wetlands and mangroves is to build up the level of certain parts with excavation of material from the remainder of the land (Jarvis and Colvin, 1981). In this way large numbers of canal estates have been constructed in south-east Queensland during the last 30 years. In many cases, scant attention was paid to the overall estuarine ecology and the consequences of large-scale habitat loss. Fortunately today, the clearing of mangroves in developed countries usually requires a detailed environmental impact study.

The main problems of such residential canal developments with regard to the ecology of estuarine fish are:

- changes to the hydrology of the estuary;
- changes to the types of habitats available;
- loss of nursery habitats;

**Fig. 9.2** Part of a residential canal estate that has replaced an intertidal mangrove area in south-east Queensland. Note the steep artificial rock walls.

- introduction of pollutants into the estuary from residential areas and marina activities.

In addition, the construction of canal estates in Australia has led to conflict with commercial and recreational fishing interests because the majority of species caught depend upon estuaries for all or part of their life cycle (Morton, 1989). Similarly it has been stated that canal developments in the Gulf of Mexico may seriously degrade coastal fisheries resources (Lindall and Trent, 1975). In upland canal estate developments in Tampa Bay, Florida, Lindall *et al.* (1973, 1975) found that in all types of canals, water stratification occurs, resulting in stagnant water with low oxygen levels. The main problem with these particular canals is that the tidal range in Florida is generally less than 1 m and adequate tidal flushing of the canals is difficult to achieve. The low oxygen levels cause the emigration or death of many organisms. Despite this, however, Lindall *et al.* (1973) recorded at least 33 species of fishes in the canals. The majority of individuals were planktivores, but most species (26) were benthic feeders. A similar trophic structure was found in the Marco Canals of Florida (Kinch, 1979), where the most abundant juvenile fishes are predominantly pelagic planktivores. However, the food and feeding habits of fish species in the canals were similar to those of fish from natural areas (Kinch, 1979).

In a comprehensive study of 34 canal estates in south-east Queensland, Morton (1989) found that the tidal canals constitute an aquatic environment that has been artificially created whilst the source rivers have been extensively modified. The canals differ from the natural estuary in their sedimentary composition, hydrology and ichthyofauna. Although the ichthyofaunas of the canals and estuaries were similar, that of the canals was less abundant. The restricted canal access, poorer water quality and clay or silt sediments probably adversely affect the influx and residence times of fishes. Substantial differences were observed in the composition of the fish faunas of the canals and undisturbed estuarine areas (Table 9.3). In undisturbed areas, gerreids and clupeids constituted less than 6% and 10% of fish numbers, but in the canals they made up 49% and 25% respectively. The dominant species in Queensland canals are thus planktivores or microbenthic carnivores such as clupeids and gerreids – similar to the situation in Florida. Another difference between estuary and canal is that macrobenthic carnivores are less abundant in canals. The lack of habitat diversity in canals and the creation of a featureless sand or mud substratum has apparently reduced the overall number of benthos and detritus feeders, many of which are important to fisheries. Morton (1989, 1992) also considers that the practice of constructing canal estates in narrow-mouthed estuaries is more detrimental to adjacent fish populations than if such developments are located on wide-open bays. Furthermore, these developments result in the loss of valuable feeding and nursery grounds for fishes of economic importance.

Well-designed canals in sandy areas with good flushing can actually increase the area of habitat for commercially important fish species. This has been documented for the Kromme estuary and marina in South Africa (Baird *et al.*, 1981; Baird and Pereyra Lago, 1992; Schumann and De Meillon, 1993). Here oceanic water from the nearby ocean rapidly penetrates the marina and there is efficient water exchange.

Considerable pollution may occur in both canal estates and marinas. In canal estates, this usually takes the form of eutrophication caused by such things as lawn fertilizer run-off. In marinas, untreated human waste may be discharged from boats, and oil, paint and antifouling compounds frequently leach into marina waters (Jenkins, 1985; Fisher *et al.*, 1987).

## 9.7  MASS MORTALITY OF FISHES

Mass mortalities of fishes, or 'fish kills' as they are popularly known, are not uncommon in estuaries. In some cases they are a result of natural phenomena, in others the proximate causes may be natural but are a result of human activities, whereas some may be a direct result of pollution. Large numbers of dead fish in an estuary are very tangible evidence of changes to the physico-chemical environment. They usually generate much publicity, but unfortunately the causes of the mortality are frequently difficult to determine with any accuracy. Mass mortalities of fishes in South African estuaries have been reviewed by Whitfield (1995), who comments that a striking feature of most fish kills is the wide range of species that are affected, despite their differing physiological tolerances.

## Naturally occurring phenomena

The most common hydrological causes are lethal combinations of one or more of the following: salinity/temperature, dissolved oxygen/temperature and dissolved oxygen/turbidity.

Fish kills in relatively undisturbed coastal lake systems of KwaZulu-Natal, where there is little pollution, show the importance of combinations of low temperatures and salinities. A large-scale mortality of fish at St Lucia in KwaZulu-Natal in 1976 was thought to have been due to a lethal combination of low salinities and low temperatures leading to osmoregulatory failure. In this instance an estimated 100 000 fish of at least 11 species died. The fish kills coincided with a period of very low salinities ( < 3.5‰) and temperatures as low as 12 °C. Fish were unable to escape to more favourable areas because of the relatively uniform conditions over much of the system (Blaber and Whitfield, 1976). Almost all the species that were affected were of marine origin and the cichlid *Oreochromis mossambicus* was unaffected. In contrast to this, a fish kill of the freshwater catfish *Clarias gariepinus* in the upper reaches of the St Lucia system was probably due to the influx of high-salinity (42‰) water into the rivers. This catfish is not tolerant of salinities above about 10‰ (Blaber, 1981b). A mass mortality of several species of fish occurred in the Kosi estuary in KwaZulu-Natal in July 1987. *Caranx sexfasciatus, Strongylura leiura, Monodactylus argenteus* and many *Ambassis* species and juvenile Mugilidae were found dead on the beach inside the estuary mouth. Examination of environmental parameters during the period involved showed that both minimum air and water temperatures reached their lowest recorded levels that year; and the period also coincided with a neap tide when water of low salinity was flowing out of the estuary mouth (Kyle, 1989).

Unusually low temperatures are thought to be responsible for periodic fish kills in Texas bays and estuaries. In the 1980s the Texas coast experienced three unusually cold weather periods (one in 1983 and two in

**Table 9.3** Comparisons of the number of fish sampled per site from the Nerang River Estuary and adjacent canal estates (modified from Morton, 1989)

| Family and Species | Estuary | Canals | Life stage* | Value† |
|---|---|---|---|---|
| Antennariidae | | | | |
| *Phrynelox striatus* | 0.0 | 0.2 | A | – |
| Atherinidae | | | | |
| *Pseudomugil signifer* | 0.0 | 25.2 | J, A | C |
| Belonidae | | | | |
| *Ablennes hians* | 0.5 | 0.0 | A | – |
| *Tylosurus macleayanus* | 2.5 | 0.0 | J, A | – |
| Callionymidae | | | | |
| *Callionymus liniceps* | 1.5 | 0.0 | A | – |
| Carangidae | | | | |
| *Caranx ignobilis* | 0.0 | 0.7 | J | R |
| *Scomberoides lysan* | 0.5 | 1.3 | J | R |
| Chandidae | | | | |
| *Ambassis jacksonensis* | 15.5 | 12.7 | J, A | – |
| *Ambassis marianus* | 9.5 | 76.8 | J, A | – |
| Clupeidae | | | | |
| *Harengula abbreviata* | 1130.5 | 342.5 | J, A | – |
| Dasyatidae | | | | |
| *Dasyatis fluviorum* | 1.0 | 0.0 | J | – |
| Engraulididae | | | | |
| *Engraulis australis* | 1.5 | 5.8 | J, A | – |
| Gerreidae | | | | |
| *Gerres ovatus* | 2251.0 | 646.8 | J, A | – |
| Gobiidae | | | | |
| *Gobiomorphus lateralis* | 0.0 | 6.3 | J, A | – |
| Haemulidae | | | | |
| *Plectorhinchus nigrus* | 1.0 | 0.0 | J | C, R |
| Hemiramphidae | | | | |
| *Arrhamphus sclerolepis* | 16.0 | 9.5 | J, A | C, R |
| *Hyporhamphus quoyi* | 1.5 | 1.5 | J, A | C, R |
| Kyphosidae | | | | |
| *Girella tricuspidata* | 0.5 | 1.5 | J, A | C, R |
| Monacanthidae | | | | |
| *Monocanthus chinensis* | 0.0 | 0.2 | J | – |
| Monodactylidae | | | | |
| *Monodactylus argenteus* | 1.0 | 2.7 | J, A | – |
| Mugilidae | | | | |
| *Liza argentea* | 43.0 | 256.3 | J, A | C |
| *Mugil cephalus* | 38.0 | 53.7 | J, A | C |
| *Mugil georgii* | 29.0 | 187.7 | J, A | C |
| *Myxus elongatus* | 0.5 | 1.5 | J | C |
| Platycephalidae | | | | |
| *Platycephalus arenarius* | 1.0 | 0.0 | J | C, R |
| *Platycephalus fuscus* | 8.0 | 1.2 | J, A | C, R |
| Pleuronectidae | | | | |
| *Pseudorhombus arsius* | 4.5 | 2.0 | J, A | C |

Table 9.3   Continued

| Family and Species | Estuary | Canals | Life stage* | Value† |
|---|---|---|---|---|
| Plotosidae | | | | |
| *Euristhmus lepturus* | 0.0 | 0.5 | A | C |
| Pomatomidae | | | | |
| *Pomatomus saltatrix* | 0.0 | 0.7 | J | C, R |
| Scatophagidae | | | | |
| *Scatophagus argus* | 0.0 | 0.7 | A | C |
| *Selenotoca multifasciata* | 0.0 | 2.2 | J, A | C |
| Sillaginidae | | | | |
| *Sillago analis* | 8.5 | 2.3 | J, A | C, R |
| *Sillago ciliata* | 148.5 | 53.7 | J, A | C, R |
| *Sillago maculata* | 16.0 | 1.7 | J | C, R |
| Soleidae | | | | |
| *Achlyopa nigra* | 0.0 | 0.5 | J | – |
| *Synaptera orientalis* | 0.0 | 0.2 | J | – |
| Sparidae | | | | |
| *Acanthopagrus australis* | 43.5 | 17.5 | J, A | C, R |
| *Rhabdosargus sarba* | 33.5 | 4.7 | J, A | C, R |
| Tetraodontidae | | | | |
| *Sphoeroides hamiltoni* | 33.5 | 3.2 | J | – |
| *Sphoeroides pleurogramma* | 3.5 | 0.5 | J, A | – |
| *Sphoeroides pleurostictus* | 31.5 | 15.0 | J, A | – |
| Terapontidae | | | | |
| *Terapon jarbua* | 0.0 | 0.2 | A | – |
| Triacanthidae | | | | |
| *Tripodichthys augustifrons* | 0.0 | 0.2 | A | – |

* A, adult; J, juvenile.
† C, commercial; R, recreational.

1989) that caused massive fish kills. Of 159 species identified, 103 were fishes, 45 were invertebrates and 11 were vertebrates other than fishes. About 14 million fishes were killed in December 1983, 11 million in February 1989 and 6 million in December 1989. These assessments are the largest in area and most comprehensive to be documented in the literature with known levels of precision (McEachron *et al.*, 1994).

Although anthropogenic influences are usually blamed for fish kills in India, a large-scale mortality of the catfish *Arius maculatus* observed in a 100 ha coastal lagoon at Therespuram was thought to have been due to high salinity, low dissolved oxygen and particularly the presence of hydrogen sulphide (Natarajan *et al.*, 1982). Toxic sulphides are also occasionally responsible for fish kills in Lake Mpungwini, part of the Kosi system of KwaZulu-Natal. This lake has an anoxic layer near the bottom that is rich in sulphides that may be brought nearer to the surface during periods of turbulence. During September 1989, thousands of fish of more than a dozen species died as a result of toxic concentrations of sulphides ($>10\,\mathrm{mg\,l^{-1}}$) being upwelled into the surface waters of the lake (Ramm, 1992).

Fish kills caused by toxic algal blooms in tropical estuaries and coastal waters are well documented in the literature. The extent to which these may be considered natural is debatable because it is believed that the sudden explosion of toxic algal populations, especially dinoflagellates (red tides), may often be a result of large amounts of nutrients (N and P) being introduced into estuaries or coastal waters by sewage disposal or agricultural run-off.

A mass mortality of fish (*Terapon puta*, *Congresox* sp. and *Pomadasys maculatus*) in Gwadar Bay, Pakistan, was caused by the toxic dinoflagellate *Prorocentrum minimum*. The relatively high temperatures of the surface water in the bay (27–29 °C), associated with high light intensity and moderate winds for most of the duration of the bloom, were factors contributing to the development of the bloom. Its maintenance was supported by the permanent availability of the phosphate and nitrogen compounds in the mixed layer of the water column (Rabbani *et al.*, 1990). In Jakarta Bay, Indonesia, red tides due to *Noctiluca scintillans* have been found throughout the year. Besides *Noctiluca* red tides, there have been blooms of diatoms dominated by *Skeletonema*, *Chaetoceros*, *Thalassiothrix* and *Coscinodiscus*. Other dinoflagellates such as *Ceratium*, *Dinophysis* and *Peridinium* also cause red tides. Mass mortalities of demersal fishes and benthic organisms in Jakarta Bay caused by these red tide organisms were first reported in July, 1986 (Adnan, 1989). A severe outbreak of red tide in Tampa Bay, Florida, in June 1970 caused massive fish kills; an estimated 3000 tonnes of various species of fish were removed from beaches around the bay (Lindall *et al.*, 1973).

### Anthropogenic sources of mortality

Most of the recorded fish kills in tropical estuaries attributable to human activities involve agricultural run-off or effluent from sugar, pulp or palm mills (pp.272–278). Many of the effects of industrial effluents are sublethal, and although very serious in the long term, have seldom been directly implicated in large fish kills. In heavily polluted areas, however, it is often difficult to determine the precise cause of fish kills. Such is the case in Visakhapatnam harbour on the east coast of India. The annual average readings of nutrients, chlorophyll and particulate organic carbon in the harbour area are much higher than those of unpolluted coastal stations, both at the surface and on the bottom. The harbour waters are polluted with nutrients and organic matter due to the discharge of a cocktail of industrial effluents and domestic sewage. Under these conditions, mass mortalities of fish occur, as well as toxic algal blooms caused by eutrophication (Satyanarayana *et al.*, 1992).

A mass mortality of sciaenid fishes in the Gulf of Nicoya, Costa Rica, in 1985 was probably due to an untraced agricultural toxin because of the diagnostic combination of tissue irregularities observed in the fish. The fact that large numbers of agrochemicals are used rather indiscriminately in the area is suggestive, but no direct evidence could be found linking the mortalities to any one compound (Szelistowski and Garita, 1989).

A massive mortality of fishes at Mengabang Lagoon on the Trengganu coast of Malaysia occurred in May 1983. This involved mainly gerreids, hemiramphids, leiognathids, *Liza* spp. and *Terapon jarbua*. The mortality was thought to have been caused by organic pollution. Very high values of ammonia and dissolved organic nitrogen were recorded in the water. Ammonia can affect the pH, and high temperatures and pH values cause more of the total ammonia to exist in the toxic un-ionized form. This is highly toxic because it passes easily through gill membranes and accumulates in the blood. The lagoon is a blind estuary that is cut off from the South China Sea for most of the year, opening only in about October when local villagers cut the sandbar to release flood waters from their paddy fields. It has no major streams feeding it and the only inflow is from rainfall or drainage from household effluent, nearby factories, and excess water from adjacent paddy fields. All of these sources are highly enriched with N or P (Shamsudin, 1988).

## 9.8 CONSERVATION OF TROPICAL ESTUARIES

Few attempts have been made to restore the ecological functions of degraded tropical estuaries. Even in developed countries, not many attempts have been made to restore habitats such as mangroves, but much attention is given to improving water quality by controlling pollution.

However, as destruction continues and fisheries yields fall, this topic is likely to increase in prominence. A number of countries have begun trials investigating the possibility of replanting mangroves. In Vietnam the destruction of vast areas of mangroves caused by overexploitation, wartime herbicide application, population pressures and pollution has been of concern to the government. Afforestation programmes using *Rhizophora apiculata* are now in force. More than 60% of bare areas previously covered by mangroves have been planted and show encouraging results (Hong, 1991). In Maharashtra State in India, large areas of mangroves have been lost and a programme of selective replanting has begun (Raddi, 1992).

In the absence of significant amounts of rehabilitation, the conservation of tropical estuaries is more important than ever. However, as is discussed in this chapter and in Chapter 8, tropical and subtropical estuaries are often subject to heavy human usage. The frequent proximity to population centres, biological diversity, high fisheries productivity and recreational value of estuaries ensure that humans are now very much part of the ecology of estuaries, which must be managed accordingly. The variety of uses to which estuaries are put leads to conflicts of interest. In this section the use of the word 'conservation' follows the definition of the IUCN (Harden Jones, 1994):

> In the case of sectors (such as agriculture, fisheries, forestry and wildlife) directly responsible for the management of living resources, conservation is that aspect of management which ensures that utilisation is sustainable and which safeguards the ecological process and genetic diversity essential for the maintenance of the resources concerned.

## Why conservation?

Developed countries with subtropical and tropical estuaries now accord great importance to the conservation of ecosystems for their heritage value and to preserve fauna and flora. This was not always the case, and only in recent years have estuaries and mangroves been considered of intrinsic value, apart from any economic importance they may have. However, in the developing world the importance of estuaries for fisheries production, as sources of raw materials, as sinks for waste and as water highways, often outweighs arguments for their conservation. Nevertheless, it is now widely realized throughout the tropics that sustainable fisheries production in estuaries is dependent on the maintenance of a functional and diverse ecosystem.

The large-scale destruction of mangroves has been discussed in section 9.5 and although the loss of mangroves has slowed in countries such as Australia and the USA, in much of the developing world, mangroves are more or less seen as obstacles to development. The average citizen views mangrove swamps as ugly, muddy, inhospitable, inaccessible and hostile environments. Based on such perceptions, there is no public outcry over the indiscriminate destruction of mangroves (Abraham, 1984). The connectivity between estuaries, mangroves and sustainable fisheries is now well established (Sasekumar, 1993) and most decision makers are aware of the importance of estuaries. Hence the problems, often viewed from an economic standpoint, now concern resolution of conflicts over whether industrial, or residential or recreational development take precedence over natural resources such as fisheries.

Unfortunately, such conflicts often involve traditional usage of estuaries versus various forms of development. In many developing countries, the estuarine environment provides local people not only with fish, but with a range of products such as fruit, medicines, thatch, timber and firewood. Loss of the environment requires that the people previously using it change their way of life. In much of Asia this is virtually impossible; in India, for example, many millions of people earn their livelihood from fisheries in mangrove estuaries. Loss of a healthy estuarine environment forces more and more already poor subsistence dwellers into greater poverty, usually with a population drift to the cities.

The costs and benefits of conserving and managing tropical estuaries and their mangrove forests have been summarized by de Leon (1994). The main cost is that declaring areas for reserves or sanctuaries means a loss of area for potential industrial development, human settlement and other uses. It may also entail the setting up of expensive legislative and enforcement structures. Areas that have been damaged may require rehabilitation through, for example, re-afforestation with mangroves, or even re-introduction of fish species.

The benefits include:

- sustainable fisheries for all sectors from subsistence to recreational;
- increasing the aesthetic value and generation of tourist income;
- protection of rare and endangered species;
- maintenance of the buffering action mangroves provide against wave action and erosion in estuaries.

## What is being conserved?

In areas of low population density, such as parts of northern Australia and the Indonesian provinces of Kalimantan and Irian Jaya, large estuarine areas remain in a relatively pristine state. The value of protecting such areas is now becoming widely recognized by governments, not only for their heritage value, but also because they have significant potential for generating income through ecotourism. The RAMSAR Convention, with its headquarters in Switzerland, has been of particular importance for the conservation of tropical estuaries. This intergovernmental convention was set up in Iran in 1971 and by 1995, 90 countries had signed up as contracting parties. The overall objective of the convention is "the conservation and wise use of wetlands by national action and international cooperation as a means of achieving sustainable development throughout the world". Estuaries fall within the

RAMSAR definition of wetlands. One of the achievements of the convention has been the listing by each of the contracting parties of wetlands of international importance in their country. Scientific and financial assistance is being offered to developing countries to help them work effectively to maintain the ecological character of their RAMSAR sites.

Tropical estuarine areas that are listed RAMSAR sites include: in Africa, the St Lucia and Kosi coastal lake systems in South Africa, the Sakumo lagoon in Ghana, the Delta du Saloum in Senegal, and parts of the lower Zaire estuary in Zaire; in Australia, parts of the Kakadu National Park in the Northern Territory, and Moreton Bay in south-east Queensland; in Asia, the Sundarbans in Bangladesh, Lake Chilka in India, the Berbak National Park in Sumatra, Indonesia, and the Red River estuary in Vietnam; in North America, the Everglades in Florida; and in South America, the Churute Mangroves in Ecuador. Unfortunately, out of the 777 (1996) listed RAMSAR sites, relatively few (about 10%) are in tropical countries, and estuaries form fewer than a quarter of these, the majority being freshwater areas. Nevertheless, more sites are being added each year and although as yet many of the sites have little legal or practical protection, there is increasing pressure on governments to develop and demonstrate effective management strategies.

Management strategies for conservation of estuaries vary considerably but have the common theme of sustainability of the resources, particularly with regard to fisheries. In the St Lucia system in South Africa the main uses are tourism and recreational fishing. In the Everglades and Kakadu National Parks, tourism is the chief use but some recreational fishing is allowed. In the Larut–Matang Forest Reserve of Malaysia, artisanal fishing and sustainable harvesting of timber are the main industries.

Fisheries form important criteria for designation of wetlands as *internationally important* under the RAMSAR convention. There are three main criteria, any one of which qualifies a site for listing with regard to fish and fisheries (Bruton, 1996):

1.  if it supports a significant proportion of indigenous fish subspecies, species or families, life history stages, species interactions and/or populations that are representative of wetland benefits and values;
2.  if it is an important source of food for fishes, spawning ground, nursery and/or migration path on which an offshore fishery or a fishery elsewhere in the catchment is dependent;
3.  if it is of special value in supporting a sustainable fishery on which a local community is dependent.

This last item is particularly significant with regard to tropical estuaries, most of which are important to local fishing communities. To quote from Bruton (1996), "The modern view of conservation is very different from that accepted at the time when the RAMSAR convention was originally promulgated. Conservation is now increasingly taken to include the cultural diversity of humans and their life styles. There is also a strong movement to recognise the importance of local communities, including indigenous peoples, and their dependence on wetlands ... it is also important to realise that the sustainable use of wetlands by traditional people has, in some cases, become an integral part of the ecology of those systems and has added value to them."

There is little doubt that attempts to conserve estuaries in the tropics must take human activities into account. For example, the Sundarbans of Bangladesh, a $4264\,km^2$ area of estuaries and mangroves, is the country's first RAMSAR site, but provides livelihood and employment to government and private sectors and to a quarter of the people living in the area, including wood-cutters, fishers, honey and wax collectors, shell collectors, timber traders and their workers, newsprint mill workers and other people engaged in the collection of non-wood minor products. The total number employed in the Sundarbans annually is estimated at more than 300 000, with twice that number for half the year (Khan and Khan, 1995). The area is capable of supporting the conservation of fish and fisheries resources with careful management, as well as wildlife resources such as tigers and birds (Sanyal, 1992) while still retaining its human usage. It possesses great potential for ecotourism, but is currently threatened by diversion of water for irrigation, construction of coastal embankments, overpopulation and large-scale collection of prawn larvae for aquaculture. The latter activity results in the destruction of larvae and juveniles of large numbers of other species (Huq, 1995).

Similarly, Lake Chilka in Orissa, India, is a RAMSAR site that also plays a very important role in the social, economic, political and cultural activities of the people living around it. It is an important fishing area and directly supports about 10 000 fishers from 122 villages who derive their livelihood from its fishes and prawns. It is also an increasingly important attraction for ecotourism with its wealth of birdlife. It is currently threatened mainly by uncontrolled expansion of aquaculture into the lake, increasing numbers of immigrant fishers, and an increasing disregard for traditional practices and conservation (Gouda, 1995).

It is not only in developing countries that people must be factored into conservation strategies. Moreton Bay in south-east Queensland, Australia, is a RAMSAR site that is extensively used by many thousands of people, both for work and for recreation (p.264). Its management is a complex process, but regardless of how it is managed, people remain a vital component. Their continued use of the Bay for such diverse activities as recreational and commercial fishing, water sports, scientific study and just messing about in boats depends upon the development of integrated management and conservation strategies.

# References

Abdus-Shahid, M., Pramanik, M.A.H., Abdul-Jabbar, M. and Ali, S. (1992) Remote sensing application to study the coastal shrimp farming area in Bangladesh. *Geocartography International*, **7**, 5–13.

Abraham, M. (1984) Public's perception of mangrove utilization and conservation, in *Workshop on Productivity of the Mangrove Ecosystem: Management Implications, Penang (Malaysia), 4–6 Oct. 1983* (eds J.E. Ong and W.K. Gong), Universiti Sains Malaysia, Penang, Malaysia 1984, pp. 164–167.

Abu-Hakima, R., El Zahar, C., Akatsu, S. and Al Abdul Elah, M. (1983) The reproductive biology of *Pomadasys argenteus* (Forskal) (family: Pomadasyidae) in Kuwaiti waters. *Technical Report, Kuwait Institute for Scientific Research, Safat, (Kuwait)*, **999**, 1–25.

Adnan, Q. (1989) Red tides due to *Noctiluca scintillans* (Macartney) Ehrenb. and mass mortality of fish in Jakarta Bay, in *Red Tides: Biology, Environmental Science, and Toxicology*. Proceedings of the first international symposium on red tides held November 10–14, 1987, in Takamatsu, Kagawa Prefecture, Japan (eds T. Okaichi, D.M. Anderson and T. Nemoto), LIPI, Jakarta, pp. 53–55.

Aguero, N.M. and Lockwood, B.A. (1986) Resource management is people management, in *The First Asian Fisheries Forum* (eds J.L. Maclean, L.B. Dizon and L.V. Hosillos), Asian Fisheries Society, Manila, Philippines, pp. 345–347.

Ahmad, M., Dahril, T. and Efizon, D. (1995) Ekologi reproduksi Ikan Terubuk (Alosa toli) di perairan Begkalis, Riau. *Jurnal Perikanan dan Kelautan, Universitas Riau, Indonesia*, **1**, 2–19.

Ahsanullah, M. (1964) Population dynamics of Hilsa in East Pakistan. *Agriculture Pakistan*, **15**, 351–365.

Albaret, J.J. (1987) Les peuplements de poissons de la Casamance (Sénégal) en période de sécheresse. *Revue d'hydrobiologie tropicale*, **20**, 291–310.

Albaret, J.J. (1994) Les poissons: biologie et peuplements, in *Environnement et Ressources aquatiques de Côte d'Ivoire, tome II – Les Milieux lagunaires* (eds J.R. Durand, Dufour, D. Guiral and Zabi), Editions de L'ORSTOM, Paris, pp. 239–279.

Albaret, J.J. and Desfossez, P. (1988) Biologie et écologie des Gerreidae (Pisces, Teleostei) en lagune Ébrié (Côte d'Ivoire). *Revue d'hydrobiologie tropicale*, **21**, 71–88.

Albaret, J.J. and Ecoutin, J.M. (1989) Communication mer–lagune: impact d'une reouverture sur l'ictyofaune de la lagune Ébrié (Côte d'Ivoire). *Revue d'hydrobiologie tropicale*, **22**, 71–81.

Albaret, J.J. and Ecoutin, J.M. (1990) Influence des saisons et des variations climatiques sur les peuplements de poissons d'une lagune tropicale en Afrique de l'Ouest. *Acta cologica*, **11**, 557–583.

Albaret, J.J. and Legendre, M. (1985) Biologie et écologie des Mugilidae en lagune Ébrié (Côte d'Ivoire) intérêt potentiel pour l'aquaculture lagunaire. *Revue d'hydrobiologie tropicale*, **18**, 281–303.

Alexander, R.McN. (1967) *Functional Design in Fishes*, Hutchinson & Co., London.

Allanson, B.R. and van Wyk, J.D. (1969) An introduction to the physics and chemistry of some lakes in northern Zululand. *Transactions of the Royal Society of South Africa*, **38**, 217–240.

Allen, G.R. (1978) A review of the archerfishes (Family Toxotidae). *Records of the West Australian Museum*, **6**, 355–377.

Allen, G.R. and Burgess, W.E. (1990) A review of the glassfishes (Chandidae) of Australia and New Guinea. *Records of the West Australian Museum, Supplement* **34**, 139–206.

Allen, G.R. and Cross, N.J. (1982) *Rainbow Fishes of Australia and Papua New Guinea*, Angus & Robertson, Sydney.

Alongi, D.M. and Sasekumar, A. (1992) Benthic communities, in *Tropical Mangrove Ecosystems* (eds A.I. Robertson and D.M. Alongi), American Geophysical Union, Washington, DC, pp. 137–171.

Anonymous (1991) *Moreton Bay Strategic Plan. Proposal for Management.* Queensland Government, Brisbane.

Apekin, V.S. and Vilenskaya, N.I. (1978) A description of the sexual cycle and the state of the gonads during the spawning migration of the striped mullet, *Mugil cephalus. Journal of Ichthyology*, **18**, 446–456.

Arakel, A.V., Hill, C.M., Piorewicz, J. and Connor, T.B. (1989) Hydro-sedimentology of the Johnstone River estuary, in *Symposium on Sediment–Water Interactions, Melbourne, Australia 16–20 February, 1987* (eds P.G. Sly and B.T. Hart), Queensland University of Technology, Brisbane, pp. 51–60.

Areces-Mallea, A.J. and Gil-Vades, A. (1980) Notas sobre la distribucion y concentracion de zinc en un area costera contaminada. *Informe científico-técnico, Instituto de Oceanología, Academia de Ciencias de Cuba*, **135**, 1–26.

Arnold, E.L. and Thompson, J.R. (1958) Offshore spawning of the striped mullet *Mugil cephalus* in the Gulf of Mexico. *Copeia*, **1958**, 130–132.

Asian Wetland News (1995) The Kuala Lumpur Statement. *Asian Wetland News*, **8**, 5–6.

Atmadja, W. and Man, A. (1994) Threats and pressures on mangroves and current management, in *Third ASEAN–Australia Symposium on Living Coastal Resources, Bangkok, Thailand, 1994* (ed C.R. Wilkinson), Australian Institute of Marine Science, Townsville, pp. 62–70.

Austin, H.M. and Austin, S.E. (1971) Juvenile fish in two Puerto Rican mangroves. *Underwater Naturalist*, **7**, 26–36.

Ayvazian, S.G. and Hyndes, G.A. (1995) Surf–zone fish asemblages in south–western Australia: do adjacent nearshore habitats and the warm Leeuwin Current influence the characteristics of the fish fauna? *Marine Biology*, **122**, 527–536.

Babu Rao, M. (1962) On the species of the genus *Setipinna* Swainson of the Godavari estuary. *Proceedings of the First All-India Congress of Zoology, 1959*, 364–369.

Bade, T. (1994) Reproduction and growth in three species of grunt (Teleostei: Haemulidae) from north Queensland waters, PhD thesis, James Cook University of North Queensland, Townsville, Queensland, Australia, 260 pp.

Baelde, P. (1990) Differences in the structures of fish assemblages in *Thalassia testudinum* beds in Guadeloupe, French West Indies, and their ecological significance. *Marine Biology*, **105**, 163–173.

Bagchi, M.M. and Ghosh, B.B. (1986) Role of tidal influence on the degree of industrial pollution caused by sulphite pulp and paper mill waste in the Hooghly

Estuary near Hazinagar. *Symposium Series of the Marine Biological Association of India*, **6**, 1316–1329.

Bagchi, M.M., Ghosh, B.B. and Majumder, S.K. (1987) Status of industrial pollution in the Hooghly Estuary due to disposal of trade effluents from Haldia oil refinery complex, in *Symposium on the Impact of Current Land Use Pattern and Water Resources Development on Riverine Fisheries* (Barrackpore, India), 25–27 Apr. 1987, Marine Biological Association of India, Cochin, p.10.

Bainbridge, V. (1960) The plankton of the inshore waters off Freetown, Sierra Leone. *Colonial Office Fishery Publications, London*, **13**, 1–48.

Bainbridge, V. (1963) The food, feeding habits and distribution of the Bonga *Ethmalosa dorsalis* (Cuvier & Valenciennes). *Journal du Conseil Permanent International pour l'Exploration de la Mer*, **28**, 270–284.

Baird, D. and Pereyra Lago, R. (1992) Nutrient status and water quality assessment of the Marina Glades canal system, Kromme Estuary, St. Francis Bay. *Water SA*, **18**, 37–42.

Baird, D., Marais, J.F.K. and Wooldridge, T. (1981) The Influence of a Marina Canal System on the Ecology of the Kromme Estuary, St. Francis Bay. *South African Journal of Zoology*, **16**, 21–34.

Balogun, J.K. (1987) Seasonal fluctuations of salinity and fish occurrence in and around Epé Lagoon, Nigeria. *African Journal of Ecology*, **25**, 55–61.

Balon, E.K. and Bruton, M.N. (1994) Fishes of the Tatinga River, Comoros, with comments on freshwater amphidromy in the goby *Sicyopterus lagocephalus*. *Ichthyological Explorations in Freshwaters*, **5**, 25–40.

Baran, E. (1995) Dynamique spatio-temporelle des peuplements de poissons estuariens en Guinée – relations avec les milieu abiotique. PhD thesis, Université de Bretagne Occidentale, France, 225 pp.

Barnes, R.S.K. (1980) *Coastal Lagoons*, Cambridge University Press, Cambridge, UK.

Barro, M. (1979) Reproduction de *Brachydeuterus auritus* Val. 1831 (Poissons, Pomadasyidae) en Côte d'Ivoire. *ORSTOM Senegal Centre de Recherches Oceanographiques de Dakar-Thiaroye, Document Scientifique*, **68**, 57–62.

Barry, B. and Posner, J.L. (1986) Suivi hydroagricôle dur barrage anti-sel de Guidel, in *L'Estuaire de la Casamance: Environnement, Pêche, Socioeconomie*, – Seminaire ISRA, Senegal, 19–24 Juin 1986 (eds L. Le Reste, A. Fontana and A. Samba), Centre de Recherche Agriculture, Djibelor, Senegal, pp. 291–306.

Barry, T.P. and Fast, A.W. (1992) Biology of the spotted scat *(Scatophagus argus)* in the Philippines. *Asian Fisheries Science*, **5**, 163–179.

Barry, T.P., Castanos, M.T., Paz-Soccoro, M., Macahilig, C. and Fast, A.W. (1993) Spawning induction in female spotted scat *(Scatophagus argus)*. *Journal of Aquaculture in the Tropics*, **8**, 121–129.

Beccari, O. (1904) *Wanderings in the Great Forests of Borneo*, Constable, London.

Beckley, L.E. (1985) Tidal exchange of ichthyoplankton in the Swartkops estuary mouth, South Africa. *South African Journal of Zoology*, **20**, 15–20.

Beckley, L.E. (1986) The ichthyoplankton assemblage of the Algoa Bay nearshore region in relation to coastal zone utilization by juvenile fish. *South African Journal of Zoology*, **21**, 244–252.

Begg, G.W. (1978) *The Estuaries of Natal – A Resource Inventory Report to the Natal Town and Regional Planning Commission*. Oceanographic Research Institute, Durban, South Africa.

Beumer, J.P. (1978) Feeding ecology of four fishes from a mangrove creek in north Queensland, Australia. *Journal of Fish Biology*, **12**, 475–490.

Beverton, R. and Holt, S. (1993) *On the Dynamics of Exploited Fish Populations*, Chapman & Hall, London. [facsimile reprint]

Bhaskar, B.R., Rao, K.S., Rao, D.P. and Prasad, Y.V.K.D. (1989) Inferior condition of mullets from polluted waters of Visakhapatnam Harbour, in *Studies on Fish Stock Assessment in Indian Waters*, Fishery Survey of India, Bombay, **2**, 120–131.

Bin Hasman, M. (1996) *A Socioeconomic Study on Terubok Fishermen in Sarawak*, Planning Report 78, Planning Branch, Department of Agriculture, Sarawak.

Blaber, S.J.M. (1973) Population size and mortality of juveniles of the marine teleost *Rhabdosargus holubi* (Pisces: Sparidae) in a closed estuary. *Marine Biology*, **21**, 219–225.

Blaber, S.J.M. (1974) Field studies of the diet of *Rhabdosargus holubi* (Steindachner) (Teleostei: Sparidae). *Journal of Zoology, London*, **173**, 407–417.

Blaber, S.J.M. (1976) The food and feeding ecology of Mugilidae in the St. Lucia Lake system. *Biological Journal of the Linnean Society*, **8**, 267–277.

Blaber, S.J.M. (1977) The feeding ecology and relative abundance of Mugilidae in Natal and Pondoland estuaries. *Biological Journal of the Linnean Society*, **9**, 259–276.

Blaber, S.J.M. (1978) The fishes of the Kosi system. *Lammergeyer*, **24**, 28–41.

Blaber, S.J.M. (1979) The biology of filter feeding teleosts in Lake St. Lucia, Zululand. *Journal of Fish Biology*, **15**, 37–59.

Blaber, S.J.M. (1980) Fishes of the Trinity Inlet system of North Queensland with notes on the ecology of fish faunas of tropical Indo–Pacific estuaries. *Australian Journal of Marine and Freshwater Research*, **31**, 137–146.

Blaber, S.J.M. (1981a) The zoogeographical affinities of estuarine fishes in south east Africa. *South African Journal of Science*, **77**, 305–307.

Blaber, S.J.M. (1981b) An unusual mass mortality of *Clarias gariepinus* in the Mkuze River at Lake St. Lucia. *Lammergeyer*, **31**, 15.

Blaber, S.J.M. (1982) The biology of *Sphyraena barracuda* in the Kosi system with notes on the Sphyraenidae of other Natal estuaries. *South African Journal of Zoology*, **17**, 171–176.

Blaber, S.J.M. (1984) The diet, food selectivity and niche of the benthic feeding fish, *Rhabdosargus sarba* (Teleostei: Sparidae) in Natal estuaries. *South African Journal of Zoology*, **19**, 241–246.

Blaber, S.J.M. (1987) Factors affecting recruitment and survival of Mugilidae in estuaries and coastal waters of the Indo–West Pacific. *American Fisheries Society Symposium*, **1**, 507–518.

Blaber, S.J.M. (1988) Fish communities in coastal lakes (Les peuplements de poissons des lacs cotieres), in *Biologie et Ecologie des Poissons d'Eau Douce* (eds C. Lévêque, M.N. Bruton and G.W. Ssentongo), ORSTOM, Paris, pp. 351–362.

Blaber, S.J.M. (1991) Deep sea, estuarine and freshwater fishes: life history strategies and ecological boundaries. *South African Journal of Aquatic Sciences*, **17**, 2–11.

Blaber, S.J.M. and Blaber, T.G. (1980) Factors affecting the distribution of juvenile estuarine and inshore fish. *Journal of Fish Biology*, **17**, 143–162.

Blaber, S.J.M. and Cyrus, D.P. (1981) A revised checklist and further notes on the fishes of the Kosi system. *Lammergeyer*, **31**, 5–14.

Blaber, S.J.M. and Cyrus, D.P. (1983) The biology of Carangidae in Natal estuaries. *Journal of Fish Biology*, **22**, 173–188.

Blaber, S.J.M. and Milton, D.A. (1990) Species composition, community structure and zoogeography of fishes of mangroves in the Solomon Islands. *Marine Biology*, **105**, 259–268.

Blaber, S.J.M. and Whitfield, A.K. (1976) Large-scale mortality of fish at St. Lucia. *South African Journal of Science*, **72**, 218.

Blaber, S.J.M. and Whitfield, A.K. (1977a) The feeding ecology of juvenile Mugilidae in south east African estuaries. *Biological Journal of the Linnean Society*, **9**, 277–284.

Blaber, S.J.M. and Whitfield, A.K. (1977b) The biology of the burrowing goby *Croilia mossambica* Smith (Teleostei: Gobiidae). *Environmental Biology of Fishes*, **1**, 197–204.

Blaber, S.J.M., Cyrus, D.P. and Whitfield, A.K. (1981) The influence of zooplankton food resources on the morphology of the estuarine clupeid *Gilchristella aestuarius*. *Environmental Biology of Fishes*, **6**, 351–355.

Blaber, S.J.M., Kure, N.F., Jackson, S. and Cyrus, D.P. (1983) The benthos of South Lake, St. Lucia, following a period of stable salinities. *South African Journal of Zoology*, **18**, 311–319.

Blaber, S.J.M., Hay, D.G., Cyrus, D.P. and Martin, T.J. (1984) The ecology of two degraded estuaries on the north coast of Natal. *South African Journal of Zoology*, **19**, 224–240.

Blaber, S.J.M., Young, J.W. and Dunning, M.C. (1985) Community structure and zoogeographic affinities of the coastal fishes of the Dampier region of–north–west Australia. *Australian Journal of Marine and Freshwater Research*, **36**, 247–266.

Blaber, S.J.M., Brewer, D.T. and Salini, J.P. (1989) Species composition and biomasses of fishes in different habitats of a tropical northern Australian estuary: their occurrence in the adjoining sea and estuarine dependence. *Estuarine Coastal and Shelf Science*, **29**, 509–531.

Blaber, S.J.M., Salini, J.P. and Brewer, D.T. (1990a) A checklist of the fishes of Albatross Bay and the Embley estuary, north eastern Gulf of Carpentaria. *CSIRO Marine Laboratories Report Series*, **210**.

Blaber, S.J.M., Brewer, D.T., Salini, J.P. and Kerr, J. (1990b) Biomasses, catch rates and patterns of abundance of demersal fishes, with particular reference to penaeid prawn predators, in a tropical bay in the Gulf of Carpentaria, Australia. *Marine Biology*, **107**, 397–408.

Blaber, S.J.M., Brewer, D.T., Salini, J.P., Kerr, J. and Conacher, C. (1992) Species composition and biomasses of fishes in tropical seagrasses at Groote Eylandt, northern Australia. *Estuarine, Coastal and Shelf Science*, **35**, 605–620.

Blaber, S.J.M., Brewer, D.T. and Salini, J.P. (1994a) Comparisons of the fish communities of tropical estuarine and inshore habitats in the Gulf of Carpentaria, northern Australia. *International Symposium Series, ECSA 22*, Olsen & Olsen, Fredensborg, pp. 363–372.

Blaber, S.J.M., Brewer, D.T. and Salini, J.P. (1994b) Dentition and diet in tropical ariid catfishes from Australia. *Environmental Biology of Fishes*, **40**, 159–174.

Blaber, S.J.M., Brewer, D.T. and Harris, A.N. (1994c) The distribution, biomass and community structure of fishes of the Gulf of Carpentaria. *Australian Journal of Marine and Freshwater Research*, **45**, 375–396.

Blaber, S.J.M., Brewer, D.T. and Salini, J.P. (1995a) Fish communities and the nursery role of the shallow inshore waters of a tropical bay in the Gulf of Carpentaria. *Estuarine Coastal and Shelf Science*, **40**, 177–193.

Blaber, S.J.M., Milton, D.A., Pang, J., Wong, P., Ong, B., Nyigo, L. and Lubim, D. (1995b) *The biology, ecology and life cycle of terubok* (Tenualosa toli) *with recommendations for the conservation, management and culture of the species. Phase 2 final report*. Ministry of Agriculture and Community Development, Kuching, Sarawak.

Blaber, S.J.M., Milton, D.A., Pang, J., Wong, P., Ong, B., Nyigo, L. and Lubim, D. (1996) The biology of the tropical shad *Tenualosa toli* from Sarawak: first evidence of protandry in the Clupeiformes? *Environmental Biology of Fishes*, **46**, 225–242.

Blaber, S.J.M., Farmer, M.F., Milton, D.A., Pang, J., Wong, P. and Ong, B. (1997) The ichthyoplankton assemblages of Sarawak and Sabah estuaries: composition, distribution and affinities. *Estuarine, Coastal and Shelf Science*, (in press).

Blay, J. jun. and Eyeson, K.N. (1982) Observations on the reproductive biology of the shad, *Ethmalosa fimbriata* (Bowdich), in the coastal waters of Cape Coast, Ghana. *Journal of Fish Biology*, **21**, 485–496.

Bleckmann, H., Waldner, I. and Schwartz, E. (1981) Frequency discrimination of the surface-feeding fish *Aplocheilus lineatus* – a prerequisite for prey location? *Journal of Comparative Physiology*, **143**, 485–490.

Boeseman, M. (1963) An annotated list of fishes from the Niger Delta. *Zoologische Verhandelingen*, **61**, 1–54.

Bok, A.H. (1984) The demographic, breeding biology and management of two mullet species (Pisces: Mugilidae) in the eastern Cape, South Africa, PhD thesis, Rhodes University, Grahamstown, South Africa, 307 pp.

Boltt, R.E. (1975) The benthos of some southern African lakes. Part V: the recovery of the benthic fauna of St Lucia lake following a period of excessively high salinity. *Transactions of the Royal Society of South Africa*, **41**, 295–323.

Bone, Q. and Marshall, N.B. (1994) *Biology of Fishes*, 2nd edn, Chapman & Hall, London.

Boto, K. and Isdale, P. (1985) Fluorescent bands in massive corals result from terrestrial fulvic acid inputs to nearshore zone. *Nature*, **315**, 396–397.

Boujard, T. and Rojas-Beltran, R. (1988) Longitudinal zonation of the fish population of the Sinnamary River (French Guiana). *Revue d'hydrobiologie tropicale*, **21**, 47–61.

Bowen, S. H. (1979) A nutritional constraint in detritivory by fishes: the stunted population of *Sarotherodon mossambicus* in Lake Sibaya, South Africa. *Ecological Monographs*, **49**, 17–31.

Braekevelt, C.R. (1985a) Fine structure of the retinal pigment epithelial region of the Archerfish *(Toxotes jaculatrix)*. *Ophthalmic Research*, **17**, 221–229.

Braekevelt, C.R. (1985b) Photoreceptor fine structure in the Archerfish *(Toxotes jaculatrix)*. *The American Journal of Anatomy*, **173**, 89–98.

Brewer, D.T. and Warburton, K. (1992) Selection of prey from a seagrass/ mangrove environment by golden lined whiting, *Sillago analis* (Whitley). *Journal of Fish Biology*, **40**, 257–271.

Brewer, D.T., Blaber, S.J.M. and Salini, J.P. (1989) The feeding biology of *Caranx bucculentus* Alleyne and Macleay (Teleostei: Carangidae) in Albatross Bay, Gulf of Carpentaria; with special reference to predation on penaeid prawns. *Australian Journal of Marine and Freshwater Research*, **40**, 657–668.

Brewer, D.T., Blaber, S.J.M. and Salini, J.P. (1991) Predation on penaeid prawns by fishes in Albatross Bay, Gulf of Carpentaria. *Marine Biology*, **109**, 231–240.

Brewer, D.T., Blaber, S.J.M., Salini, J.P. and Milton, D.A. (1994) Aspects of the biology of *Caranx bucculentus* (Teleostei: Carangidae) from the Gulf of Carpentaria. *Australian Journal of Marine and Freshwater Research*, **45**, 413–428.

Brewer, D.T., Blaber, S.J.M. and Salini, J.P. (1995) Feeding ecology of predatory fishes from Groote Eylandt, with special reference to predation on penaeid prawns. *Estuarine Coastal and Shelf Science*, **40**, 577–600.

Briggs, J.C. (1974) *Marine Zoogeography*, McGraw–Hill, New York.

Brouns, J.J.W.M. and Heijs, F.M.L. (1985) Tropical seagrass systems in Papua New Guinea. A general account of the environment, marine flora and fauna. *Koninklijke Nederlandse Akademie van Wetenskappen, Proceedings Series C, Biological and Medical*, **88**, 145–182.

Browder, J.A., McClellan, D.B., Harper, D.E., Kandrashoff, M.G. and Kandrashoff, W. (1993) A major developmental defect observed in several Biscayne Bay, Florida, fish species. *Environmental Biology of Fishes*, **37**, 181–188.

Brown, M.E. (ed.) (1957) *The Physiology of Fishes*, Academic Press, London.

Bruton, M.N. (1980) An outline of the ecology of Lake Sibaya, with emphasis on the vertebrate communities, in *Studies on the Ecology of Maputaland* (eds M.N. Bruton and K.H. Cooper), Rhodes University and The Natal Branch of the Wildlife Society of Southern Africa, Durban, pp. 382–407.

Bruton, M.N. (1985) The effects of suspensoids on fish. *Hydrobiologia*, **125**, 221–241.

Bruton, M.N. (1996) Draft resolution on adoption of specific criteria for identifying wetlands of international importance based on fish and fisheries and guidelines for their application, in *6th Meeting of the Conference of the Contracting Parties to RAMSAR, Brisbane, Australia, 19–27 March 1996*, Document 6.17, RAMSAR, Brisbane, 7 pp.

Bruton, M.N. and Kok, H.M. (1980) The freshwater fishes of Maputaland, in *Studies on the Ecology of Maputaland* (eds M.N. Bruton, and K.H. Cooper), Rhodes University and The Natal Branch of the Wildlife Society of Southern Africa, Durban, pp. 210–244.

Bulpin, T.V. (1966) *Natal and the Zulu Country*. Books of Africa, Cape Town.

Burren, K.R., Nittim, R. and Hossain, S. (1981) Trial dredging at the port of Chittagong, Bangladesh, in *Australian Conference on Coastal and Ocean Engineering 1981: Offshore Structures, Perth (Australia) 25 Nov. 1981*, National Committee on Coastal and Ocean Engineering Australia, Melbourne, pp. 60–61.

Burton, J.D. and Liss, P.D. (eds) (1976) *Estuarine Chemistry*, Academic Press, London.

Caldwell, D.K., Ogren, L.H. and Giovannoli, L. (1959) Systematic and ecological notes on some fishes collected in the vicinity of Tortuguero, Caribbean coast of Costa Rica. *Revista de Biologia Tropical*, **7**, 7–33.

Campos, J.A., Burgos, B. and Gamboa, C. (1984) Effect of shrimp trawling on the commercial ichthyofauna of the Gulf of Nicoya, Costa Rica. *Revista de Biologia Tropical*, **32**, 203–207.

Carr, W.E.S. and Adams, C.A. (1973) Food habits of juvenile marine fishes occupying seagrass beds in the estuarine zone near Crystal River, Florida. *Transactions of the American Fisheries Society*, **3**, 511–540.

Cervigón, F. (1969) Las especies de los géneros Anchovia y Anchoa (Pisces: Engraulidae) de Venezuela y áreas adyacentes del Mar Caribe y Atlántico hasta 23S. *Memoirs Sociedad Ciencias Naturales La Salle*, **29**, 193–246.

Cervigón, F. (1985) The ichthyofauna of the Orinoco estuarine water delta in the west Atlantic coast, Caribbean, in *Fish Community Ecology in Estuaries and Coastal Lagoons: Towards an Ecosystem Integration* (ed. A. Yáñez-Arancibia), UNAM Press, Mexico City, pp. 57–78.

Cervigón, F., Cipriani, R., Fischer, W., Garibaldi, L., Hendrickx, M. and 5 others (1992) *Fichas FAO de Identificación de Especies para los Fines de la Pesca. Guía de Campo de las Especies Comerciales Marinas y de Aguas Salobres de la Costa Septentrional de Sur América*. FAO, Rome.

Chan, E.H. and Chua, T.E. (1979) The food and feeding habits of greenback grey mullet, *Liza subviridis* (Valenciennes), from different habitats and at various stages of growth. *Journal of Fish Biology*, **15**, 165–171.

Chan, S.T.H. and Yeung, W.S.B. (1983) Sex control and sex reversal in fish under natural conditions, in *Fish Physiology*, Vol. IXB (eds W.S. Hoare, D.J. Randall and E.M. Donaldson), Academic Press, New York, pp. 171–222.

Chandra, R. (1962) A preliminary account of the distribution and abundance of fish larvae in the Hooghly estuary. *Indian Journal of Fisheries*, **9**, 48–70.

Chandrasekaran, V.S. and Natarajan, R. (1992) Small-scale fishery of Pichavaram mangrove swamp, southeast India. *Naga, the ICLARM Quarterly*, **15**, 41–43.

Chapman, V.J. (1976) *Mangrove Vegetation*, J. Cramer, Vaduz.

Charles, R. (1975) Aspects of the biology of the Mojarra Eucinostomus gula (Quoy & Gaimard), in the Biscayne Bay, Florida, MSc thesis, University of Miami, USA, 218 pp.

Charles-Dominique, E. (1982) Exposé synoptique des données biologiques sur l'Ethmalose (*Ethmalosa fimbriata*, S. Bowdich, 1825). *Revue d'hydrobiologie tropicale*, **15**, 283–404.

Chavance, P., Hernández, D.F., Yáñez-Arancibia, A. and Linares, F.A. (1984) Ecología, biología y dinámica de las poblaciones de *Bairdiella chrysoura* en la Laguna de Terminós, sur del Golfo de México (Pisces: Scienidae). *Annales del Instituto de Ciencias del Mar y Limnologia, Universidad Nacional Autónoma de México*, **11**, 123–162.

Chavance, P., Yáñez-Arancibia, A., Flores, D., Lara-Dominguez, A. and Amezcua, F. (1986) Ecology, biology and population dynamics of *Archosargus rhomboidalis* (Pisces, Sparidae) in a tropical lagoon system, southern Gulf of Mexico. *Annales del Instituto de Ciencias del Mar y Limnologia, Universidad Nacional Autónoma de México*, **13**, 11–30.

Cheong, L. and Yeng, L. (1987) Status of sea bass *(Lates calcarifer)* culture in Singapore. *ACIAR Proceedings*, **20**, 65–68.

Chong, V.C. and Sasekumar, A. (1994) Status of mangrove fisheries in the ASEAN region, in *Third ASEAN–Australia Symposium on Living Coastal Resources, Bangkok, Thailand, 1994* (ed. C.R. Wilkinson), Australian Institute of Marine Science, Townsville, pp. 56–61.

Chong, V.C., Sasekumar, A., Leh, M.U.C. and D'Cruz, R. (1990) The fish and prawn communities of a Malaysian coastal mangrove system, with comparisons to adjacent mud flats and inshore waters. *Estuarine, Coastal and Shelf Science*, 31, 703–722.

Choy, S.K. (1993) The commercial and artisanal fisheries of the Larut Matang district of Perak. *Proceedings of a Workshop on Mangrove Fisheries and Connections*, August 26–30, 1991, Ipoh, Malaysia (ed. A. Sasekumar), Ministry of Science, Technology & Environment, Kuala Lumpur, pp. 27–40.

Chua, Thia-Eng (1973) An ecological study of the Ponggol estuary in Singapore. *Hydrobiologia*, **43**, 505–533.

Clayton, D.A. (1993) Mudskippers. *Oceanography and Marine Biology An Annual Review*, **31**, 507–577.

Coates, D. (1987) Observations on the biology of Tarpon, *Megalops cyprinoides* (Broussonet) (Pisces: Megalopidae), in the Sepik River, northern Papua New Guinea. *Australian Journal of Marine and Freshwater Research*, **38**, 529–535.

Coates, D. (1990) Aspects of the biology of the perchlet *Ambassis interrupta* Bleeker (Pisces: Ambassidae) in the Sepik River, Papua New Guinea. *Australian Journal of Marine and Freshwater Research*, **41**, 267–274.

Coates, D. (1991) Biology of fork–tailed catfishes from the Sepik River, Papua New Guinea. *Environmental Biology of Fishes*, **31**, 55–74.

Coetzee, D.J. (1981) Analysis of the gut contents of the needlefish, *Hyporhamphus knysnaensis* (Smith), from Rondevlei, southern Cape. *South African Journal of Zoology*, **16**, 14–20.

Cogger, H.G. (1988) *Reptiles and Amphibians of Australia*, Reed, Sydney.

Collette, B.B. (1974) The garfishes (Hemiramphidae) of Australia and New Zealand. *Records of the Australian Museum*, **29**, 11–105.

Commonwealth of Australia (1990) *Ecologically Sustainable Development*. Catalogue No. 90. 13867, Australian Government Publishing Service, Canberra.

Compagno, L.J.V. (1984) *FAO Species Catalogue 4, Sharks of the World. Part II – Carcharhiniformes. FAO Fisheries Synopsis*, Vol. 4, Part 2.

Conand, F. (1991) Biology and phenology of *Amblygaster sirm* (Clupeidae) in New Caledonia, a sardine of the coral environment. *Bulletin of Marine Science*, **48**, 137–149.

Connell, D.W. (1974) A kerosene-like taint in the sea mullet *Mugil cephalus* (Linnaeus) I. Composition and environmental occurrence of the tainting substance. *Australian Journal of Marine and Freshwater Research*, **25**, 7–24.

Cordell, J. (1974) The lunar-tide fishing cycle in northeastern Brazil. *Ethnology*, **13**, 379–392.

Cordell, J. (1978) Swamp dwellers of Bahia. *Natural History*, **1978**, 62–74.

Cordell, J. (1989) Social marginality and sea tenure in Bahia, in *A Sea of Small Boats* (ed. J. Cordell), Cultural Survival Inc., Cambridge, MA, pp. 125–151.

Cordone, A.J. and Kelly, D.W. (1961) The influences of inorganic sediment on the aquatic life of streams. *California Fish and Game*, **47**, 189–228.

Costa, M.J. and Elliott, M. (1991) Fish usage and feeding in two industrialised estuaries – the Tagus, Portugal, and the Forth, Scotland. *International Symposium Series*, ECSA 19, Olsen & Olsen, Fredensborg, pp. 289–297.

Coull, B.C., Greenwood, J.G., Fielder, D.R. and Coull, B.A. (1995) Subtropical Australian juvenile fish eat meiofauna: experiments with winter whiting *Sillago maculata* and observations on other species. *Marine Ecology Progress Series*, **125**, 13–19.

Cushing, D.H. (1975) *Marine Ecology and Fisheries*, Cambridge University Press, Cambridge, UK.

Cyrus, D.P. (1988a) Episodic events and estuaries: effects of cyclonic flushing on the benthic fauna and diet of *Solea bleekeri* (Teleostei) in Lake St Lucia on the south-eastern coast of Africa. *Journal of Fish Biology*, **33**, 1–7.

Cyrus, D.P. (1988b) Turbidity and other physical factors in Natal estuarine systems. Part 1: selected estuaries. *Journal of the Limnological Society of Southern Africa*, **14**, 60–71.

Cyrus, D.P. (1988c) Turbidity and other physical factors in Natal estuarine systems. Part 2: estuarine lakes. *Journal of the Limnological Society of Southern Africa*, **14**, 72–81.

Cyrus, D.P. (1991) The reproductive biology of *Solea bleekeri* (Teleostei) in Lake St Lucia on the south-east coast of Africa. *South African Journal of Marine Science*, **10**, 45–51.

Cyrus, D.P. and Blaber, S.J.M. (1982) Mouthpart structure and function and the feeding mechanisms of *Gerres* (Teleostei). *South African Journal of Zoology*, **17**, 117–121.

Cyrus, D.P. and Blaber, S.J.M. (1983) The food and feeding ecology of Gerreidae Bleeker 1859, in the estuaries of Natal. *Journal of Fish Biology*, **22**, 373–393.

Cyrus, D.P. and Blaber, S.J.M. (1984a) Predation and sources of mortality of Gerreidae Bleeker 1859, in Natal estuaries. *Lammergeyer*, **32**, 14–20.

Cyrus, D.P. and Blaber, S.J.M. (1984b) The feeding of Gerreidae in the Kosi system with special reference to their seasonal diet. *Lammergeyer*, **32**, 35–49.

Cyrus, D.P. and Blaber, S.J.M. (1984c) The reproductive biology of *Gerres* (Teleostei) Bleeker 1859, in Natal estuaries. *Journal of Fish Biology*, **24**, 491–504.

Cyrus, D.P. and Blaber, S.J.M. (1987a) The influence of turbidity on juvenile marine fishes in estuaries. Part 1, field studies. *Journal of Experimental Marine Biology and Ecology*, **109**, 53–70.

Cyrus, D.P. and Blaber, S.J.M. (1987b) The influence of turbidity on juvenile marine fishes in estuaries. Part 2, laboratory studies. *Journal of Experimental Marine Biology and Ecology*, **109**, 71–91.

Cyrus, D.P. and Blaber, S.J.M. (1988) The potential effects of dredging activities and increased silt load on the St Lucia system, with special reference to turbidity and the estuarine fauna. *Water SA*, **14**, 43–48.

Cyrus, D.P. and Blaber, S.J.M. (1992) Turbidity and salinity in a tropical northern Australian estuary and their influence on fish distribution. *Estuarine, Coastal and Shelf Science*, **35**, 545–563.

Dall, W., Hill, B.J., Rothlisberg, P.C. and Staples, D.J. (1990) The biology of the Penaeidae. *Advances in Marine Biology*, **27**, 1–489.

Darnell, R.M. (1961) Trophic spectrum of an estuarine community based on studies of Lake Pontchartrain, Louisiana. *Ecology*, **42**, 553–568.

Datta, N.C., Bandyopadhyay, B.K. and Barman, S.S. (1984) On the food of an euryhaline perch *Scatophagus argus* (Cuv. and Val.) and the scope of its culture in fresh water. *International Journal of the Academy of Ichthyology, Modinagar*, **5**, 121–124.

Davis, T.L.O. (1984) Estimation of fecundity in barramundi, *Lates calcarifer* (Bloch), using an automatic particle counter. *Australian Journal of Marine and Freshwater Research*, **35**, 111–118.

Davis, T.L.O. (1985) The food of barramundi, *Lates calcarifer* (Bloch), in coastal and inland waters of Van Dieman Gulf and the Gulf of Carpentaria, Australia. *Journal of Fish Biology*, **26**, 669–682.

Davis, T.L.O. (1987) Biology of wildstock *Lates calcarifer* in northern Australia. *ACIAR Proceedings*, **20**, 22–29.

Dawson, C.E. (1981) Notes on four pipefishes (Syngnathidae) from the Persian Gulf. *Copeia*, **1981**, 87–95.

Day, J.H. (1951) The ecology of South African estuaries. Part I. General considerations. *Transactions of the Royal Society of South Africa*, **33**, 53–91.

Day, J.H. (1974) The ecology of Morrumbene estuary, Moçambique. *Transactions of the Royal Society of South Africa*, **41**, 43–97.

Day, J.H. (1981) The nature, origin and classification of estuaries, in *Estuarine Ecology with Particular Reference to Southern Africa* (ed. J.H. Day), A.A. Balkema, Cape Town, pp. 1–6.

Day, J.H. and Grindley, J.R. (1981) The estuarine ecosystem and environmental constraints, in *Estuarine Ecology with Particular Reference to Southern Africa* (ed. J.H. Day), A.A. Balkema, Cape Town, pp. 345–372.

Day, J.H., Millard, N.A.H. and Broekhuysen, G.J. (1954) The ecology of South African estuaries. Part IV. The St Lucia System. *Transactions of the Royal Society of South Africa*, **34**, 129–156.

Day, J.H., Blaber, S.J.M. and Wallace, J.H. (1981) Estuarine fishes, in *Estuarine*

*Ecology with Particular Reference to Southern Africa* (ed. J.H. Day) A.A. Balkema, Cape Town, pp. 97–221.

Day, J.W. and Yáñez–Arancibia, A. (1985) Coastal lagoons and estuaries as an environment for nekton, in *Fish Community Ecology in Estuaries and Coastal Lagoons: Towards an Ecosystem Integration* (ed. A. Yáñez-Arancibia), UNAM Press, Mexico City, pp. 17–34.

Day, J.W. jun., Hall, C.A.S., Kemp, W.M. and Yáñez-Arancibia, A. (1989) *Estuarine Ecology*, John Wiley & Sons, New York.

De, D.K. and Datta, N.C. (1990) Age, growth, length–weight relationship and relative condition in hilsa, *Tenualosa ilisha* (Hamilton) from the Hooghly estuarine system. *Indian Journal of Fisheries*, **37**, 199–209.

De Kock, A.C. and Lord, D.A. (1988) Kinetics of the uptake of and elimination of polychlorinated biphenyls by an estuarine fish species *(Rhabdosargus holubi)* after aqueous exposure. *Chemosphere*, **17**, 2381–2390.

De Martini, E.E. and Fountain, R.K. (1981) Ovarian cycling frequency and batch frequency in the queenfish, *Seriphus politus*: attributes representative of serial spawning fishes. *Fisheries Bulletin, US*, **79**, 547–559.

De Silva, S.S. (1986) Reproductive biology of *Oreochromis mossambicus* populations of man-made lakes in Sri Lanka: a comparative study. *Aquaculture and Fisheries Management*, **17**, 31–47.

De Silva S.S. and Silva, E.I.L. (1979) Biology of young grey mullet, *Mugil cephalus* L., populations in a coastal lagoon in Sri Lanka. *Journal of Fish Biology*, **15**, 9–20.

De Sylva, D.P. (1963) Systematics and life history of the Great Barracuda, *Sphyraena barracuda* (Walbaum). *Studies in Tropical Oceanography*, **1**, 1–179.

De Sylva, D.P. (1973) Barracudas (Pisces: Sphyraenidae) of the Indian Ocean and adjacent seas – a preliminary review of their systematics and ecology. *Journal of the Marine Biological Association of India*, **15**, 74–94.

De Sylva, D.P. (1985) Nektonic food webs in estuaries, in *Fish Community Ecology in Estuaries and Coastal Lagoons: Towards an Ecosystem Integration* (ed. A. Yáñez-Arancibia), UNAM Press, Mexico City, pp. 233–246.

Deegan, L.A. and Thompson, B.A. (1985) The ecology of fish communities in the Mississippi River deltaic plain, in *Fish Community Ecology in Estuaries and Coastal Lagoons: Towards an Ecosystem Integration* (ed A. Yáñez-Arancibia), UNAM Press, Mexico City, pp. 35–56.

Dekhnik, T.V. (1953) Reproduction of *Mugil cephalus* in the Black Sea. *Comptes Rendus (Doklady) de l'Academie des Sciences de l'URSS*, **93**, 201–204.

Dentzau, M.W. and Chittenden, M.E. jun. (1990) Reproduction, movements, and apparent population dynamics of the Atlantic threadfin *Polydactylus octonemus* in the Gulf of Mexico. *Fishery Bulletin US*, **88**, 439–462.

Dill, L.M. (1977) Refraction and the spitting behavior of the Archerfish *(Toxotes chatareus)*. *Behavioral Ecology and Sociobiology*, **2**, 169–184.

Dionne, M. and Folt, C.L. (1991) An experimental analysis of macrophyte growth forms as fish foraging habitat. *Canadian Journal of Fisheries and Aquatic Sciences*, **48**, 123–131.

Diouf, P.S., Kebe, L., Le Reste, L., Bousso, T., Diadhiou, H.D. and Gaye, A.B. (1991) Plan d'action forestier – pêche et aquaculture continentales. *CRODT, FAO, Ministère du Développement et de l'Hydraulique*, **1**, 1–268.

Doan, K.H. (1941) Relation of sauger catch to turbidity in Lake Erie. *Ohio Science*, **41**, 449–452.

Dodd, J.M. (1983) Reproduction in cartilaginous fishes (Chondrichthyes), in *Fish*

*Physiology*, Vol. IXA (eds W.S. Hoare, D.J. Randall and E.M. Donaldson), Academic Press, New York, pp. 31–95.

Dublin-Green, C.O. (1990) The foraminiferal fauna of the Bonny Estuary. A baseline study. *Technical paper of the Nigerian Institute of Oceanography and Marine Research*, **64**, 1–27.

Dufour, P. (1987) Region 3: Afrique occidentale: Lagunes du Nigeria, in *African Wetlands and Shallow Water Bodies. Zones Humides et Lacs Peu Profonds d'Afrique* (eds M.J. Burgis, and J.J. Symoens), 1987, no. 211, pp. 196–200.

Dunn, I.G. (1982) *The hilsa fishery of Bangladesh, 1982: an investigation of its present status with an evaluation of current data.* A report prepared for the Fisheries Advisory Service, Planning, Processing and Appraisal Project, Field Document 2. FAO, Rome, pp. 1–70.

Durand, J.R. and Chantraine, J.M. (1982) L'environnement climatique des lagunes Ivoriennes. *Revue d'hydrobiologie tropicale*, **15**, 85–113.

Durand, J.R., Amon Kothias, J.B., Ecoutin, J.M., Gerlotto, F., Hiee Daré, J.P. and Laé, R. (1978) Statistiques de pêche en Lagune Ébrié (Côte d'Ivoire): 1976 et 1977. *Centre de Recherches Oceanographiques Abidjan, Documents Scientifique*, **67**, 114.

Dutrieux, E. (1991) Study of the ecological functioning of the Mahakam delta (East Kalamantan, Indonesia). *Estuarine, Coastal and Shelf Science*, **32**, 415–420.

Dyer, K.R. (1972) Sedimentation in estuaries, in *The Estuarine Environment* (eds R.S.K. Barnes and J. Green), Applied Science Publishers Ltd, London, pp. 10–32.

Dyer, K.R. (1973) *Estuaries: A Physical Introduction*, John Wiley & Sons, London.

Écoutin, J.M. and Delahaye, M. (1989) Les sennes tournantes de Vridi (Lagune Ébrié, Côte d'Ivoire). *Centre de Rechereches Océanographiques Abidjan, Documents Scientifique*, 17, 59–77.

Eisma, D., van der Gaast, S.J., Martin, J.M. and Thomas, A.J. (1978) Suspended matter and bottom deposits of the Orinoco Delta: turbidity, mineralogy and elementary composition. *Netherlands Journal of Sea Research*, **12**, 224–251.

Ekman, S. (1953) *Zoogeography of the Sea*, Sidgwick & Jackson, London.

Elliott, M. and Dewailly, F. (1995) The structure and components of European estuarine fish assemblages. *Netherlands Journal of Aquatic Sciences*, **29**, 397–417.

Elliott, M. and Taylor, C.J.L. (1989) The structure and functioning of an estuarine/marine fish community in the Forth estuary, Scotland. *Polish Academy of Sciences, Institute of Oceanology, Proceedings of the 21st EMBS*, 227–240.

Elshoud, G.C.A. and Koomen, P. (1985) A biomechanical analysis of spitting in Archer Fishes (Pisces, Perciformes, Toxotidae). *Zoomorphology*, **105**, 240–252.

Elst – *see* van der Elst.

Fagade, S.O. and Olaniyan, C.I.O. (1972) The biology of the West African shad, *Ethmalosa fimbriata* (Bowdich) in the Lagos Lagoon. *Journal of Fish Biology*, **4**, 519–533.

Fagade, S.O. and Olaniyan, C.I.O. (1973) The food and feeding interrelationship of the fishes in the Lagos Lagoon. *Journal of Fish Biology*, **5**, 205–227.

Fagade, S.O. and Olaniyan, C.I.O. (1974) Seasonal distribution of the fish fauna of the Lagos Lagoon. *Bulletin de L'IFAN*, **34A**, 244–252.

Fagbami, A.A., Udo, E.J. and Odu, C.T.I. (1988) Vegetation damage in an oil field in the Niger Delta of Nigeria. *Journal of Tropical Ecology*, **4**, 61–75.

FAO (1995) *FAO Yearbook of Fishery Statistics 1993*, FAO Fisheries Series 44, FAO, Rome.

Fernandez, I. and Devaraj, M. (1989) Reproductive biology of the gold spotted grenadier anchovy, *Coilia dussumieri* (Cuvier & Valenciennes), along the northwest coast of India. *Indian Journal of Fisheries*, **36**, 11–18.

Findlay, I.W.O. (1978) Marine biology of the Sierra Leone River Estuary. 1. The physical environment. *Bulletin of the Institute of Marine Biology and Oceanography, Fourah Bay College, Sierra Leone*, **3**, 48–64.

Fischer, W. and Bianchi, G. (eds) (1984) *FAO Fisheries Identification Sheets for Fishery Purposes. Western Indian Ocean (fishing area 51)*, FAO, Rome.

Fischer, W. and Whitehead, P.J.P. (1974) *FAO Fisheries Identification Sheets for Fishery Purposes. Eastern Indian Ocean (fishing area 57) and Western Central Pacific (fishing area 71)*, FAO, Rome.

Fisher, J.S., Perdue, R.R., Overton, M.F., Sobsey, M.D. and Sill, B.L. (1987) Comparison of water quality at two recreational marinas during a peak-use period, in *Working Papers of the North Carolina State University Sea Grant Program 1987*, pp. 1–53.

Fisher, W.L., McGowen, J.H., Brown, L.F. and Groat, C.G. (1972) *Environmental Geologic Atlas of the Texas Coastal Zone – Galveston–Houston Area*, University of Texas, Bureau of Economic Geology, Houston, TX.

Flores-Coto, C., Ocana Luna, A., Luna Calvo, A. and Zavala Garcia, F. (1988) Abundancia de algunas especies de anchoas en la Laguna de Terminos (Mexico), estimada a traves de la captura de huevos. *Annales del Instituto de Ciencias del Mar y Limnologia, Universidad Nacional Autónoma de México*, **15**, 125–134.

Fontana, A. and Leguen, J.C. (1969) Étude de la maturité sexuelle et de la fecondité de *Pseudotolithes (fonticulus) elongatus*. *Cahiers ORSTOM Serie Oceanographie*, **7**, 9–19.

Forbes, A.T. and Cyrus, D.P. (1992) Impact of a major cyclone on a southeast African estuarine lake system. *Netherlands Journal of Sea Research*, **30**, 265–272.

Forbes, A.T. and Cyrus, D.P. (1993) Biological effects of salinity gradient reversals in a southeast African estuarine lake. *Netherlands Journal of Aquatic Ecology*, **27**, 483–488.

Fryer, G., Greenwood, P.H. and Trewavas, E. (1955) Scale–eating habits of African cichlid fishes. *Nature, London*, **175**, 1089–1090.

Gamboa, B.R., Garcia, A.G., Benitey, J.A. and Okuda, T. (1971) Estudio de las condiciones hidrográficas y quimicas en el agua de la Laguna de Lacarigua. *Boletin del Instituto Oceanografico de Venezuela, Universidad de Oriente*, **10**, 55–72.

García, C.B. and Solano, O.D. (1995) *Tarpon atlanticus* in Colombia: a big fish in trouble. *Naga, The ICLARM Quarterly*, **18** (3), 47–49.

Garcia, S. (1975) Los recursos pesqueros regionales de Tuxpan, Veracruz a Tampico, Tamps, y su posible industrialización. *Instituto Nacional de Pesca (Mexico) Boletin Informativo*.

Gargantiel, E.J. (1982) Programme on cage culture of finfish of the Southeast Asian Fisheries Development Center Aquaculture Department, in *Report of the training course on small scale pen and cage culture for finfish, Los Banos, Laguna, Philippines, 26–31 October 1981 and Aberdeen, Hong Kong, 1–13 November 1981* (eds R.D. Guerrero and V. Soesanto) FAO–UNDP South China Sea Fisheries Development and Coordination Programme, Manila, Philippines, pp. 191–195.

Garratt, P.A. (1993) Spawning of riverbream, *Acanthopagrus berda*, in Kosi estuary. *South African Journal of Zoology*, **28**, 26–31.

Garrett, R.N. (1987) Reproduction in Queensland barramundi (*Lates calcarifer*). *ACIAR Proceedings*, **20**, 38–43.

Gee, J.M. (1989) An ecological and economic review of meiofauna as food for fish. *Zoological Journal of the Linnean Society,* **96**, 243–261.

Gerking, S.D. (1994) *Feeding Ecology of Fish,* Academic Press, San Diego.

Ghiselin, M.T. (1969) The evolution of hermaphroditism among animals. *The Quarterly Review of Biology,* **44**, 189–208.

Ghosh, A., Das, K.M., Chattopadhyay, G.N., Chakraborti, P.K., Ghosh, A., Naskar, K.R., Das, R.K. and Mondal, S.K. (1987) Waste recycling through fish culture in estuarine wetlands – a productive method of land use, in *Symposium on the Impact of Current Land Use Pattern and Water Resources Development on Riverine Fisheries (Barrackpore, India), 25–27 Apr. 1987,* Marine Biological Association of India, Cochin, p. 78.

Ghosh, T.K. and Konar, S.K. (1993) Mass mortality of fish in the River Mathabhanga-Churni, West Bengal. *Environmental Ecology,* **11**, 833–838.

Giannini, R. and Paiva-Filho, A.M. (1990) Aspectos bioecologicos de *Stellifer rastrifer* (Perciformes: Sciaenidae) na Baia de Santos, SP. *Boletim do Instituto Oceanografico de Sao Paulo,* **38**, 57–67.

Gilbert, C.R. and Kelso, D.P. (1971) Fishes of the Tortuguero area, Carribbean Costa Rica. *Bulletin of the Florida State Museum Biological Sciences,* **16**, 1–54.

Gilmore, G. and Trent, L. (1974) Abundance of benthic macroinvertebrates in natural and altered estuarine areas. *National Marine Fisheries Service Report SSRF,* **677**, 1–13.

Gloerfelt-Tarp, T. and Kailola, P.J. (1984) *Trawled Fishes of Southern Indonesia and Northwestern Australia,* Australian Development Assistance Bureau, Canberra.

Goodwin, J.M. and Finucane, J.H. (1985) Reproductive biology of blue runner (*Caranx crysos*) from the eastern Gulf of Mexico. *Northeast Gulf Science,* **7**, 139–146.

Gouda, R. (1995) Environmental problems and management strategy of a coastal wetland of Orissa, Chilka Lake, on the east coast of India. *Abstract No. 102, Workshop 1, International Conference on Wetlands and Development, 8–14 October 1995, Selangor, Malaysia.* Asian Wetland Bureau, Kuala Lumpur.

Green, J. (1968) *The Biology of Estuarine Animals,* Sidgwick & Jackson, London.

Grey, D.L. (1987) An overview of *Lates calcarifer* in Australia and Asia. *ACIAR Proceedings,* **20**, 15–21.

Griffin, R.K. (1995) Recreational fishing surveys in the Northern Territory – 1978–1993, in *Recreational Fishing: What's the Catch?* (ed. D.A. Hancock), Australian Government Publishing Service, Canberra, pp. 113–119.

Guggisberg, C.A.W. (1972) *Crocodiles. Their Natural History, Folklore and Conservation,* Purnell, Cape Town.

Gunn, J.S. (1990) A revision of selected genera of the family Carangidae (Pisces) from Australian waters. *Records of the Australian Museum,* Supplement **12**, 1–77.

Gunter, G. (1967) Some relationships of estuaries to the fisheries of the Gulf of Mexico, in *Estuaries* (ed. G.H. Lauff), American Association for the Advancement of Science, Washington, DC, pp. 621–638.

Haines, A.K. (1979) An ecological survey of fish of the lower Purari River system, Papua New Guinea. *Department of Minerals and Energy, Papua New Guinea, Purari River Hydroelectric Scheme Environmental Studies,* **6**, 1–102.

Halder, G.C., Haroon, A.K.Y., Khan, M.A.A. and Tsai, C.F. (1991) Fish nursery ground investigation of the Feni River estuary. *Bangladesh Journal of Zoology,* **19**, 85–94.

Harden Jones, F.R. (1984) A view from the ocean, in *Mechanisms of Migration in*

*Fishes* (eds J.D. McGleave, G.P. Arnold, J.J. Dodson, and W.H. Neill) Plenum, London, pp. 1–26.

Harden Jones, F.R. (1994) *Fisheries Ecologically Sustainable Development: Terms and Concepts*, IASOS, University of Tasmania, Hobart.

Hardy, G.S. (1981) New records of pufferfishes (Family Tetraodontidae) from Australia and New Zealand, with notes on *Sphoeroides pachygaster* (Muller & Troschel) and *Lagocephalus sceleratus* (Gmelin). *Records of the National Museum of New Zealand*, **1**, 311–316.

Harrel, R.C. and Hall, M.A. (1991) Macrobenthic community structure before and after pollution abatement in the Neches River estuary (Texas). *Hydrobiologia*, **211**, 241–252.

Harris, S.A. (1996) Larval fish assemblages of selected estuarine and coastal systems in KwaZulu-Natal, South Africa, PhD thesis, University of Zululand, South Africa, 208 pp.

Harris, S.A., Cyrus, D.P. and Forbes, A.T. (1995) The larval fish assemblage at the mouth of the Kosi estuary, KwaZulu-Natal, South Africa. *South African Journal of Marine Science*, **16**, 333–350.

Harris, S.A. and Cyrus, D.P. (1995) Occurrence of larval fishes in the St Lucia estuary, KwaZulu-Natal, South Africa. *South African Journal of Marine Science*, **16**, 351–364.

Harrison, T.D. (1991) A note on the diet and feeding selectivity of juvenile river-bream, *Acanthopagrus berda* (Forskal, 1775), in a subtropical mangrove creek. *South African Journal of Zoology*, **26**, 36–42.

Harrison, T.D. and Whitfield, A.K. (1995) Fish community structure in three temporarily open/closed estuaries on the Natal coast. *Ichthyological Bulletin of the J.L.B. Smith Institute*, **64**, 1–80.

Hay, D.G. (1985) The macrobentos of the St Lucia narrows, MSc thesis, University of Natal, Pietermaritzburg, KwaZulu-Natal, South Africa, 162 pp.

Hayes, M.O. (1967) Hurricanes as geological agents, south Texas coast. *Bulletin of the American Society of Petroleum Geology*, **51**, 937–942.

Hecht, T. and van der Lingen, C.D. (1992) Turbidity-induced changes in feeding strategies of fish in estuaries. *South African Journal of Zoology*, **27**, 95–107.

Hedgpeth, J.W. (1967) Ecological aspects of the Laguna Madre, a hypersaline estuary, in *Estuaries* (ed. G.H. Lauff), American Association for the Advancement of Science, Washington, DC, pp. 408–419.

Hedgpeth, J.W. (1982) Estuarine dependence and colonization. *Atlantica*, **5**, 57–58.

Helfrich, P. and Allen, P.M. (1975) Observations on the spawning of mullet, *Crenimugil crenilabis* (Forskal), at Eniwetak, Marshall Islands. *Micronesica*, **11**, 219–225.

Hernandez, R.R.A. and Calderon, M.G. (1974) *Inventario preliminar de la flora y fauna acuática de la Bahía de Jiquilisco*. Ministerio de Agricultura y Granadería, Dirección General de Recursos Naturales Renovables, Servicio de Recursos Pesqueros, El Salvador.

Herrera, A. and Charles, A.T. (1994) Costa Rican coastlines: mangroves, reefs, fisheries and people, in *Coastal Zone Canada – 94, Cooperation in the Coastal Zone. Conference Proceedings. Volume 2* (eds P.G. Wells and P.J. Ricketts), Coastal Zone Canada Association, Dartmouth, Nova Scotia, pp. 612–624.

Hiatt, R.W. (1944) Food-chains and the food cycle in the Hawaiian fish ponds – Part 1. The food and feeding habits of mullet *(Mugil cephalus)*, milkfish *(Chanos chanos)*, and the ten-pounder *(Elops machnata)*. *Transactions of the American Fisheries Society*, **74**, 250–261.

Hildebrand, S.F. (1946) A descriptive catalog of the shorefishes of Peru. *Bulletin of the US National Museum,* **189,** 1–530.

Hill, B.J. (1969) Bathymetry and possible origin of Lakes Sibaya, Nhlange and Sifungwe in Zululand. *Transactions of the Royal Society of South Africa,* **38,** 205–216.

Hill, B.J., Blaber, S.J.M. and Boltt, R.E. (1975) The limnology of Lagoa Poelela, Moçambique. *Transactions of the Royal Society of South Africa,* **41,** 263–271.

Hoar, W.S. and Randall, D.J. (eds) (1969) *Fish Physiology,* Volume III, *Reproduction and Growth, Biolumiscence, Pigments, and Poisons,* Academic Press, London.

Hoar, W.S., Randall, D.J. and Donaldson, E.M. (eds) (1983a) *Fish Physiology,* Volume IX, *Reproduction Part A, Endocrine Tissues and Hormones,* Academic Press, London.

Hoar, W.S., Randall, D.J. and Donaldson, E.M. (eds) (1983b) *Fish Physiology,* Volume IX, *Reproduction Part B, Behaviour and Fertility Control,* Academic Press, London.

Hoese, D.F. (1986) Gobiidae, in *Smith's Sea Fishes* (eds M.M. Smith and P.C. Heemstra), Macmillan, Johannesburg, South Africa, pp. 774–807.

Hoese, D.F., Larson, H.K. and Llewellyn, L.C. (1980) Family Eleotridae Gudgeons, in *Freshwater Fishes of South-eastern Australia* (ed. R.M. McDowall), A.H. & A.W. Reed, Sydney, Australia, pp.169–185.

Hoese, H.D. (1966) Ectoparasitism by juvenile sea catfish, *Galeichthys felis. Copeia,* **1966,** 880–881.

Hoin-Radkovsky, I., Bleckmann, H. and Schwartz, E. (1984) Determination of source distance in the surface-feeding fish *Pantodon buchholzi* Pantodontidae. *Animal Behaviour,* **32,** 840–851.

Holt, G.J., Godbout, R. and Arnold, C.R. (1981) Effect of temperature and salinity on egg hatching and larval survival of Red Drum *Sciaenops ocellata. Fishery Bulletin, US,* **79,** 569–573.

Holt, G.J., Holt, S.A. and Arnold, C.R. (1985) Diel periodicity of spawning in sciaenids. *Marine Ecology Progress Series,* **27,** 1–7.

Hong, P.N. (1991) Status of mangrove ecosystems in Vietnam: some management considerations, in *Workshop for Establishing a Global Network of Mangrove Genetic Resource Centres for Adaptation to Sea Level Rise, Madras (India), 15–19 Jan. 1991* (eds S.V. Deshmukh and R. Mahalingam), Centre for Adaptation to Sea Level Rise, Madras, pp. 53–63.

Hora, S.L. (1938) A preliminary note on the spawning grounds and bionomics of the so-called Indian Shad, *Hilsa ilisha* (Ham.), in the River Ganges. *Records of the Indian Museum,* **40,** 147–158.

Hora, S.L. and Nair, K.K. (1940) Further observations on the bionomics and fishery of the Indian Shad, *Hilsa ilisha* (Ham.), in Bengal waters. *Records of the Indian Museum,* **42,** 35–50.

Horst – *see* van der Horst.

Houde, E.D. (1989) Comparative growth and energetics of marine fish larvae, in *Third ICES Symposium on the Early Life History of Fish, Bergen (Norway), 3–5 Oct. 1988* (eds J.H.S. Blaxter, J.C. Gamble and H. von Westernhagen), *Rapports et Procès-verbaux des Réunions, Conseil International pour l'Exploration de la Mer,* **191,** p.479.

Houde, E.D., Almatar, S., Leak, J.C. and Dowd, C.E. (1986) Ichthyoplankton abundance and diversity in the western Arabian Gulf. *Kuwait Bulletin of Marine Science,* **8,** 107–393.

Hum, K.S., Thanh, N.C. and Lee, T.L. (1981) Palm oil waste treatment study in Malaysia. *Effluent Water Treatment Journal*, **21**, 452–455.

Huq, A.K.M.F. (1995) The Sundarbans mangrove swamp. *Abstract No. 139, Workshop 1, International Conference on Wetlands and Development, 8–14 October 1995, Selangor, Malaysia*, Asian Wetland Bureau, Kuala Lumpur.

Husain, Z. and Sufi, M.S.K. (1962a) *Hilsa ilisha* (Ham.) and fish ladders at Ghulam Mohammed Barrage on the River Indus, West Pakistan. *Agriculture Pakistan*, **13**, 335–345.

Husain, Z. and Sufi, M.S.K. (1962b) Biological and economic effects of barrages on *Hilsa ilisha* (Ham.) and its fisheries on the Indus. *Agriculture Pakistan*, **13**, 346–349.

Hutchings, P. and Saenger, P. (1987) *Ecology of Mangroves*, University of Queensland Press, St Lucia, Australia.

Ikusemiju, K. and Olaniyan, C.I.O. (1977) The food and feeding habits of the catfishes, *Chrysichthys walkeri* (Gunther), *Chrysichthys filamentosus* (Boulenger) and *Chrysichthys nigrodigitatus* (Lacépède) in the Lekki Lagoon, Nigeria. *Journal of Fish Biology*, **10**, 105–112.

Ikusemiju, K., Oki, A.A. and Graham-Douglas, M. (1983) On the biology of an estuarine population of the clupeid *Pellonula afzeliusi* (Johnels) in Lagos Lagoon, Nigeria. *Hydrobiologia*, **102**, 55–59.

INDERENA (1974) *Estadísticas, actividad pesquera en Colombia*. Instituto para el Desarollo de los Recursos Renovables Naturales. Oficina de Paneación, Ministerio de Agricultura, Colombia. (In Spanish)

Infante-Guevara, C. (1992) Mount Pinatubo's effects on Philippine fisheries. *Naga, The ICLARM Quarterly*, **15**(1), 8–10.

Inger, R.E. (1955) Ecological notes on the fish fauna of a coastal drainage of North Borneo. *Fieldiana, Zoology*, **37**, 47–90.

Jafri, S.I.H. and Melvin, G.M. (1988) An annotated bibliography (1803–1987) of the Indian Shad, *Tenualosa ilisha* (Ham.) (Clupeidae: Teleostei). Manuscript Report no. 178e, International Development Research Centre, Ottawa.

James, P.S.B.R., Rengaswamy, V.S., Raju, A., Mottanraj, G. and Gandhi, V. (1983) Induced spawning and larval rearing of the grey mullet *Liza macrolepis* (Smith). *Indian Journal of Fisheries*, **30**, 185–202.

Janekarn, V. and Boonruang, P. (1986) Composition and occurrence of fish larvae in mangrove areas along the east coast of Phuket Island, western peninsular, Thailand. *Phuket Marine Biological Center Research Bulletin*, **44**, 1–22.

Janekarn, V. and Kiørboe, T. (1991a) The distribution of fish larvae along the Andaman coast of Thailand. *Phuket Marine Biological Center Research Bulletin*, **56**, 41–61.

Janekarn, V. and Kiørboe, T. (1991b) Temporal and spatial distribution of fish larvae and their environmental biology in Phang-Nga bay. *Phuket Marine Biological Center Research Bulletin*, **56**, 23–40.

Jarvis, R.M. and Colvin, M.F. (1981) Some aspects of canal estate construction in south-east Queensland, in *Australian Conference on Coastal and Ocean Engineering 1981: Offshore Structures, Perth (Australia) 25 Nov. 1981*, National Committee on Coastal and Ocean Engineering Australia, Melbourne, pp. 142–143.

Jenkins, W.A. (1985) Environmental effects of pollutants associated with marina development: a synthesis of existing research, in *The Fate and Effects of Pollutants: A Symposium (College Park, MD) (USA), 26–27 Apr. 1985*, Technical Report, Maryland University Sea Grant Program, pp. 63–64.

Jeyaseelan, M.J.P. and Krushnamurthy, K. (1980) Role of mangrove forests of

Pichavaram as fish nurseries. *Proceedings of the Indian National Sciences Academy,* **46B**, 48–53.

Jhingran, V.G. (1991) *Fish and Fisheries of India,* 3rd edn, Hindustan Publishing Corporation, Delhi.

Jhingran, V.G. and Gopalakrishnan, V. (1973) Estuarine fisheries resources of India in relation to adjacent seas. *Journal of Marine Biological Association of India,* **15**, 323–334.

Jhingran, V.G. and Natarajan, A.V. (1969) A study of the fisheries and fish populations of the Chilka Lake during the period 1957–1965. *Journal of the Inland Fisheries Society of India,* **1**, 49–126.

Jobling, M. (1995) *Environmental Biology of Fishes,* Chapman & Hall, London.

Johannes, R.E. (1978) Reproductive strategies of coastal marine fishes in the tropics. *Environmental Biology of Fishes,* **1**, 65–84.

John, D.M. and Lawson, G.W. (1990) A review of mangrove and coastal ecosystems in West Africa and their possible relationships. *Estuarine Coastal and Shelf Science,* **31**, 505–518.

Johnson, C.R. (1973) Behaviour of the Australian crocodiles, *Crocodylus johnstoni* and *C. porosus. Zoological Journal of the Linnean Society,* **52**, 315–336.

Johnson, S.A. (1981) Estuarine dredge and fill activities: a review of impacts. *Environmental Management,* **5**, 427–440.

Jones, R.L., Kelley, D.W. and Owen, L.W. (1963) Delta fish and wildlife protection study. Resources and Agriculture, California, Sacramento, CA, Report Number 2, 73 pp.

Jones, S. and Menon, P.M.G. (1952) Observations on the development and systematics of the fishes of the genus *Coilia* Gray. *Journal of the Zoological Society of India,* **4**, 17–36.

Jones, S. and Sujansingani, K.H. (1951) The Hilsa of the Chilka Lake. *Journal of the Bombay Natural History Society,* **50**, 264–280.

Jones, S. and Sujansingani, K.H. (1954) Fish and fisheries of the Chilka Lake with statistics of fish catches for the years 1948–1950. *Indian Journal of Fisheries,* **1**, 256–344.

Joseph, K.O. and Srivastava, J.P. (1993) Mercury in the Ennore estuary and in fish from Madras coastal waters. *Journal of Environmental Biology,* **14**, 55–62.

Juanes, F. (1994) What determines prey size selectivity in piscivorous fishes, in *Theory and Application in Fish Feeding Ecology* (eds D.J. Stouder, K.L. Fresh and R.J. Feller), Belle W. Baruch Library in Marine Sciences no. 18, University of South Carolina Press, Columbia, SC, pp. 79–100.

Kader, M.A., Bhuiyan, A.L., Manzur, A.R.M.M. and Khuda, I. (1988) The reproductive biology of *Gobioides rubicundus* (Ham. Buch.) in the Karnaphuli River estuary, Chittagong. *Indian Journal of Fisheries,* **35**, 239–250.

Kagwade, P.V. (1976) Sexuality in *Polydactylus indicus* (Shaw). *Indian Journal of Fisheries,* **21**, 323–329.

Kagwade, P.V. (1988) Present status of polynemid fishery in India, in *National Symposium on Research and Development in Marine Fisheries, Mandapam Camp (India), 16 Sep. 1987,* CMFRI Special Publication, **40**, p. 14.

Kailola, P.J. (1990) The catfish family Ariidae (Teleostei) in New Guinea and Australia: relationships, systematics and zoogeography, PhD thesis, University of Adelaide, Australia, 525 pp.

Kailola, P.J., Williams, M.J., Stewart, P.C., Reichelt, R.E., McNee, A. and Grieve, C. (1993) *Australian Fisheries Resources,* Bureau of Resource Sciences and the Fisheries Research and Development Corporation, Canberra.

Kapetsky, J.M. (1981) Some considerations for the management of coastal lagoon and estuarine fisheries. *FAO Fisheries Technical Paper*, **218**, 1–47.

Kare, B. (1995) A decline of the barramundi (*Lates calcarifer*) (Bloch) fishery in the Western Province, with a review on the research and fishery. Unpublished manuscript, Department of Fisheries and Marine Resources, Papua New Guinea.

Käse, R.H. and Bleckmann, H. (1987) Prey localization by surface wave ray-tracing: fish track bugs like oceanographers track storms. *Experientia*, **43**, 290–292.

Kawasaki, T. (1980) Fundamental relations among the selections of life-history in marine teleosts. *Bulletin of the Japanese Society of Scientific Fisheries*, **46**, 289–293.

Kayal, S. and Connell, D.W. (1995) Polycyclic aromatic hydrocarbons in biota from the Brisbane River estuary, Australia. *Estuarine Coastal and Shelf Science*, **40**, 475–493.

Khalid, A.R. and Wan–Mustafa, W.A. (1992) External benefits of environmental regulation: resource recovery and the utilisation of effluents. *Environmentalist*, **12**, 277–285.

Khan, A. and Khan, Z.R. (1995) Wetland biodiversity: the socio-economic and conservation perspective in Bangladesh. *Asian Wetland News*, **8**, 4–6.

Khoo, H.K. (1989) The fisheries in the Matang and Merbok mangrove ecosystems. *Proceedings of the 12th Annual Seminar of the Malaysian Society of Marine Sciences*, 147–169.

Killam, K.A., Hochberg, R.J. and Rzemien, E.C. (1992) Synthesis of basic life histories of Tampa Bay species. *Tampa Bay National Estuary Program Report* – 10–92, pp. 1–255.

Kinch, J.C. (1979) Trophic habits of the juvenile fishes within artificial waterways – Macro Island, Florida. *Contributions in Marine Science*, **22**, 77–90.

Kishi, Y. (1978) Egg size difference among three populations of the goby, *Tridentiger obscurus*. *Japanese Journal of Ichthyology*, **24**, 278–280

Klein, R., Loya, Y., Gvirtzman, G., Isdale, P.J. and Susic, M. (1990) Seasonal rainfall in the Sinai Desert during the late Quaternary inferred from fluorescent bands in fossil corals. *Nature*, **345**, 145–147.

Konchina, Y.V. (1978) Some data on the biology of grunts (Family Pomadasyidae). *Journal of Ichthyology*, **17**, 548–558.

Kowtal, G.V. (1976) Studies of the juvenile fish stock of Chilka Lake. *Indian Journal of Fisheries*, **23**, 31–40.

Krishnamurthy, K. and Jeyaseelan, M.J. (1981) Early life history of fishes from Pichavaram mangrove ecosystem of India. *Rapports et Procès-verbaux des Réunions, Conseil Permanent International pour l'exploration de la Mer*, **178**, 416–423.

Kumar, P., Samal, R.C. and Das, N.K. (1989) Remote sensing study on the coastal process of Orissa coast, in *Proceedings of the Third National Conference on Dock and Harbour Engineering, Suratkal (India), 6–9 Dec. 1989*, Orissa Remote Sensing Centre, Bubaneswar, India, pp. 673–679.

Kuo, C.M., Shedhdah, Z.H. and Milisen, K.K. (1973) A preliminary report on the development, growth and survival of laboratory reared larvae of the grey mullet, *Mugil cephalus* L. *Journal of Fish Biology*, **5**, 459–470.

Kurup, B.M. and Samuel, C.T. (1983) Observations on the spawning biology of *Liza parsia* (Hamilton–Buchanan) in the Cochin Estuary. *Mahagasar*, **16**, 371–380.

Kurup, B.M. and Samuel, C.T. (1991a) Spawning biology of *Gerres filamentosus*

Cuvier in the Cochin estuary. *Fisheries Technology Journal, Kochi University, India*, **28**, 19–24.

Kurup, B.M. and Samuel, C.T. (1991b) Observation on the spawning biology of *Nibea albida* (Cuvier) in the Cochin estuary. *Journal of the Marine Biological Association of India*, **33**, 99–106.

Kuwamura, T. (1986) Sex of the mouthbrooder, with reference to ancestral parental behavior and reproductive success, in *Indo–Pacific Fish Biology, Proceedings of the Second International Conference on Indo–Pacific Fishes, Tokyo (Japan), 29 July – 3 August 1985* (eds T. Uyeno, R. Arai, T. Taniuchi and K. Matsuura), Ichthyological Society of Japan, Tokyo, pp. 946–947.

Kwei, E.A. (1977) Biological, chemical and hydrological characters of coastal lagoons of Ghana, West Africa. *Hydrobiologia*, **56**, 157–174.

Kyle, R. (1988) Aspects of the ecology and exploitation of the fishes of the Kosi Bay system, KwaZulu, South Africa, PhD thesis, University of Natal, Pietermaritzburg, KwaZulu–Natal, South Africa, 271 pp.

Kyle, R. (1989) A mass mortality of fish at Kosi Bay. *Lammergeyer*, **40**, 39.

Lacerda, L.D., Carvalho, C.E.V., Rezende, C.E. and Pfeiffer, W.C. (1993) Mercury in sediments from the Paraiba do Sul River continental shelf, S.E. Brazil. *Marine Pollution Bulletin*, 26, 220–222.

Lal, B. (1987) Impact of Farraka Barrage on the hydrological changes and productivity potential of Hooghly estuary, in *Symposium on the Impact of Current Land Use Pattern and Water Resources Development on Riverine Fisheries (Barrackpore, India), 25–27 Apr. 1987*, p. 9.

Lal, P.N. (1984) Environmental implications of coastal development in Fiji. *Ambio*, 13, 316–321.

Langlois, T.H. (1941) Two processes operating for the reduction in abundance or elimination of fish species from certain types of water. *Transactions of the Sixth North American Wildlife Conference*, pp. 189–201.

Larkum, A.W.D., McComb, A.J. and Shepherd, S.A. (eds) (1989) *Biology of Seagrasses – A Treatise on the Biology of Seagrasses with Special Reference to the Australian Region*, Elsevier, Amsterdam.

Laserre, G. (1979) Bilan de la situation des peches: aux Pangalanes Est. *(Zone Tamatave–Andevovanto) au Lac Anony (region Fort Dauphin)*. Perspective et Amenagement. Consultant's Report to MAG/76/002.

Lasiak, T.A. (1984) Structural aspects of the surf zone fish assemblage at King's Beach, Algoa Bay, South Africa: short term fluctuations. *Estuarine, Coastal and Shelf Science*, 18, 347–360.

Lasiak, T.A. (1986) Juveniles, food and the surf zone habitat: implications for teleost nursery areas. *South African Journal of Zoology*, 21, 51–56.

Last, P.R. and Stevens, J.D. (1994) *Sharks and Rays of Australia*, CSIRO, Hobart, Australia.

Latiff, M.A., Weber, W. and Kean, L.A. (1976) *Demersal fish resources in Malaysian waters – 9. Second trawl survey off the coast of Sarawak*. Ministry of Agriculture, Kuala Lumpur, Malaysia.

Law, A.T. and Singh, A. (1987) Distribution of mercury in the Kelang Estuary. *Pertanika*, 10, 175–181.

Law, A.T. and Singh, A. (1988) Heavy metals in fishes in the Kelang estuary, Malaysia. *Malayan Nature Journal*, 41, 505–513.

Le Reste, L. (1986) Consequences d'un barrage écluse anti–sel sur l'environnement aquatique et la pêche, in *L'Estuaire de la Casamance: Environnement, Pêche, Socioeconomie – Seminaire ISRA, Senegal, 19–24 Juin 1986* (eds L. Le Reste, A.

Fontana and A. Samba), Centre de Recherche Oceanographie, Dakar, pp. 307–316.

Lea, R., Grey, D. and Griffin, R. (1987) Utilisation and wildstock management of the barramundi (*Lates calcarifer*) in the Northern Territory. *ACIAR Proceedings*, **20**, 82–86.

Lee, C.L., Peerzada, N. and Guinea, M. (1993) Control of aquatic plants in aquaculture using silver scat, *Selenotoca multifasciata*. *Journal of Applied Aquaculture*, **2**, 77–83.

Lee, C.S. and Menu, B. (1981) Effects of salinity on egg development and hatching in grey mullet *Mugil cephalus* L. *Journal of Fish Biology*, **19**, 179–188.

Leis, J.M. (1991) The pelagic stage of reef fishes: the larval biology of coral reef fishes, in *The Ecology of Fishes on Coral Reefs* (ed. P.F. Sale), Academic Press, San Diego, pp. 183–230.

Leis, J.M. and Rennis D.S. (1983) *The Larvae of Indo–Pacific Coral Reef Fishes*, University of Hawaii Press, Honolulu.

Lenanton, R.C.J. and Hodgkin, E.P. (1985) Life history strategies of fish in some temperate Australian estuaries, in *Fish Community Ecology in Estuaries and Coastal Lagoons: Towards an Ecosystem Integration* (ed. A. Yáñez-Arancibia), UNAM Press, Mexico City, pp. 277–284.

Lenanton, R.C.J. and Potter, I.C. (1987) Contribution of estuaries to commercial fisheries in temperate Western Australia and the concept of estuarine dependence. *Estuaries*, **10**, 28–35.

León, R.A. and Racedo, J.B. (1985) Composition of fish communities in the lagoon and estuarine complex of Cartagena Bay, Cienaga de Tesca and Cienaga Grande de Santa Marta, Colombian Caribbean, in *Fish Community Ecology in Estuaries and Coastal Lagoons: Towards an Ecosystem Integration* (ed. A. Yáñez-Arancibia), UNAM Press, Mexico City, pp. 535–556.

de Leon, R.O.D. (1994) Benefits and costs of managed mangroves, in *Third ASEAN–Australia Symposium on Living Coastal Resources, Bangkok, Thailand, 1994* (ed. C.R. Wilkinson), Australian Institute of Marine Science, Townsville, pp. 84–86.

Lewis, A.D. (1990) Tropical South Pacific tuna baitfisheries. *ACIAR Proceedings*, **30**, 10–21.

Liao, I.C. (1975) Experiments on the induced breeding of the grey mullet in Taiwan from 1963–1973. *Aquaculture*, **6**, 31–58.

Lin, Hui Chen and Dunson, W.A. (1993) The effect of salinity on the acute toxicity of cadmium to the tropical, estuarine, hermaphroditic fish, *Rivulus marmoratus*: a comparison of Cd, Cu, and Zn tolerance with *Fundulus heteroclitus*. *Archives of Environmental Contamination and Toxicology*, **25**, 41–47.

Lindall, W.N., jun., and Trent, L. (1975) Housing development canals in the coastal zone of the Gulf of Mexico: ecological consequences, regulations and recommendations. *Marine Fisheries Review*, **37**, 19–24.

Lindall, W.N. jun., Hall, J.R. and Saloman, C.H. (1973) Fishes, macroinvertebrates, and hydrological conditions of upland canals in Tampa Bay, Florida. *Fishery Bulletin, US*, **71**, 155–163.

Lindall, W.N., jun., Fable, W.A., jun. and Collins, L.A. (1975) Additional studies of the fishes, macroinvertebrates, and hydrological conditions of upland canals in Tampa Bay, Florida. *Fishery Bulletin, US*, **73**, 81–85.

Little, M.C., Reay, P.J. and Grove, S.J. (1988a) Distribution gradients of ichthyoplankton in an East African Mangrove creek. *Estuarine Coastal and Shelf Science*, **26**, 669–677.

Little, M.C., Reay, P.J. and Grove, S.J. (1988b) The fish community of an East African mangrove creek. *Journal of Fish Biology*, **32**, 729–747.

Livingston, R.J. (1982) Trophic organization of fishes in a coastal seagrass system. *Marine Ecology Progress Series*, **7**, 1–2.

Loftus, W.F. (1987) Possible establishment of the Mayan cichlid, *Cichlasoma urophthalmus* (Guenther) (Pisces: Cichlidae), in Everglades National Park, Florida. *Florida Science*, **50**, 1–6.

Longhurst, A.R. (1957) The food of the demersal fish of a West African estuary. *Journal of Animal Ecology*, **26**, 369–387.

Longhurst, A.R. (1960) A summary of the food of West African demersal fish. *Bulletin de l'IFAN, Series A*, **22**, 267–282.

Longhurst, A.R. (1962) A review of the oceanography of the Gulf of Guinea. *Bulletin de l'IFAN, Series A*, **24**, 633–663.

Longhurst, A.R. (1963) The bionomics of the fishery resources of the eastern tropical Atlantic. *Colonial Office Fishery Publications, London*, **20**, 1–66.

Longhurst, A.R. (1983) Benthic–pelagic coupling and export of organic carbon from a tropical Atlantic continental shelf – Sierra Leone. *Estuarine, Coastal and Shelf Science*, **17**, 261–285.

Longhurst, A.R. and Pauly, D. (1987) *Ecology of Tropical Oceans*, Academic Press, San Diego.

Lowe-McConnell, R.H. (1962) The fishes of the British Guiana continental shelf, Atlantic coast of South America, with notes of their natural history. *Journal of the Linnean Society (Zoology)*, **44**, 669–700.

Lowe-McConnell, R.H. (1966) The sciaenid fishes of British Guiana. *Bulletin of Marine Science*, **16**, 20–57.

Lowe-McConnell, R.H. (1987) *Ecological Studies in Tropical Fish Communities*, Cambridge University Press, Cambridge.

Lüling, K.H. (1963) The archer fish. *Scientific American*, **209**, 100–109.

Luther, G. (1962) The food habits of *Liza macrolepis* (Smith) and *Mugil cephalus* L. (Mugilidae). *Indian Journal of Fisheries*, **9**, 604–626.

Lyle, J.M. (1984) Mercury concentrations in four carcharhinid and three hammerhead sharks from coastal waters of the Northern Territory. *Australian Journal of Marine and Freshwater Research*, **35**, 441–451.

Lyle, J.M. (1986) Mercury and selenium concentrations in sharks from northern Australian waters. *Australian Journal of Marine and Freshwater Research*, **37**, 309–321.

Lyons, J. and Schneider, D.W. (1990) Factors influencing fish distribution and community structure in a small coastal river in southwestern Costa Rica. *Hydrobiologia*, **203**, 1–14.

McCluney, W.R. (1975) Radiometry of water turbidity measurements. *Journal of the Water Pollution Control Federation*, **47**, 252–266.

McClusky, D.S. (1971) *Ecology of Estuaries*, Heinemann Educational Books Ltd, London.

McDowall, R.M. (1980) *Freshwater Fishes of South-eastern Australia*, A.H. & A.W. Reed, Sydney.

McDowall, R.M. (1988) *Diadromy in Fishes: Migrations between Freshwater and Marine Environments*, Croom Helm, London.

McEachron, L.W., Matlock, G.C., Bryan, C.E., Unger, P., Cody, T.J. and Martin, J.H. (1994) Winter mass mortality of animals in Texas bays. *Northeast Gulf Science*, **13**, 121–138.

McHugh, J.L. (1967) Estuarine nekton, in *Estuaries* (ed. G.H. Lauff), American

Association for the Advancement of Science, Washington, DC, pp. 581–620.

MacNae, W. (1968) A general account of the fauna and flora of mangrove swamps and forests in the Indo-West-Pacific region. *Advances in Marine Biology*, **6**, 73–270.

McPherson, B.F. and Miller, R.L. (1987) The vertical attenuation of light in Charlotte Harbor, a shallow, subtropical estuary, south-western Florida. *Estuarine Coastal and Shelf Science*, **25**, 721–737.

Majid, D.S. (1992) *Coastal Fisheries Management in Malaysia*, Department of Fisheries, Kuala Lumpur.

Major, P.F. (1973) Scale-feeding behavior of the leatherjacket *Scomberoides lysan* and two species of the genus *Oligoplites* (Pisces: Carangidae). *Copeia*, **1973**, 151–154.

Marr, J.C. (1982) The realities of fishery management in the Southeast Asia region, in *Theory and Management of Tropical Fisheries* (eds D. Pauly and G.I. Murphy), ICLARM/CSIRO, Manila, pp. 299–307.

Marten, G.G. and Polovina, J.J. (1982) A comparative study of fish yields from various tropical ecosystems, in *Theory and Management of Tropical Fisheries* (eds D. Pauly and G.I. Murphy), ICLARM/CSIRO, Manila, pp. 255–285.

Martin, T.J. (1988) Interaction of salinity and temperature as a mechanism for spatial separation of three co–existing species of Ambassidae (Cuvier) (Teleostei) in estuaries on the south east coast of Africa. *Journal of Fish Biology*, **33** (Supplement A), 9–15.

Martin, T.J. and Blaber, S.J.M. (1983) The feeding ecology of Ambassidae (Osteichthyes: Perciformes) in Natal estuaries. *South African Journal of Zoology*, **18**, 353–362.

Martin, T.J., Cyrus, D.P. and Forbes, A.T. (1992) Episodic events: the effects of cyclonic flushing on the ichthyoplankton of St Lucia estuary on the southeast coast of Africa. *Netherlands Journal of Sea Research*, **30**, 273–278.

Martin, T.J., Brewer, D.T. and Blaber, S.J.M. (1995) Factors affecting distribution and abundance of small demersal fishes in the Gulf of Carpentaria, Australia. *Marine and Freshwater Research*, **46**, 909–920.

Martinez-Palacios, C.A. and Ross, L.G. (1992) The reproductive biology and growth of the Central American cichlid *Cichlasoma urophthalmus* (Guenther). *Journal of Applied Ichthyology*, **8**, 1–4.

Matlock, G.C. (1987) The role of hurricanes in determining year-class strength of red drum. *Contributions in Marine Science*, **30**, 39–47.

Maurer, D., Epifanio, C., Dean, H., Howe, S., Vargas, J., Dittel, A. and Murillo, M. (1984) Benthic invertebrates of a tropical estuary: Gulf of Nicoya, Costa Rica. *Journal of Natural History*, **18**, 47–61.

Medway, Lord (1978) *The Wild Mammals of Malaya (Peninsula Malaysia) and Singapore*, 2nd edn, Oxford University Press, Kuala Lumpur, Malaysia.

Melville-Smith, R. and Baird, D. (1980) Abundance, distribution and species composition of fish larvae in the Swartkops estuary. *South African Journal of Zoology*, **15**, 72–78.

Melvin, G.D. (1984) Investigation of the Hilsa fishery of Bangladesh. *FAO Field Document 5, May 1984*.

Menon, A.G.K. (1977) A systematic mongraph of the Tongue Soles of the genus *Cynoglossus* Hamilton–Buchanan (Pisces: Cynoglossidae). *Smithsonian Contributions to Zoology*, **238**, 1–129.

de Merona, B. (1987) Aspectos ecologicos da ictiofauna no Baixo Tocantins. *Acta Amazonica*, **16**, 109–124.

Millard, N.A.H. and Broekhuysen, J. (1970) The ecology of South African estuaries part X. St. Lucia: a second report. *Zoologia Africana*, 5, 277–307.

Miller, J.M., Crowder, L.B. and Moser, M.L. (1985) Migration and utilization of estuarine nurseries by juvenile fishes: an evolutionary perspective. *Contribution in Marine Science*, 27, 338–352.

Milton, D.A. and Blaber, S.J.M. (1991) Maturation, spawning seasonality and proximal spawning stimuli of six species of tuna baitfish in Solomon Islands. *Fisheries Bulletin U.S.*, 89, 221–237.

Milward, N.E. (1974) Studies on the taxonomy, ecology and physiology of Queensland mudskippers. PhD thesis, University of Queensland, Australia, 376 pp.

Mines, A.N., Smith, I.R. and Pauly, D. (1986) An overview of the fisheries of San Miguel Bay, Philippines, in *The First Asian Fisheries Forum* (eds J.L. Maclean, L.B. Dizon and L.V. Hosillos), Asian Fisheries Society, Manila, Philippines, pp. 385–388.

Moazzam, M. and Rizvi, S.H.N. (1980) Fish entrapment in the seawater intake of power plant at Karachi coast. *Environmental Biology of Fishes*, 5, 49–57.

Mohanty, S.K. (1975) The breeding of economic fishes of the Chilka lake – a review. *Bulletin of the Department of Marine Science, University of Cochin*, 7, 543–559.

Mohsin, A.K.M., Said, M.Z.B.M., Ambak, M.A.B., Hayase, M.N. and Sekioka, M. (1988) Marine catch by experimental trawlings in the south-western portion of the South China Sea. *Faculty of Fisheries and Marine Science, Universiti Pertanian Malaysia, Occasional Publication*, 4, 113–120.

Moore, P.G. (1973) The kelp fauna of northeast Britain. II. Multivariate classification: turbidity as an ecological factor. *Journal of Experimental Marine Biology and Ecology*, 13, 127–163.

Moore, R. (1979) Natural sex inversion in the giant perch *(Lates calcarifer)*. *Australian Journal of Marine and Freshwater Research*, 30, 803–813.

Moore, R. (1980) Reproduction and migration in *Lates calcarifer* (Bloch). PhD thesis, University of London, 407 pp.

Moore, R. (1982) Spawning and early life history of barramundi, *Lates calcarifer* (Bloch), in Papua New Guinea. *Australian Journal of Marine and Freshwater Research*, 33, 663–670.

Moore, W.S. and Todd, J.F. (1993) Radium isotopes in the Orinoco estuary and eastern Caribbean Sea. *Journal of Geophysical Research*, 98, 2233–2244.

Moriarty, D.J.W. (1976) Quantitative studies on bacteria and algae in the food of the mullet *Mugil cephalus* L. and the prawn *Metapenaeus bennettae* (Racek and Dall). *Journal of Experimental Marine Biology and Ecology*, 22, 131–143.

Morton, R.M. (1989) Hydrology and fish fauna of canal developments in an intensively modified Australian estuary. *Estuarine, Coastal and Shelf Science*, 28, 43–58.

Morton, R.M. (1992) Fish assemblages in residential canal developments near the mouth of a subtropical Queensland estuary. *Australian Journal of Marine and Freshwater Research*, 43, 1359–1371.

Morton, R.M. (1993) Enhancement of estuarine habitats in association with development, in *Sustainable Fisheries through Sustaining Fish Habitat* (ed. D.A. Hancock), Australian Society for Fish Biology Workshop, 12–13 August, 1992, Bureau of Resource Sciences, Canberra, pp. 64–69.

Mukherjee, A.K. (1971) Food-habits of the water birds of the Sundarban 24 – Parganas District, West Bengal, India. *Journal of the Bombay Natural History Society*, 68, 37–64.

Mukhopadhyay, M.K., Vass, K.K., Bagchi, M.M. and Mitra, P. (1995) Environmental impact on breeding biology and fisheries of *Polynemus paradiseus* in Hooghly–Matlah estuarine system. *Environmental Ecology,* **13**, 395–399.

Munro, I.S.R. (1967) *The Fishes of New Guinea,* Department of Agriculture, Stock and Fisheries, Port Moresby.

Murphy, M.D. and Taylor, R.G. (1990) Reproduction, growth, and mortality of red drum *Sciaenops ocellatus* in Florida waters. *Fisheries Bulletin, US,* **88**, 531–542.

Myers, G.S. (1949) Salt tolerance of fresh–water fish groups in relation to zoogeographical problems. *Bijdragen tot de Dierkunde,* **28**, 315–322.

Myers, G.S. (1952) How the shooting apparatus of the archer fish was discovered. *Aquarium Journal,* **23**, 210–214.

Nair, P.V. (1958) Seasonal changes in the gonad of *Hilsa ilisha* (Ham.). *Philippine Journal of Science,* **87**, 255–276.

Nammalwar, P. (1992) Field bioassay in Cooum and Adyar estuaries for environmental management, in *Tropical Ecosystems: Ecology and Management* (eds K.P. Singh and J.S. Singh), Wiley Eastern Limited, New Delhi, pp. 359–370.

Nance, J.M. (1991) Effects of oil/gas field produced water on the macrobenthic community in a small gradient estuary. *Hydrobiologia,* **220**, 189–204.

Nandan, S.B. and Abdul-Azis, P.K. (1995) Pollution indicators of coconut husk retting areas in the kayals of Kerala. *International Journal of Environmental Studies A,* **47**, 19–25.

Natarajan, P., Ramadhas, V. and Ramanathan, N. (1982) A case report of mass mortality of marine catfish. *Science and Culture,* **48**, 182–183.

Neira, F.J. and Potter, I.C. (1992) Movement of larval fishes through the entrance channel of a seasonally open estuary in western Australia. *Estuarine, Coastal and Shelf Science,* **35**, 213–224.

Neira, F.J. and Potter, I.C. (1994) The larval fish assemblage of the Nornalup–Walpole estuary, a permanently open estuary on the southern coast of Western Australia. *Australian Journal of Marine and Freshwater Research,* **45**, 1193–1207.

Neira, F.J., Potter, I.C. and Bradley, J.S. (1992) Seasonal and spatial changes in the larval fish fauna within a large temperate Australian estuary. *Marine Biology,* **112**, 1–16.

Nelson, J.S. (1984) *Fishes of the World,* 2nd edn, Wiley-Interscience, New York.

Nemoto, T. (1971) La pesca en el lago de Maracaibo. Projecto de Investigacion y Desarollo Pesquero MAC–PNUD. *FAO Technical Paper,* **24**, 1–56.

Nikijuluw, V.P.H. and Naamin, N. (1994) Current and future community based fishery management in Indonesia. *Indonesian Agricultural Research and Development Journal,* **16**, 19–23.

Norcross, B.L. and Shaw, R.F. (1984) Oceanic and estuarine transport of fish eggs and larvae: a review. *Transactions of the American Fisheries Society,* **113**, 153–165.

Nordlie, F.G. (1981) Feeding and reproductive biology of eleotrid fishes in a tropical estuary. *Journal of Fish Biology,* **18**, 97–110.

Nordlie, F.G. and Kelso, D.P. (1975) Trophic relationships in a tropical estuary. *Revista de Biologia Tropical,* **23**, 77–99.

Norman, J.R. and Greenwood, P.H. (1975) *A History of Fishes,* 2nd edn, Ernest Benn, London.

Nuruzzaman, A.K.M. (1987) Impact of Hilsa fisheries on the rural economy of Bangladesh. *BARC – CIRDAP jointly sponsored workshop, 10–11 June 1987,* Bangladesh Agricultural Research Council, pp. 1–16.

Odum, W.E. (1970) Utilization of the direct grazing and plant detritus food chains by the striped mullet *Mugil cephalus*, in *Marine Food Chains* (ed. J.H. Steele), Oliver & Boyd, London, pp. 222–240.

Oliveira, A.M. and Kjerfve, B. (1993) Environmental responses of a tropical lagoon system to hydrological variability: Mundau-Manguaba, Brazil. *Estuarine Coastal and Shelf Science*, **37**, 575–591.

Ong, J.E. (1982) Aquaculture, forestry and conservation of Malaysian mangroves. *Ambio*, **11**, 252–257.

Ong, J.E. (1995) The ecology of mangrove conservation and management. *Proceedings of the Asia–Pacific Symposium on Mangrove Ecosystems 1–3 September, 1993 (Hong Kong)* (eds Wong, Yuk Shan and N.F.Y. Tam), Hong Kong University of Science and Technology, Hong Kong, p. 29 (abstract).

Ong, T.L. and Sasekumar, A. (1984) The trophic relationship of fishes in the shallow waters adjoining a mangrove swamp, in *Proceedings of the Asian Symposium on Mangrove Environment Research and Management, Kuala Lumpur, 25–29 August, 1980* (eds E. Soepadmo, A.N. Rao and D.J. Macintosh), University of Malaya and UNESCO, pp. 453–469.

Ossom, E.M. and Rhykerd, C.L. (1985) The mangrove swamps of Southern Nigeria, II: problems and prospects. *Proceedings of the Indiana Academy of Sciences*, **95**, 500.

Palmer, H.D. and Goss, M.G. (1979) *Ocean Dumping and Marine Pollution*, Dowden, Hutchinson and Ross, Pittsburgh, PA.

Panda, K.K., Lenka, M. and Panda, B.B. (1990) Monitoring and assessment of mercury pollution in the vicinity of a chloralkali plant. 1. Distribution, availability and genotoxicity of sediment mercury in the Rushikulya Estuary, India. *Science of the Total Environment*, **96**, 281–296.

Pandian, T.J. (1969) Feeding habits of the fish *Megalops cyprinoides* Broussonet, in the Cooum backwaters, Madras. *Journal of the Bombay Natural History Society*, **65**, 569–580.

Panigrahi, A.K. and Konar, S.K. (1989) Impact of different industrial effluents on Hooghly Estuary ecosystem with reference to oil refinery at Haldia, West Bengal. *Environmental Ecology*, **7**, 57–61.

Panvisavas, S., Agamanon, P., Arthorn–Thurasook, T. and Khatikarn, K. (1991) Mangrove deforestation and uses in Ban Don Bay, Thailand. *ICLARM Conference Proceedings*, **22**, 223–230.

Paranagua, M.N. and Eskinazi-Leça, E. (1985) Ecology of a northern tropical estuary in Brazil and technological perspectives in fishculture, in *Fish Community Ecology in Estuaries and Coastal Lagoons: Towards an Ecosystem Integration* (ed. A. Yáñez-Arancibia), UNAM Press, Mexico City, pp. 595–614.

Pathiratne, A. and Costa, H.H. (1984) Morphological and histological changes in the gonads of the estuarine cichlid fish *Etroplis suratensis* (Bloch) during gonadal development. *Mahagasar*, **17**, 211–220.

Pati, S. (1983) Growth changes in relation to food habits of silver pomfret, *Pampus argenteus* (Euphrasen). *Indian Journal of Animal Sciences*, **53**, 53–56.

Patnaik, S. (1973) Some aspects of the fishery and biology of the Chilka Khuranti, *Rhabdosargus sarba* (Forskal). *Journal of the Inland Fisheries Society of India*, **5**, 102–114.

Pauly, D. (1976) The biology, fishery and potential for aquaculture of *Tilapia melanotheron* in a small West African lagoon. *Aquaculture*, **7**, 33–49.

Pauly, D. (1985) Ecology of coastal and estuarine fishes in Southeast Asia: a Philippine case study, in *Fish Community Ecology in Estuaries and Coastal*

*Lagoons: Towards an Ecosystem Integration* (ed. A. Yáñez-Arancibia), UNAM Press, Mexico City, pp. 499–514.

Pauly, D. (1988) Fisheries research and the demersal fisheries of southeast Asia, in *Fish Population Dynamics* (ed. J.A. Gulland), John Wiley, London, pp. 329–348.

Pauly, D. (1994) *On the Sex of Fish and the Gender of Scientists*, Chapman & Hall, London.

Pauly, D. and Murphy, G.I. (eds) (1982) *Theory and Management of Tropical Fisheries*, ICLARM, Manila and CSIRO, Australia.

Pauly, D. and Palomares, M.L. (1987) Shrimp consumption by fish in Kuwait waters: a methodology, preliminary results and their implications for management and research. *Kuwait Bulletin of Marine Science*, 9, 101–125.

Payne, A.I. (1976) The relative abundance and feeding habits of the Grey Mullet species occurring in an estuary in Sierra Leone, West Africa. *Marine Biology*, 35, 277–286.

Percival, M. and Womersley, J.S. (1975) Floristics and ecology of the mangrove vegetation of Papua New Guinea. *Botany Bulletin*, 8, Department of Forestry, Division of Botany, Lae, Papua New Guinea.

Perkins, E.J. (1974) *The Biology of Estuaries and Coastal Waters*, Academic Press, London.

Perry, J.E. (1989) *The Impact of the September 1987 Floods on the Estuaries of Natal/Kwazulu; a Hydro–photographic Perspective*. CSIR Research Report 640, CSIR, Stellenbosch.

Peters, K.M. and McMichael, R.H. (1987) Early life history of the Red Drum *Sciaenops ocellatus* (Pisces Sciaenidae), in Tampa Bay, Florida. *Estuaries*, 10, 92–107.

Phillips, D.J.H. (1994) Pollution and environmental control, in *Marine Biology* (eds L.S. Hammond and R.N. Synnot), Longman Cheshire, Melbourne, pp. 405–418.

Phillips, P.C. (1981) Diversity and fish community structure in a Central American mangrove embayment. *Revista de Biologia Tropical*, 29, 227–236.

Pillay, S.R. and Rosa, H. (1963) Synopsis of biological data on Hilsa *Hilsa ilisha* (Hamilton) 1822. *FAO Fisheries Synopsis*, 25, 1–60.

Pillay, T.V.R. (1958) Biology of Hilsa, *Hilsa ilisha* (Ham.) of the River Hooghly. *Indian Journal of Fisheries*, 5, 201–257.

Pillay, T.V.R. (1967a) Estuarine fisheries of the Indian Ocean coastal zone, in *Estuaries* (ed. G.H. Lauff), American Association for the Advancement of Science, Washington, DC, pp. 647–657.

Pillay, T.V.R. (1967b) Estuarine fisheries of West Africa, in *Estuaries* (ed. G.H. Lauff), American Association for the Advancement of Science, Washington, DC, pp. 639–646.

Pinto, L. (1988) Population dynamics and community structure of fish in the mangroves of Pagbilao, Philippines. *Journal of Fish Biology*, 33, 35–43.

Pitcher, T.J. and Hart, P.J.B. (1983) *Fisheries Ecology*, Croom Helm, London.

Piumsombun, S. (1993) *The Socioeconomic Feasibility of Introducing Fishery Right System in the Coastal Waters of Thailand*. Department of Fisheries, Bangkok.

Poiner, I.R., Walker, D.I. and Coles, R.G. (1989) Regional studies – seagrasses of tropical Australia, in *Biology of Seagrasses* (eds A.W.D. Larkum, A.J. McComb and S.A. Shepherd), Elsevier, Amsterdam, pp. 279–303.

Pollock, B.R. (1982) Spawning period and growth of yellowfin bream, *Acanthopagrus australis* (Guenther), in Moreton Bay, Australia. *Journal of Fish Biology*, 21, 349–355.

Pollock, B.R. (1984) Relations between migration, reproduction and nutrition in yellowfish bream *Acanthopagrus australis*. *Marine Ecology Progress Series*, **19**, 17–23.

Pollock, B.R. (1985) The reproductive cycle of yellowfin bream, *Acanthopagrus australis* (Günther), with particular reference to protandrous sex inversion. *Journal of Fish Biology*, **26**, 301–311.

Pollock, B.R., Weng, H. and Morton, R.M. (1983) The seasonal occurrence of postlarval stages of yellowfin bream, *Acanthopagrus australis* (Guenther), and some factors affecting their movement into an estuary. *Journal of Fish Biology*, **22**, 409–415.

Pomeroy, R.S. (1995) Community-based and co-management institutions for sustainable coastal fisheries management in southeast Asia. *Ocean and Coastal Management*, **27**, 143–162.

Pomfret, J.R., Elliott, M., O'Reilly, M.G. and Phillips, S. (1991) Spatial and temporal patterns in the fish communities in two UK North Sea estuaries. *International Symposium Series, ECSA 19*, Olsen & Olsen, Fredensborg, pp. 277–284.

Pooley, A.C. (1982) Further observation on the Nile crocodile *Crocodylus niloticus* in the St Lucia system, in *St Lucia Research Review, February, 1982* (ed. R.H. Taylor), Natal Parks, Game and Fish Preservation Board, Pietermaritzburg, Natal, South Africa, pp. 144–161.

Poore, G.C.B., Just, J. and Cohen, B.F. (1994) Composition and diversity of Crustacea, Isopoda of the southeastern Australian continental shelf. *Deep Sea Research, Part A, Oceanographic Research Papers*, **41**, 677–693.

Potter, I.C., Claridge, P.N. and Warwick, R.M. (1986) Consistency of seasonal changes in an estuarine fish assemblage. *Marine Ecology Progress Series*, **32**, 217–228.

Potter, I.C., Beckley, L.E., Whitfield, A.K. and Lenanton, R.C.J. (1990) Comparisons between the roles played by estuaries in the life cycles of fishes in temperate western Australia and southern Africa. *Environmental Biolgy of Fishes*, **28**, 143–178.

Potts, G.W. and Wootton, R.J. (1984) *Fish Reproduction: Strategies and Tactics*, Academic Press, London.

Powell, G.V.N., Fourqurean, J.W., Kenworthy, W.J. and Zieman, J.C. (1991) Bird colonies cause seagrass enrichment in a subtropical estuary: observational and experimental evidence. *Estuarine Coastal and Shelf Science*, **32**, 567–579.

Prabhakara Rao, A.V. (1968) Observations on the food and feeding habits of *Gerres oyena* (Forskal) and *G. filamentosus* Cuvier from the Pulicat Lake, with notes on the food of allied species. *Journal of the Marine Biological Association of India*, **10**, 332–346.

Primavera, J.H. (1995) Mangroves and brackishwater pond culture in the Philippines. *Proceedings of the Asia–Pacific Symposium on Mangrove Ecosystems 1–3 September, 1993 (Hong Kong)* (eds Wong, Yuk Shan and N.F.Y. Tam), Hong Kong University of Science and Technology, Hong Kong, p. 62 (abstract).

Pritchard, D.W. (1967) What is an estuary, physical viewpoint, in *Estuaries* (ed. G.H. Lauff), American Association for the Advancement of Science, Washington, DC, pp. 3–5.

Qasim, S.Z. (1970) Some problems related to the food chain in a tropical estuary, in *Marine Food Chains* (ed. J.H. Steele), Oliver & Boyd, Edinburgh, pp. 45–51.

Quddus, M.M.A. (1982) Two types of Hilsa ilisha and their population biology in Bangladesh waters. PhD thesis, University of Tokyo, 180 pp.

Quinn, N.J. and Kojis, B. L. (1987) The influence of diel cycle, tidal direction and trawl alignment on beam trawl catches in an equatorial estuary. *Environmental Biolgy of Fishes*, **19**, 297–308.

Quinn, R.H. (1987) Analysis of fisheries logbook information from the Gulf of Carpentaria, Australia. *ACIAR Proceedings*, **20**, 92–95.

Quinn, R.H. (1992) *Fisheries Resources of the Moreton Bay Region.* Queensland Fish Management Authority, Brisbane.

Rabbani, M.M., Rehman, A.-U. and Harms, C.E. (1990) Mass mortality of fishes caused by dinoflagellate bloom in Gwadar Bay, southwestern Pakistan, in *Toxic Marine Phytoplankton* (eds E. Graneli, B. Sundstroem, L. Edler and D.M. Anderson), National Institute of Oceanography, Karachi, pp. 209–214.

Raddi, A.G. (1992) Afforestation of mangrove wetlands, in *Tropical Ecosystems: Ecology and Management* (eds K.P. Singh and J.S. Singh), Wiley Eastern Limited, New Delhi, pp. 295–300.

Raja, B.T.A. (1985) Current knowledge of the biology and fishery of Hilsa Shad, *Hilsa ilisha* (Ham. Buch.) of upper Bay of Bengal. *Internal Report, Bay of Bengal Project Document, Colombo, Sri Lanka.*

Ramesh, S. (1987) Ecology of the human pathogen, *E. coli* in Porto Novo coastal environs, in *A Special Collection of Papers to Felicitate Dr. S.Z. Qasim on his Sixtieth Birthday* (eds R. Natarajan, T.S.S. Rao, B.N. Desai, G. Narayana Swamy and S.R. Bhat), Centre for Advanced Studies in Marine Biology, Annamalai University, Porto Novo, Tamil Nadu, pp. 67–84.

Ramirez, C.A.R., Szelistowski, W.A. and Lopez, S.M.I. (1989) Spawning pattern and larvae recruitment in Gulf of Nicoya anchovies (Pisces: Engraulidae). *Revista de Biologia Tropical*, 37, 55–62.

Ramm, A.E.L. (1992) Aspects of the biogeochemistry of sulphur in Lake Mpungwini, southern Africa. *Estuarine Coastal and Shelf Science*, **34**, 253–261.

Rankin, J.C. and Jensen, F.B. (eds) (1993) *Fish Ecophysiology*, Chapman & Hall, London.

Rao, D.P., Bhaskar, B.R., Rao, K.S., Prasad, Y.V.K.D., Rao, N.S. and Rao, T.N.V.V. (1990) Haematological effects in fishes from complex polluted waters of Visakhapatnam Harbour. *Marine Environmental Research*, 30, 217–231.

Rawlinson, N.J.F., Milton, D.A., Blaber S.J.M., Sesewa, A. and Sharma, S. (1995) The subsistence fishery of Fiji. *ACIAR Monograph*, **35**, 1–138.

Reinboth, R. (1980) Can sex inversion be environmentally induced. *Biology of Reproduction*, **22**, 49–59.

Renfro, W.C. (1960) Salinity relations of some fishes in the Aransas River, Texas. *Tulane Studies in Zoology*, **8**, 83–91.

Ricker, W.E. (1958) Handbook of computations for biological statistics of fish populations. *Bulletin of the Fisheries Research Board of Canada*, **119**, 1–300.

Rimmer, M.A. and Merrick, J.R. (1982) A review of reproduction and development in the fork-tailed catfishes (Ariidae). *Proceedings of the Linnean Society of New South Wales*, **107**, 41–50.

Risebrough, R.W. and Anderson, D. (1975) Some effects of DDE and PCB on mallards and their eggs. *Journal of Wildlife Management*, **39**, 508–513.

Robblee, M.B. and Zieman, J.C. (1984) Diel variation in the fish fauna of a tropical seagrass feeding ground. *Bulletin of Marine Science*, **34**, 335–345.

Roberts, T.R. (1970) Scale-eating American characoid fishes, with special reference to *Probolodus heterostomus. Proceedings of the Californian Academy of Science*, **28**, 383–390.

Roberts, T.R. (1978) An ichthyological survey of the Fly River in Papua New

Guinea with descriptions of new species. *Smithsonian Contributions to Zoology*, **281**, 1–72.

Robertson, A.I. (1988) Abundance, diet and predators of juvenile banana prawns, *Penaeus maerguiensis*, in a tropical mangrove estuary. *Australian Journal of Marine and Freshwater Research*, **39**, 467–478.

Robertson, A.I. and Alongi, D.M. (eds) (1992) *Tropical Mangrove Ecosystems*, American Geophysical Union, Washington, DC.

Robertson, A.I. and Blaber, S.J.M. (1992) Plankton, epibenthos and fish communities, in *Tropical Mangrove Ecosystems* (eds A.I. Robertson and D.M. Alongi), American Geophysical Union, Washington, DC, pp. 173–224.

Robertson, A.I. and Duke, N.C. (1987) Mangroves as nursery sites: comparisons of the abundance and species composition of fish and crustaceans in mangroves and other nearshore habitats in tropical Australia. *Marine Biology*, **96**, 193–205.

Robertson, A.I. and Duke, N.C. (1990) Mangrove fish communities in tropical Australia: spatial and temporal patterns in densities, biomass and community structure. *Marine Biology*, **104**, 369–379.

Robertson, A.I. and Klumpp, D.W. (1983) Feeding habits of the southern Australian garfish *Hyporhamphus melanochir*: a diurnal herbivore and nocturnal carnivore. *Marine Ecology Progress Series*, **10**, 197–201.

Robertson, A.I., Daniel, P.A. and Dixon, P. (1988) Zooplankton dynamics in mangrove and other nearshore habitats in tropical Australia. *Marine Ecology Progress Series*, **43**, 139–150.

Robertson, A.I., Daniel, P.A. and Dixon, P. (1991) Mangrove forest structure and productivity in the Fly River estuary, Papua New Guinea. *Marine Biology*, **111**, 147–155.

Robertson, A.I., Alongi, D.M. and Boto, K.G. (1992) Food chains and carbon fluxes, in *Tropical Mangrove Ecosystems* (eds A.I. Robertson and D.M. Alongi), American Geophysical Union, Washington, DC, pp. 293–326.

Rogers, P.P. and Gould, R.R. (1995) Resource allocation – a management perspective, in *Recreational Fishing: What's the Catch?* (ed. D.A. Hancock), Australian Government Publishing Service, Canberra, pp. 195–201.

Ross, R.M. (1990) The evolution of sex-change mechanisms in fishes. *Environmental Biology of Fishes*, **29**, 81–93.

Ross, S.T., McMichael, R.H. and Ruple, D.L. (1987) Seasonal and diel variation in the standing crop of fishes and macroinvertebrates from a Gulf of Mexico surf zone. *Estuarine, Coastal and Shelf Science*, **25**, 391–412.

Russell, D.J. and Garrett, R.N. (1983) Use by juvenile barramundi, *Lates calcarifer* (Bloch) and other fishes of temporary supralittoral habitats in a tropical estuary in northern Australia. *Australian Journal of Marine and Freshwater Research*, **34**, 805–811.

Ryan, P.A. (1991) The success of the Gobiidae in tropical Pacific insular streams. *New Zealand Journal of Zoology*, **18**, 25–30.

Sadovy, Y. and Shapiro, D.Y. (1987) Criteria for the diagnosis of hermaphroditism in fishes. *Copeia*, **1987**, 136–156.

Saila, S.B. (1975) Some aspects of fish production and cropping in estuarine systems, in *Estuarine Research: Chemistry, Biology and the Estuarine System*, Volume 1 (ed. L.E. Cronin), Academic Press, New York, pp. 473–493.

Sainsbury, K.J., Kailola, P.J. and Leyland, G.G. (1985) *Continental Shelf Fishes of Northern and North-western Australia*, CSIRO Division of Fisheries Research, Hobart, Australia.

Sale, P.F. (ed.) (1991) *The Ecology of Fishes on Coral Reefs*, Academic Press, San Diego.

Salini, J.P. and Shaklee, J.B. (1988) Genetic structure of barramundi *(Lates calcarifer)* stocks from northern Australia. *Australian Journal of Marine and Freshwater Research*, **39**, 317–329.

Salini, J.P., Blaber, S.J.M. and Brewer, D.T. (1990) Diets of piscivorous fishes in a tropical Australian estuary with particular reference to predation on penaeid prawns. *Marine Biology*, **105**, 363–374.

Salini, J.P., Blaber, S.J.M. and Brewer, D.T. (1992) Diets of sharks from estuaries and nearshore waters of the northeastern Gulf of Carpentaria. *Australian Journal of Marine and Freshwater Research*, **43**, 87–96.

Salini, J.P., Blaber, S.J.M. and Brewer, D.T. (1994) Diets of trawled predatory fish of the Gulf of Carpentaria with particular reference to prawn predators. *Australian Journal of Marine and Freshwater Research*, **45**, 397–412.

Samarakoon, J.I. (1983) Breeding patterns of the indigenous cichlids *Etroplus suratensis* and *Etroplus maculatus* in an estuary in Sri Lanka. *Mahagasar*, **16**, 357–362.

Samarakoon, J.I. (1991) The ecological history of Negombo lagoon: an example of the rapid assessment of management issues. *ICLARM Conference Proceedings*, **22**, 83–88.

Samuelian, J. and O'Connor, J.M. (1985) Structure–activity relationships and accumulation of PCB congeners in estuarine fishes: a field study. *Estuaries*, **8** (2B), 93A (abstract no. 239).

Sanders, H.L. (1968) Marine benthic diversity: a comparative study. *American Naturalist*, **102**, 243–282.

Sanyal, P. (1992) Sunderbans mangrove: wildlife potential and conservation, in *Tropical Ecosystems: Ecology and Management* (eds K.P. Singh and J.S. Singh), Wiley Eastern Limited, New Delhi, pp. 309–313.

Sarkar, A.K. (1993) *Dynamic behaviour of a detritus–based food chain model of Sundarban Estuary, India*. Interdisciplinary Conference on Natural Resources Modeling and Analysis (Rome Italy, 1993), Ministerio Agricoltora e Foreste, Rome (summary only).

Sarkar, S.K. (1979) Further studies on seasonal and spatial variations of salinity in Chilka Lake. *Journal of the Inland Fisheries Society of India*, **11**, 1–9.

Sarker, A.L., Al–Daham, N.K. and Bhatti, M.N. (1980) Food habits of the mudskipper, *Pseudapocryptes dentatus* (Val.). *Journal of Fish Biology*, **17**, 635–639.

Sarmani, S., Pauzi-Abdullah, Md., Baba, I. and Abdul Majid, A. (1992) Inventory of heavy metals and organic micropollutants in an urban water catchment drainage basin, in *Sediment Water Interactions*, International Symposium on sediment/water interactions, Uppsala, Sweden (ed. B.T. Hart and P.G. Sly), pp. 669–674.

Sarojini, K.K. (1954) The food and feeding habits of the grey mullets: *Mugil parsia* Hamilton and *M. speigleri* Bleeker. *Indian Journal of Fisheries*, **1**, 67–93.

Sasekumar, A. (1980) *The present state of mangrove ecosystems in Southeast Asia and the impact of pollution. Malaysia*. FAO/UNDP South China Sea Fisheries Development and Coordination Programme Report, Manila, Philippines, pp. 1–21.

Sasekumar, A. (1993) A review of mangrove–fisheries connections, in *Proceedings of a Workshop on Mangrove Fisheries and Connections*, August 26–30, 1991, Ipoh, Malaysia (ed. A. Sasekumar), Ministry of Science, Technology and Environment, Kuala Lumpur, pp. 47–50.

Sasekumar, A. and Wilkinson, C.R. (1994) Compatible and incompatible uses of mangroves in ASEAN, in *Third ASEAN–Australia Symposium on Living Coastal Resources, Bangkok, Thailand, 1994* (ed. C.R. Wilkinson), Australian Institute of Marine Science, Townsville, pp. 77–83.

Sasekumar, A., Ong, T.L. and Thong, K.L. (1984) Predation of mangrove fauna by marine fishes, in *Proceedings of the Asian Symposium on Mangrove Environment Research and Management, Kuala Lumpur, 25–29 August, 1980,* University of Malaya and UNESCO, (eds E. Soepadmo, A.N. Rao and D.J. Macintosh), pp. 378–384.

Sasekumar, A., Chong, V.C., Lim, K.H. and Singh, H. (1994a) The fish community of Matang waters, in *Proceedings Third ASEAN Australia Symposium on Coastal Living Resources,* vol. 1: *Status reviews* (eds S. Sudara, C.R. Wilkinson and L.M. Chou), Chulalongkorn University, Bangkok, pp. 457–464.

Sasekumar, A., Chong, V.C. and Singh, H. (1994b) Physical and chemical characteristics of the Matang mangrove waters, in *Proceedings Third ASEAN Australia Symposium on Coastal Living Resources,* vol. 2: *Research Papers* (eds S. Sudara, C.R. Wilkinson and L.M. Chou), Chulalongkorn University, Bangkok, pp. 446–453.

Sathyashree, P.K., Sitarami Reddy, P. and Natarajan, R. (1987) Maturity and spawning of *Osteomugil speigleri* (Bleeker) in Porto Novo waters. *Journal of the Marine Biological Association of India,* 23, 1–6.

Satyanarayana, D., Sahu, S.D. and Panigrahy, P.K. (1992) Physico-chemical characteristics in the coastal environment of Visakhapatnam – A case study. *Journal of the Marine Biological Association of India,* 34, 103–109.

Saucier, M.H. and Baltz, D.M. (1993) Spawning site selection by spotted seatrout, *Cynoscion nebulosus,* and black drum, *Pogonias cromis,* in Louisiana. *Environmental Biology of Fishes,* 36, 257–272.

Schumann, E.H. and De Meillon, L. (1993) Hydrology of the St Francis Bay marina, South Africa. *Transactions of the Royal Society of South Africa,* 48, 323–338.

Scoffin, T.P., Tudhope, A.W. and Brown, B.E. (1989) Fluorescent and skeletal density banding in *Porites lutea* from Papua New Guinea and Indonesia. *Coral Reefs,* 7, 169–178.

Shameem, A. (1992) Comparative studies on the feeding habits of marine and estuarine carangids. *Journal of the Marine Biological Association of India,* 34, 262–268.

Shamsudin, L. (1988) Water quality and mass fish mortality at Mengabang lagoon, Trengganu, Malaysia. *Malayan Nature Journal,* 41, 515–527.

Shaw, G.R. and Connell, D.W. (1980) Relationships between steric factors and bioconcentration of polychlorinated biphenyls (PBB's) by the sea mullet (*Mugil cephalus* Linnaeus). *Chemosphere,* 9, 731–743.

Sheridan, P.F. and Trimm, D.L. (1983) Summer foods of Texas coastal fishes relative to age and habitat. *Fisheries Bulletin, U.S.,* 81, 643–647.

Sheridan, P.F., Trimm, D.L. and Baker, B.M. (1984) Reproduction and food habits of seven species of northern Gulf of Mexico fishes. *Contributions to Marine Science, University of Texas,* 27, 175–204.

Sien, C.L. (1979) Coastal changes, marine pollution and the fishing industry in Singapore, in *Economics of Aquaculture, Sea Fishing and Coastal Resource Use in Asia.* Proceedings of the second biennial meeting of the Agricultural Economics Society of Southeast Asia, November 3–6, 1977, Tigbauan, Iloilo, Philippines (eds A.R. Librero and W.L. Collier), Agricultural Economics Society of SE Asia, Iloilo, pp. 333–344.

Silva, E.I.L. and De Silva, S.S. (1981) Aspects of the biology of grey mullet, *Mugil cephalus* L., adult populations of a coastal lagoon in Sri Lanka. *Journal of Fish Biology*, **19**, 1–10.

Singh, H. and Sasekumar, A. (1994) Distribution and abundance of marine catfish (fam: Ariidae) in the Matang mangrove waters, in *Proceedings Third ASEAN Australia Symposium on Coastal Living Resources*, vol. 1: *Status Reviews* (eds S. Sudara, C.R. Wilkinson and L.M. Chou), Chulalongkorn University, Bangkok, pp. 464–471.

Sivalingam, P.M. and Thavaraj, S. (1978) Efficiency of canal biodegradation of palm oil sludge and its pollution effects. *Bulletin of the Japanese Society of Scientific Fisheries*, **44**, 1229–1237.

Smith, H.M. (1936) The archer fish. *Natural History*, **38**, 3–11.

Smith, H.M. (1945) The freshwater fishes of Siam, or Thailand. *Bulletin of the United States National Museum*, **188**, 1–622.

Smith, M.M. and Heemstra, P.C. (1986) *Smith's Sea Fishes*, Macmillan, Johannesburg.

Smith, R.L., Salini, J.P. and Blaber, S.J.M. (1991) Food intake and growth in the Moses perch, *Lutjanus russelli* (Bleeker), with reference to predation on penaeid prawns. *Journal of Fish Biology*, **38**, 897–903.

Smith, R.L., Salini, J.P. and Blaber, S.J.M. (1992) Food intake and growth in the blue-spotted trevally, *Caranx bucculentus*, with reference to predation on penaeid prawns. *Journal of Fish Biology*, **40**, 315–324.

Smith, T.J., III, Hudson, J.H., Robblee, M.B., Powell, G.V.N. and Isdale, P.J. (1989) Freshwater flow from the Everglades to Florida Bay: a historical reconstruction based on fluorescent banding in the coral *Solenastrea bournoni*. *Bulletin of Marine Science*, **44**, 274–282.

Sorenson, D.L., McCarthy, M.W., Middlebrooks, E.J. and Porcella, D.B. (1977) Suspended and dissolved solids effects on freshwater biota: a review. *US Environmental Protection Agency Report*, EPA-600/3-77-042.

Spitzer, P.R., Risebrough, R.W., Grier, J.W. and Sindelar, C.R. (1977) Eggshell thickness–pollutant relationships among North American Ospreys, in *Transactions of the North American Osprey Research Conference* (ed. J. Ogden), US National Park Service, Washington, DC, pp. 13–20.

Srinivasan, M. and Mahajan, B.A. (1989) Mercury pollution in an estuarine region and its effect on a coastal population. *International Journal of Environmental Studies*, **35**, 63–69.

St Mary, C.M. (1994) Sex allocation in a simultaneous hermaphrodite, the blue-banded goby *(Lythrypnus dalli)*: the effects of body size and behavioral gender and the consequences for reproduction. *Behavioural Ecology*, **5**, 304–313.

Staples, D.J. (1980) Ecology of juvenile and adolescent banana prawns, *Penaeus merguiensis*, in a mangrove estuary and adjacent offshore area of the Gulf of Carpentaria. I. Immigration and settlement of postlarvae. *Australian Journal of Marine and Freshwater Research*, **31**, 635–652.

Steinke, T.D. and Ward, C.J. (1989) Some effects of the cyclones Domoina and Imboa on mangrove communities in the St Lucia Estuary. *South African Journal of Botany*, **55**, 340–348.

Subani, W. and Wahyono, M.M. (1987) Kerusakan ekosistem perairan pantai dan dampaknya terhadap sumberdaya perikanan di pantai selatan Bali, barat dan timur Lombok dan Teluk Jakarta. *Jurnal Penelitian Perikanan Laut*, **42**, 53–70. (In Indonesian)

Subrahmanyam, C.B. (1985) Fish communities of a bay estuarine–marsh system in north Florida, in *Fish Community Ecology in Estuaries and Coastal Lagoons: Towards an Ecosystem Integration* (ed. A. Yáñez-Arancibia), UNAM Press, Mexico City, pp. 191–206.

Sulaiman, H. (1994) Status of traditional marine tenure in Malaysia, in *Traditional Marine Tenure and Sustainable Management of Marine Resources in Asia and the Pacific* (eds G.R. South, D. Goulet, S. Tuqiri M. and Church), University of the South Pacific, Suva, Fiji, p. 160.

Suparta, M.H., Blackshaw, A.W. and Capra, M.F. (1984) Sex reversal and the gonads of the yellowfin bream, *Acanthopagrus australis* Gunther, in *Proceedings of the 10th International Congress on Animal Reproduction and Artificial Insemination, Urbana, IL (USA), 10–14 June 1984*, p. 44.

Sutherland, J.P. (1980) Dynamics of the epibenthic community on roots of the mangrove *Rhizophora mangle*, at Bahia de Buche, Venezuela. *Marine Biology*, **58**, 75–84.

Swanson, C. (1996) Early development of milkfish: effects of salinity on embryonic and larval metabolism, yolk absorption and growth. *Journal of Fish Biology*, **48**, 405–421.

Swenson, W.A. (1978) Influence of turbidity on fish abundance in western Lake Superior. *Research Report, US Environmental Protection Agency, Duluth*, pp. 1–84.

Szelistowski, W.A. (1989) Scale-feeding in juvenile marine catfishes (Pisces: Ariidae). *Copeia*, **1989**, 517–519.

Szelistowski, W.A. and Garita, J. (1989) Mass mortality of Sciaenid fishes in the Gulf of Nicoya, Costa Rica. *Fishery Bulletin, U.S.*, **87**, 363–365.

Tampi, P.R.S. (1959) The ecological and fisheries characteristics of a salt water lagoon near Mandapam. *Journal of the Marine Biological Association of India*, **1**, 113–130.

Tang, Y.A. (1964) Induced spawning of striped mullet by hormone injection. *Japanese Journal of Ichthyology*, **12**, 23–28.

Taylor, R.H. (1982) St Lucia estuary: the aquatic environment, in *St Lucia Research Review, February, 1982* (ed. R.H. Taylor), Natal Parks, Game and Fish Preservation Board, Pietermaritzburg, Natal, South Africa, pp. 42–56.

Taylor, W.R. (1964) Fishes of Arnhem Land. *Records of the American –Australian Scientific Expedition to Arnhem Land*, **4**, 45–307.

Thangaraja, M. (1984) Laboratory reared fish eggs and larvae and subsequent stages from plankton of Vellar estuary, Porto Novo: 2. The flatfish, *Cynoglossus puncticeps* (Richardson). *Mahagasar*, **17**, 103–111.

Thayer, G.W., Colby, D.R. and Hettler, W.F. jun. (1987) Utilization of the red mangrove prop root habitiat by fishes in south Florida. *Marine Ecology – Progress Series*, **35**, 25–38.

Thayib, S.S., Kunarso, D.H. and Ruyitno (1991) A study of microbial indicators in Segara Anakan, Java, and its adjacent waters. *ICLARM Conference Proceedings*, **22**, 431.

Thomson, J.M. (1959) Some aspects of the ecology of Lake Macquarie, NSW with regard to an alleged depletion of fish. IX. The fishes and their food. *Australian Journal of Marine and Freshwater Research*, **10**, 365–374.

Thomson, J.M. (1963) Synopsis of biological data on the grey mullet *Mugil cephalus* Linnaeus, 1758. *CSIRO Fisheries and Oceanographic Synopsis 1.*

Thomson, J.M. (1966) The grey mullets. *Oceanography and Marine Biology An Annual Review*, **4**, 301–335.

Thresher, R.E. (1984) *Reproduction in Feef Rishes*. T.F.H. Publications, Neptune City, NJ.

Trawavas, E. (1977) The sciaenid fishes (croakers or drums) of the Indo–Pacific. *Transactions of the Zoological Society of London*, **33**, 253–541.

Tucker, J.W. jun. and Faulkner, B.E. (1987) Voluntary spawning patterns of captive spotted seatrout. *Northeast Gulf Science*, **9**, 59–63.

Turnbull, D.A. and Lewis, J.B. (1981) Pollution ecology of a small tropical estuary in Barbados, West Indies. 1. Water quality characteristics. *Reports of the Marine Science Center, McGill University*, **35**, 1–55.

Ugwumba, O.A. (1989) Distribution and growth pattern of the ten-pounder *Elops lacerta* (Val.) in the freshwater, estuarine and marine environments of Lagos, Nigeria. *Archiv für Hydrobiologie*, **115**, 451–462.

UNDP/UNESCO (1991) *Final report of the integrated multidisciplinary survey and research programme of the Ranong mangrove ecosystem*, UNDP/UNESCO regional project research and its application to the management of the mangroves of Asia and the Pacific (RAS/86/120), Bangkok, Thailand.

van der Elst, R. (1981) *A Guide to the Common Sea Fishes of Southern Africa*, C. Struik, Cape Town.

van der Elst, R.P. and Penney, A. (1995) Strategies for data collection in marine recreational and commercial linefisheries of South Africa, in *Recreational Fishing: What's the Catch?* (ed. D.A. Hancock), Australian Government Publishing Service, Canberra, pp. 31–41.

van der Horst, G. and Erasmus, T. (1978) The breeding cycle of male *Liza dumerili* (Teleostei: Mugilidae) in the mouth of the Swartkops Estuary. *Zoologica Africana*, **13**, 259–273.

Vargas, J.A. (1995) The Gulf of Nicoya estuary, Costa Rica – past, present and future cooperative research. *Helgoländer Meeresuntersuchungen*, **49**, 821–828.

Vega-Cendejas, M.E., Hernandez, M. and Arreguin-Sanchez, F. (1994) Trophic interrelations in a beach seine fishery from the northwestern coast of the Yucatan Peninsula, Mexico. *Journal of Fish Biology*, **44**, 647–659.

Venkatesan, V. (1969) *A preliminary study of the estuaries and backwaters in south Arcot District, Tamil Nadu (South India). Part II. Fisheries.* First All-India symposium on estuarine biology, Tambaram, Madras. Central Marine Fisheries Research Institute, Madras, pp. 1–9.

Verdeaux, F. (1979) La pêche lagunaire en Côte d'Ivoire. Contexte sociologique et formes d'exploitation du milieu naturel. *Communication au Séminaire UNESCO sur les ecosystems côtiers, Dakar, June 1979*, UNESCO, Paris, pp. 1–7.

Wallace, J.H. (1975a) The estuarine fishes of the east coast of South Africa. I. Species composition and length distribution in the estuarine and marine environments. II. Seasonal abundance and migrations. *Investigational Report, Oceanographic Research Institute, Durban*, **40**, 1–72.

Wallace, J.H. (1975b) The estuarine fishes of the east coast of South Africa. III. Reproduction. *Investigational Report, Oceanographic Research Institute, Durban*, **41**, 1– 51.

Wallace, J.H. and van der Elst, R.P. (1975) The estuarine fishes of the east coast of South Africa. IV. Occurrence of juveniles in estuaries. V. Ecology, estuarine dependence and status. *Investigational Report, Oceanographic Research Institute, Durban*, **42**, 1–63.

Warburton, K. (1978) Community structure, abundance and diversity of fish in a Mexican coastal lagoon system. *Estuarine and Coastal Marine Science*, **7**, 497–519.

Warburton, K. (1979) Growth and production of some important species of fish in a Mexican coastal lagoon system. *Journal of Fish Biology*, **14**, 449–464.

Warburton, K. and Blaber, S.J.M. (1992) Patterns of recruitment and resources use in a shallow water fish assemblage in Moreton Bay, Queensland. *Marine Ecology Progress Series*, **90**, 113–126.

Ward, C.J. (1982) Aspects of the ecology and distribution of submerged macrophytes and shoreline vegetation of Lake St Lucia, in *St Lucia Research Review, February, 1982* (ed. R.H. Taylor), Natal Parks, Game and Fish Preservation Board, Pietermaritzburg, Natal, South Africa, pp. 77–87.

Ward, J.A. and Samarakoon, J.I. (1981) Reproductive tactics of the Asian cichlids of the genus *Etroplus* in Sri Lanka. *Environmental Biology of Fishes*, **6**, 95–103.

Ward, J.A. and Wyman, R.L. (1977) Ethology and ecology of cichlid fishes of the genus *Etroplus* in Sri Lanka: preliminary findings. *Environmental Biology of Fishes*, **2**, 137–145.

Warlen, S.M. and Burke, J.S. (1990) Immigration of larvae of fall/winter spawning marine fishes into a North Carolina estuary. *Estuaries*, **13**, 453–461.

Warner, R.R. (1978) The evolution of hermaphroditism and unisexuality in aquatic and terrestrial vertebrates, in *Contrasts in Behavior* (eds E.S. Reese and F.J. Lighter), Wiley Interscience, New York, pp. 78–95.

Warner, R.R. (1988) Sex change in fishes: hypotheses, evidence, and objections. *Environmental Biology of Fishes*, **22**, 81–90.

Watts, J.C.D. (1958) The hydrology of a tropical West African estuary. *Bulletin de l'IFAN, Series A*, **20**, 697–752.

Weidner, D. (1992) Costa Rican fisheries, 1990. *National Marine Fisheries Service, Office of International Affairs, NMFS/FIA2/92–31*, pp.1–15.

Weinstein, M.P. and Heck, K.L. (1979) Ichthyofauna of seagrass meadows along the Caribbean coast of Panama and in the Gulf of Mexico: composition, structure and community ecology. *Marine Biology*, **50**, 97–107.

Welcomme, R.L. (1972) An evaluation of the acadja method of fishing as practised in the coastal lagoons of Dahomey (West Africa). *Journal of Fish Biology*, **4**, 39–55.

Welcomme, R.L. (1979) *Fisheries Ecology of Floodplain Rivers*, Longmans, London.

Whitehead, P.J.P. (1973) The clupeoid fishes of the Guianas. *Bulletin of the British Museum of Natural History (Zoology)*, **5**, 1–227.

Whitehead, P.J.P. (1985) *FAO Species Catalogue 7, Clupeoid Fishes of the World. Part I – Chirocentridae, Clupeidae and Pristigasteridae*. FAO Fisheries Synopsis 125, volume 7, Part 1. FAO, Rome.

Whitehead, P.J.P., Nelson, G.J. and Wongratana, T. (1988) *FAO Species Catalogue 7, Clupeoid Fishes of the World. Part II – Engraulididae*. FAO Fisheries Synopsis 125, volume 7, Part 2. FAO, Rome.

Whitfield, A.K. (1980a) A quantitative study of the trophic relationships within the fish community of the Mhlanga estuary, South Africa. *Estuarine and Coastal Marine Science*, **10**, 417–435.

Whitfield, A.K. (1980b) Distribution of fishes in the Mhlanga estuary in relation to food resources. *South African Journal of Zoology*, **15**, 159–165.

Whitfield, A.K. (1980c) Factors influencing the recruitment of juvenile fishes into the Mhlanga estuary. *South African Journal of Zoology*, **15**, 166–169.

Whitfield, A.K. (1980d) A checklist of fish species recorded from Maputaland estuarine systems, in *Studies on the Ecology of Maputaland* (eds M.N. Bruton and K.H. Cooper), Rhodes University and the Natal Branch of the Wildlife Society of Southern Africa, Durban, pp. 204–209.

Whitfield, A.K. (1989a) Ichthyoplankton in a southern African surf zone: nursery area for the postlarvae of estuarine associated fish species? *Estuarine, Coastal and Shelf Science*, **29**, 533–547.

Whitfield, A.K. (1989b) Fish larval composition, abundance and seasonality in a southern African estuarine lake. *South African Journal of Zoology*, **24**, 217–224.

Whitfield, A.K. (1990) Life–history styles of fishes in South African estuaries. *Environmental Biology of Fishes*, **28**, 295–308.

Whitfield, A.K. (1994a) An estuary–association classification for the fishes of southern Africa. *South African Journal of Science*, **90**, 411–417.

Whitfield, A.K. (1994b) Fish species diversity in southern African estuarine systems: an evolutionary perspective. *Environmental Biology of Fishes*, **40**, 37–48.

Whitfield, A.K. (1995) Mass mortalities of fish in South African estuaries. *South African Journal of Aquatic Science*, **21**, 29–34.

Whitfield, A.K. (1996) A review of estuarine ichthyology in South Africa over the past 50 years. *Transactions of the Royal Society of South Africa*, (in press)

Whitfield, A.K. and Blaber, S.J.M. (1978a) Feeding ecology of piscivorous birds at Lake St. Lucia. Part 1: diving birds. *Ostrich*, **49**, 185–198.

Whitfield, A.K. and Blaber, S.J.M. (1978b) Scale–eating habits of the marine teleost *Terapon jarbua* (Forskal). *Journal of Fish Biology*, **12**, 61–70.

Whitfield, A.K. and Blaber, S.J.M. (1978c) Resource segregation among iliophagous fish in Lake St. Lucia, Zululand. *Environmental Biology of Fishes*, **3**, 293–296.

Whitfield, A.K. and Blaber, S.J.M. (1978d) Food and feeding ecology of piscivorous fishes in Lake St. Lucia, Zululand. *Journal of Fish Biology*, **13**, 675–691.

Whitfield, A.K. and Blaber, S.J.M. (1978e) Distribution, movements and fecundity of Mugilidae at Lake St. Lucia. *Lammergeyer*, **26**, 53–62.

Whitfield, A.K. and Blaber, S.J.M. (1979a) The distribution of the freshwater cichlid *Sarotherodon mossambicus* (Peters) in estuarine systems. *Environmental Biology of Fishes*, **4**, 77–81.

Whitfield, A.K. and Blaber, S.J.M. (1979b) Feeding ecology of piscivorous birds at Lake St. Lucia. Part 2: wading birds. *Ostrich.*,**50**, 1–9.

Whitfield, A.K. and Blaber, S.J.M. (1979c) Feeding ecology of piscivorous birds at Lake St. Lucia. Part 3: swimming birds. *Ostrich*, **50**, 10–20.

Whitfield, A.K. and Blaber, S.J.M. (1979d) Predation on grey mullet *(Mugil cephalus* L.) by *Crocodylus niloticus* at St. Lucia, South Africa. *Copeia*, **1979**, 266–269.

Whitfield, A.K. and Blaber, S.J.M. (1979e) The diet of *Atilax paludinosus* (water mongoose) at St. Lucia, South Africa. *Mammalia*, **44**, 315–318.

Whitfield, A.K. and Wooldridge, T.H. (1994) Changes in freshwater supplies to southern African estuaries: some theoretical and practical considerations, in *Changes in Fluxes in Estuaries: Implications from Science to Management* (eds K.R. Dyer and R.J. Orth), Olsen & Olsen, Fredensborg.

Whitfield, A.K., Blaber, S.J.M. and Cyrus, D.P. (1981) Salinity ranges of some southern African fish species occurring in estuaries. *South African Journal of Zoology*, **16**, 151–155.

Wickbom, L. (1992) Studies of tidal choking and water exchange in Negombo lagoon, Sri Lanka. *Fisheries Development Series no. 67, National Swedish Board of Fisheries*, 1–22.

Wijeyaratne, M.J.S. and Costa, H.H. (1987a) On management of the finfish fishery of the Negombo lagoon, Sri Lanka. *Indian Journal of Fisheries*, **34**, 41–47.

Wijeyaratne, M.J.S. and Costa, H.H. (1987b) The food, feeding and reproduction in an estuarine population of green–back mullet, *Liza tade* (Forsskål) in the Negombo Lagoon, Sri Lanka. *Ophelia*, **27**, 171–180.

Wijeyaratne, M.J.S. and Costa, H.H. (1987c) The food, feeding and reproduction of the Borneo mullet *Liza macrolepis* (Smith), in a coastal estuary in Sri Lanka. *Indian Journal of Fisheries*, **34**, 283–291.

Wijeyaratne, M.J.S. and Costa, H.H. (1988) The food, fecundity and gonadal maturity of *Valamugil cunnesius* (Pisces: Mugilidae) in the Negombo lagoon, Sri Lanka. *Indian Journal of Fisheries*, **35**, 71–77.

Williams, F. (1965) Further notes on the biology of east African pelagic fish of the families Carangidae and Sphyraenidae. *East African Agricultural and Forestry Journal*, **31**, 141–168.

Williams, N.V. (1962) The seasonal distribution of the teleost fish fauna in Lagos Harbour, creek and lagoon in relation to salt tolerance, MSC thesis, University of Wales, Bangor, 125 pp. (unpaginated)

Williams, V.R. and Clarke, T.A. (1983) Reproduction, growth and other aspects of the biology of the gold–spot herring *Herklotsichthys quadrimaculatus* (Clupeidae), a recent introduction to Hawaii. *Fisheries Bulletin, US*, **81**, 587–597.

Wilson, C.A. and Nieland, D.L. (1994) Reproductive biology of red drum, *Sciaenops ocellatus*, from the neritic waters of the northern Gulf of Mexico. *Fisheries Bulletin, US*, **92**, 841–850.

Winemiller, K.O. and Leslie, M.A. (1992) Fish assemblages across a complex, tropical freshwater/marine ecotone. *Environmental Biology of Fishes*, **34**, 29–50.

Wiseman, K.A. and Sowman, M.R. (1992) An evaluation of the potential for restoring degraded estuaries in South Africa. *Water S.A.*, **18**, 13–19.

Wolanski, E. (1986) An evaporation driven salinity maximum zone in Australian tropical estuaries. *Estuarine, Coastal and Shelf Science*, **22**, 415–424.

Wolanski, E. and Ridd, P.V. (1986) Tidal mixing and trapping in mangrove swamps. *Estuarine, Coastal and Shelf Science*, **23**, 759–771.

Wong, Yuk Shan and Tam, N.F.Y. (eds) (1995) *Proceedings of the Asia–Pacific Symposium on Mangrove Ecosystems 1–3 September, 1993, (Hong Kong)*, Hong Kong University of Science and Technology, Hong Kong.

Woodland, D.J. (1984) Siganidae, in *FAO Species Identification Sheets for Fishery Purposes, Western Indian Ocean* (eds W. Fischer and G. Bianchi), FAO, Rome.

Woodland, D.J. (1990) Revision of the fish family Siganidae with descriptions of two new species and comments on distribution and biology. *Indo–Pacific Fishes*, **19**, 1–136.

Wootton, R.J. (1990) *Ecology of Teleost Fishes*, Chapman & Hall, London.

Wright, C.I. and Mason, T.R. (1993) Management and sediment dynamics of the St. Lucia Estuary mouth, Zululand, South Africa. *Environmental Geology*, **22**, 227–241.

Yamamoto, T. (1969) Sex differentiation, in *Fish Physiology*, Vol. III (eds W.S. Hoare and D.J. Randall), Academic Press, New York, pp. 117–175.

Yáñez-Arancibia, A. (1978) Taxonomy, ecology and structure of fish communities in coastal lagoons with ephemeral inlets on the Pacific coast of Mexico. *Instituto de Ciencias del Mar y Limnologia, Universidad Nacional Autónoma de México, Publicaciones Especiales*, **2**, 1–306.

Yáñez-Arancibia, A. and Lara Dominguez, A.L. (1983) Dinámica ambiental de la Boca de Estero Pargo y estructura de sus comunidades de peces en cambios estacionales ciclos de 24 hrs (Laguna de Términos, sur del Golfo de México). *Annales del Instituto de Ciencias del Mar y Limnologia, Universidad Nacional Autónoma de México*, **10**, 85–116.

Yáñez-Arancibia, A. and Sánchez-Gil, P. (1986) The demersal fishes of the southern Gulf of Mexico. 1. Environmental characterization, ecology and evaluation of species, populations and communities. *Instituto de Ciencias del Mar y Limnologia, Universidad Nacional Autónoma de México, Publicaciones Especiales*, **9**, 1–230.

Yáñez-Arancibia, A., Linares, F.A. and Day, J.W. jun. (1980) Fish community structure and function in Terminos lagoon, a tropical estuary in the southern Gulf of Mexico, in *Estuarine Perspectives* (ed. V. Kennedy), Academic Press, New York, pp. 465–482.

Yáñez-Arancibia, A., Lara-Dominguez, A.L., Sanchez-Gil, P., Vargas Maldonado, I., García Abad, M. De La C., Álvarez-Guillén, H., Tapia García, M., Flores Hernández, D. and Amezcua Linares, F. (1985) Ecology and evaluation of fish community in coastal ecosystems: estuary–shelf interrelationships in the southern Gulf of Mexico, in *Fish Community Ecology in Estuaries and Coastal Lagoons: Towards an Ecosystem Integration* (ed A. Yáñez-Arancibia), UNAM Press, Mexico City, pp. 475–498.

Yáñez-Arancibia, A., Lara-Dominguez, A.L., Rojas-Galaviz, J.L., Sanchez-Gil, P., Day, J.W. and Madden, C.J. (1988) Seasonal biomass and biodiversity of estuarine fishes coupled with tropical habitat heterogeneity (southern Gulf of Mexico). *Journal of Fish Biology*, **33** (Suppl. A), 191–200.

Yap, Y.N. (1995) The biology of sciaenid fishes in the Matang mangrove waters, MSc. thesis, University of Malaya, Kuala Lumpur, Malaysia, 172 pp.

Yap, Y.N., Sasekumar, A. and Chong, V.C. (1994) Sciaenid fishes of the Matang mangrove waters, in *Proceedings Third ASEAN Australia Symposium on Coastal Living Resources*, vol. 2: *Research Papers*, (eds S. Sudara, C.R. Wilkinson and L.M. Chou), Chulalongkorn University, Bangkok, pp. 472–476.

Yau, K.H. (1991) Water quality of Brunei Bay and estuary. *ICLARM Conference Proceedings*, **22**, 189–193.

Young, P.C., Leis, J.M. and Hausfeld, J.F. (1986) Seasonal and spatial distribution of fish larvae in waters over the North West Continental Shelf of Western Australia. *Marine Ecology Progress Series*, **31**, 209–222.

Zavala Garcia, F., Flores Coto, C. and Mendez Vargas, M.L. (1988) Desarrollo y distribucion larvaria de *Gobiosoma robustum* Ginsburg (Pisces: Gobiidae), Laguna de Terminos, Campeche. *Annales del Instituto de Ciencias del Mar y Limnologia, Universidad Nacional Autónoma de México*, **15**, 237–244.

Zhang-Renzhai (1987) The development of fertilized egg and larvae of spotted grunt. *Journal of the Fisheries Society of China, Qingdao*, **11**, 241–246.

# Taxonomic index

Ablennes, *see* Belonidae
Abudefduf, *see* Pomacentridae
Acanthopagrus, *see* Sparidae
Acanthuridae 56
  *Acanthurus grammoptilus* 80
Acentrogobius, *see* Gobiidae
Achirus, *see* Soleidae
Achlyopa, *see* Soleidae
Aegiceras 16
  *see also* Mangrove
Aetobatus, *see* Myliobatididae
Agonostomus, *see* Mugilidae
Albula vulpes 104
Alepes, *see* Carangidae
Alosa, *see* Clupeidae
Ambassidae 62, 63, 64, 73, 143,
  287
Ambassis gymnocephalus 52, 62, 63,
  64, 65, 126, 127, 138, 139,
  172
  *Ambassis interruptus* 172
  *Ambassis jacksonensis* 288
  *Ambassis kopsii* 63
  *Ambassis marianus* 288
  *Ambassis nalua* 65, 217
  *Ambassis natalensis* 54, 56, 125,
  126, 127, 129, 172, 178–
  179, 224
  *Ambassis productus* 56, 121, 126,
  127, 129, 172, 178–179,
  183, 224
  *Parambassis macrolepis* 63, 181
  *Parambassis wolffi* 63
Amniataba, *see* Teraponidae
Amoya, *see* Gobiidae
Anampses, *see* Labridae
Anchoa, *see* Engraulididae
Anchovia, *see* Engraulididae

Anchoviella, *see* Engraulididae
Anguilla marmorata 60
Anisotremus, *see* Haemulidae
Anodontostoma, *see* Clupeidae
Apogon, *see* Apogonidae
Apogonidae 76, 116, 230
  *Apogon albimaculosus* 78
  *Apogon hyalosoma* 74, 116
  *Apogon quadrifasciatus* 80
  *Apogon ruppelli* 67, 120, 197
  *Siphamia roseigaster* 74
Archerfish, *see* Toxotidae
Archosargus, *see* Sparidae
Ardea, *see* Ardeidae
Ardeidae
  *Ardea cinerea* 226–227
  *Ardea goliath* 226–227
  *Egretta alba* 226–227
  *Egretta garzetta* 226–227
Argyrosomus, *see* Sciaenidae
Ariidae 46, 60, 62, 63, 89, 158–
  159, 196, 198, 203, 206, 221
  *Ariopsis seemani* 150
  *Arius argyropleuron* 67, 68, 71,
  72, 73, 160–162, 206
  *Arius armiger* 160–162
  *Arius bilineatus* 160–162,
  *Arius caelatus* 62
  *Arius couma* 94
  *Arius felis* 96, 97, 98, 101, 150,
  203–204
  *Arius graeffei* 143, 144, 160–162
  *Arius grandicassis* 94, 95
  *Arius herzbergii* 94
  *Arius heudeloti* 84, 88, 203–204
  *Arius latiscutatus* 84, 88, 126
  *Arius leptaspis* 68, 73, 74, 143,
  144, 160–162, 203

# Geographic index

# Subject index